Haibin Ning
Thermoplastic Composites

Also of interest

Thermoplastic Elastomers.
At a Glance
Günter Scholz, Manuela Gehringer, 2021
ISBN 978-3-11-073983-1, e-ISBN 978-3-11-073984-8
e-ISBN (EPUB) 978-3-11-073998-5

Superabsorbent Polymers.
Chemical Design, Processing and Applications
Edited by Sandra Van Vlierberghe, Arn Mignon, 2021
ISBN 978-1-5015-1910-9, e-ISBN 978-1-5015-1911-6,
e-ISBN (EPUB) 978-1-5015-1171-4

Sustainability of Polymeric Materials
Edited by Valentina Marturano, Veronica Ambrogi,
Pierfrancesco Cerruti, 2020
ISBN 978-3-11-059093-7, e-ISBN 978-3-11-059058-6,
e-ISBN (EPUB) 978-3-11-059069-2

Inorganic and Organometallic Polymers
Narendra Pal Singh Chauhan, Narendra Singh Chundawat, 2019
ISBN 978-1-5015-1866-9, e-ISBN 978-1-5015-1460-9,
e-ISBN (EPUB) 978-1-5015-1479-1

Chemistry of High-Energy Materials
Thomas M. Klapötke, 2019
ISBN 978-3-11-062438-0, e-ISBN 978-3-11-062457-1,
e-ISBN (EPUB) 978-3-11-062469-4

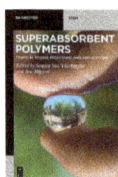

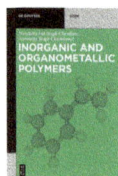

Haibin Ning

Thermoplastic Composites

Principles and Applications

DE GRUYTER

Author
Haibin Ning, PhD
Materials Processing and Applications Development Center
Department of Materials Science and Engineering
The University of Alabama at Birmingham
501 Bldg, 501 12th St S
Birmingham, AL 35294
USA
ning@uab.edu

ISBN 978-1-5015-1903-1
e-ISBN (PDF) 978-1-5015-1905-5
e-ISBN (EPUB) 978-1-5015-1161-5

Library of Congress Control Number: 2021940736

Bibliographic information published by the Deutsche Nationalbibliothek
The Deutsche Nationalbibliothek lists this publication in the Deutsche Nationalbibliografie;
detailed bibliographic data are available on the Internet at http://dnb.dnb.de.

© 2022 Walter de Gruyter GmbH, Berlin/Boston
Cover image: icestylecg/iStock/Getty Images Plus
Typesetting: Integra Software Services Pvt. Ltd.
Printing and binding: CPI books GmbH, Leck

www.degruyter.com

Dedicated to my beloved father

Preface

The popularity of thermoplastic composites as lightweight high-performance engineering materials has been steadily growing in the last several decades. The thermoplastic composite has been used more and more as an alternative material to traditional metallic materials and thermoset composites in various applications that require superior mechanical proprieties, low density, great toughness, high production rate, cost-effectiveness, and so on.

There are a number of books available on general polymer matrix composites or thermoset composites, but rarely are there any comprehensive books about thermoplastic composites. A major objective of this book is to fill the gap by summarizing the advancement of knowledge in the thermoplastic composite from its initial development and providing a broad overview on various aspects of this advanced structural material, including its constituents, pre-impregnation techniques, molding processes, additive manufacturing, characterization methods, mechanics, and recycling. It is intended to raise the interest about the thermoplastic composite among students and engineers and deliver valuable information on the principles of thermoplastic composites and their constituents. Examples and case studies on the use of thermoplastic composites are also provided in the book to bring their practical applications to readers.

I have been fortunate to receive help and support from numerous people, to whom I would like to express my gratitude, in alphabetic order: J. Andrews, C. Barham, S. Bartus, B. Bittmann-Hennes, A. Catledge, K.K. Chawla, X. Deng, P. Guha, A. Hassen, G. Husman, M. Janney, G.M. Janowski, N. Lu, D. Lerew, Q. Liu, Yantong Liu, V. Merchant, J. Papo, S. Pillay, T. Pillay, F. Samalot, S. Sen, J.C. Serrano, B. Thattai, R. Thompson, L. Townsend, C. Ulven, K. Ulven, U.K. Vaidya, H. Wang, M.R. Wright, L. Zeng, J. Zhang, and L. Zhou. Special thanks are due to my family members for their continuous encouragement and support. Last but not least, I want to express my most sincere gratitude and appreciation to my parents for their guidance and unconditional love.

Haibin Ning
Birmingham, AL, USA
April 2021

https://doi.org/10.1515/9781501519055-202

Contents

Chapter 4
Processing of thermoplastic composites —— 129

Chapter 1
Introduction

A composite material, or a composite, is a mixture of two or more constituent materials that remain physically distinguishable but have properties that cannot be achieved only by any individual constituent material. Matrix and reinforcement are the main constituent materials in the composite. The matrix surrounds the reinforcement and forms a continuous phase in the composite. It protects the reinforcement from damage and transfers the load to the reinforcement. On the other hand, the reinforcement possesses excellent strength and modulus and contributes to the majority of the load-bearing capability for the composite.

It is believed that the first composite material made by mankind was wattle and daub, a mixture of sticks and mud, for building and construction use, dated back to more than 6,000 years ago. Since then, different forms of natural materials have been used to create various composite materials throughout human civilization, including papier-mâché (paper strips glued together with adhesive) and wood. Modern composite materials with significantly improved performance only started to take shape in the early 1900s after synthetic polymers and fibers were invented. From then on, composites have been gaining significant interests in various industrial sectors, and new markets have been continuously established ranging from products used in daily life to advanced niche applications.

The matrix in the composite can be different types of materials, such as polymer, metal, ceramic, and carbonaceous materials. Based on the matrix material type, there are four categories of composite materials: polymer matrix composite (PMC), metal matrix composite (MMC), ceramic matrix composite (CMC), and carbon and graphite matrix composite (CGMC). MMC, CMC, and CGMC have been mainly used in advanced niche applications that require high performance at extreme temperatures. Their raw material cost and processing cost are considerably high. On the other hand, PMCs generally possess superior specific strength and modulus at a relatively low cost although their performance is limited at elevated temperatures. PMCs occupy about 90% of the composite market share.

PMCs have polymeric materials, or polymers, as the matrix material. The polymer is an organic material consists of repeating units known as monomers. The monomer is made of light elements such as hydrogen, carbon, and nitrogen; therefore, it possesses a low density, typically ranging from 0.9 to 1.5 g/cm^3. The repeated units are called molecular chains, polymer chains, or molecules, which can be arranged in different forms. Based on the arrangement, there are three types of polymers: thermoplastics, thermosets, and elastomers. Each of the polymers has a unique structure and distinct properties.

Figure 1.1 shows the structure of these three types of polymers. Thermosets have a cross-linked structure that their molecular chains are linked via certain molecules.

https://doi.org/10.1515/9781501519055-001

If the repeat unit in the molecular chains or monomer is considered to be one "snake," the analogy for the cross-linked structure of the polymer chains in the thermoset is snakes holding the head or tail of the other snakes nearby (Fig. 1.1a), forming a giant network structure.

Thermoplastics have a series of molecular chains that are not cross-linked (Fig. 1.1b). The analogy for the structure of thermoplastics is a series of "snake chains" formed by one snake holding the tail of another snake. One "snake chain" (molecular chain) might cross over other "snake chains" (molecular chains) around it, but the chains are not linked at intersections. The "snake chains" in the thermoplastic have the freedom of sliding past each other when adequate loading or heat is applied. This unique structure results in improved toughness in thermoplastics compared to thermosets. It also allows the thermoplastic to be reheated and reshaped for easy recycling. However, this structure might cause some issues under certain conditions. For example, the thermoplastic or its composites have dimensional instability issues when heated or applied with a constant stress (creep). It is noted that the analogy of snake chains will be used throughout the book to describe some unique characteristics of thermoplastics and thermoplastic composites.

Elastomers have a structure that is between the structures of thermosets and thermoplastics. Their molecular chains have a certain extent of cross-linking but some molecular chain sections have similar structures to thermoplastics. Figure 1.1c shows the partially cross-linked structure of the elastomer.

Since the development of the first fully synthetic thermoplastic in the 1930s, the thermoplastic has become a crucial and versatile material in our daily life as well as science and engineering fields. The thermoplastic is lightweight and tough and has been widely used as the packaging films, toy components, grocery bags, beverage containers, electronic carriers, and so on.

The thermoplastic has gained substantial attention as a matrix material for composites. When reinforcements, such as fibers, are added to the thermoplastic, its mechanical properties can be significantly improved. The composite material based on thermoplastics is called thermoplastic matrix composites, or thermoplastic composites, one type of polymer matrix composite. The other type of PMCs is thermoset matrix composites (or thermoset composites). Figure 1.2 shows different composites, including thermoplastic composites, and their general characteristics. Carbonaceous matrix composites are not included in the figure. The carbonaceous matrix composite has similar properties with CMC, such as excellent modulus and low density; however, it is subjected to oxidation at elevated temperatures in the presence of oxygen.

Both the thermoplastic matrix and the reinforcement contribute to the properties of the thermoplastic composite. The matrix provides attributes, such as low density, high toughness, corrosion resistance, and recyclability, to the thermoplastic composite while the reinforcement offers high strength and high modulus. Compared

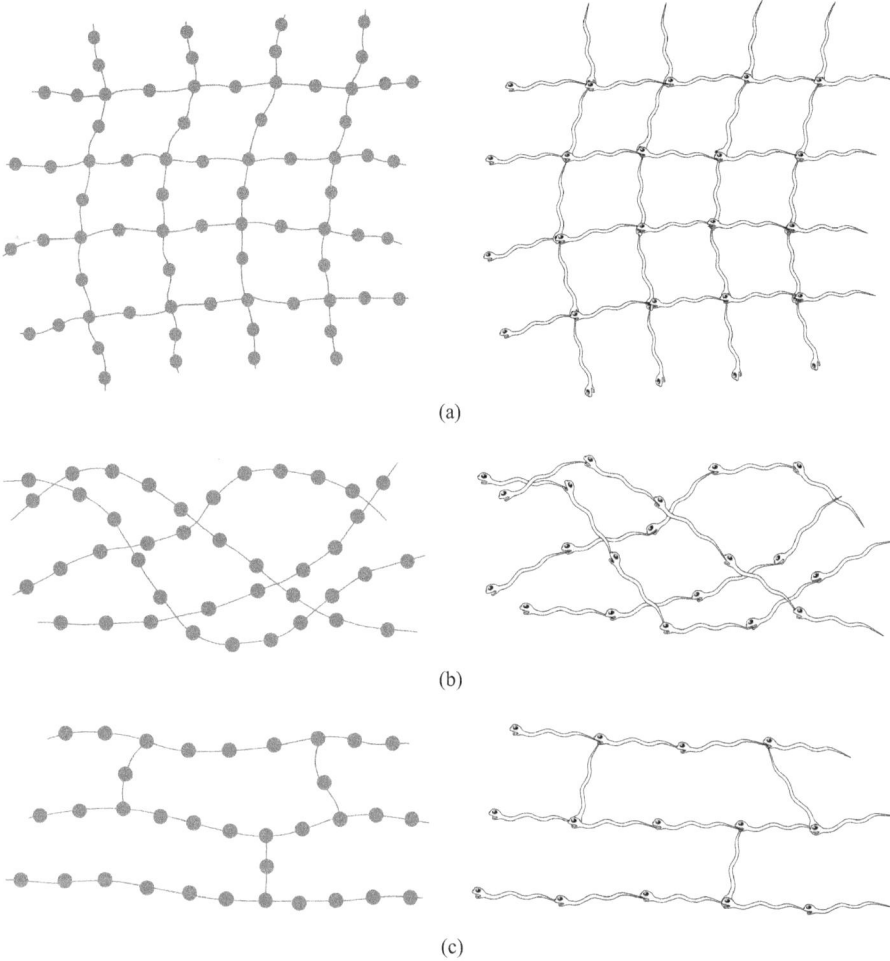

(a)

(b)

(c)

Fig. 1.1: The molecular chain structure in (a) thermosets, (b) thermoplastics, and (c) elastomers. Individual snakes and snake chains are used as analogies to monomers and polymer chains, respectively, in order to show the difference in the structure of each polymer at the molecular level.

to traditional metallic materials, the thermoplastic composite has the following advantages:
1. High specific strength and modulus
2. Good corrosion and chemical resistance
3. Low coefficient of thermal expansion
4. Design flexibility
5. Overmolding potential
6. Near-net shape and reduced secondary operations

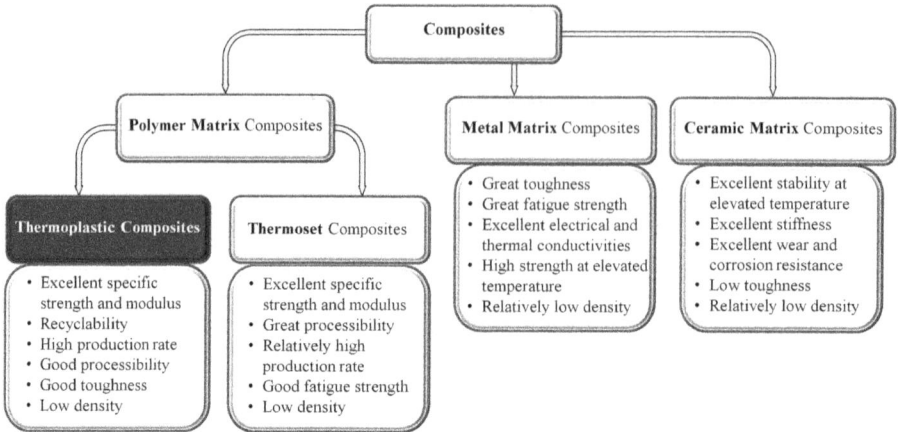

Fig. 1.2: Classification of composites based on matrix materials, including thermoplastic composites, and general characteristics of each composite material.

The thermoplastic composite also has advantageous characteristics over thermoset matrix composites. For example, the thermoplastic generally has complete polymerization and requires no time for the polymerization reaction to happen during processing as opposed to necessary chemical reaction time in processing thermosets or thermoset composites. Therefore, the production rate for thermoplastic or thermoplastic composite products is much higher compared to that for thermoset or thermoset composite products. In addition, the thermoplastic or its composites have an unlimited shelf life. The advantages that thermoplastic composites have over thermoset matrix composites are listed below:

1. Unlimited shelf life
2. High production rate
3. Recyclability
4. Low emission of volatile organic compound
5. Excellent toughness and impact resistance
6. Various joining options
7. Wide range of mechanical properties

The superior material properties of the thermoplastic composite have enabled its increasing use in a variety of industrial sectors, including aviation, automotive, construction, energy, marine, defense, sporting goods, and electrical equipment. Figure 1.3 shows some typical use of the thermoplastic composites, including truck bed liner in automotive application (Fig. 1.3a), compressed gas tank in energy application (Fig. 1.3b), leading edge component for aviation application (Fig. 1.3c), and floor board for construction application (Fig. 1.3d). Because of the increasing interest from those industrial sectors, the global thermoplastic composite market has been projected to grow with a compound annual growth rate of 5.2% [1].

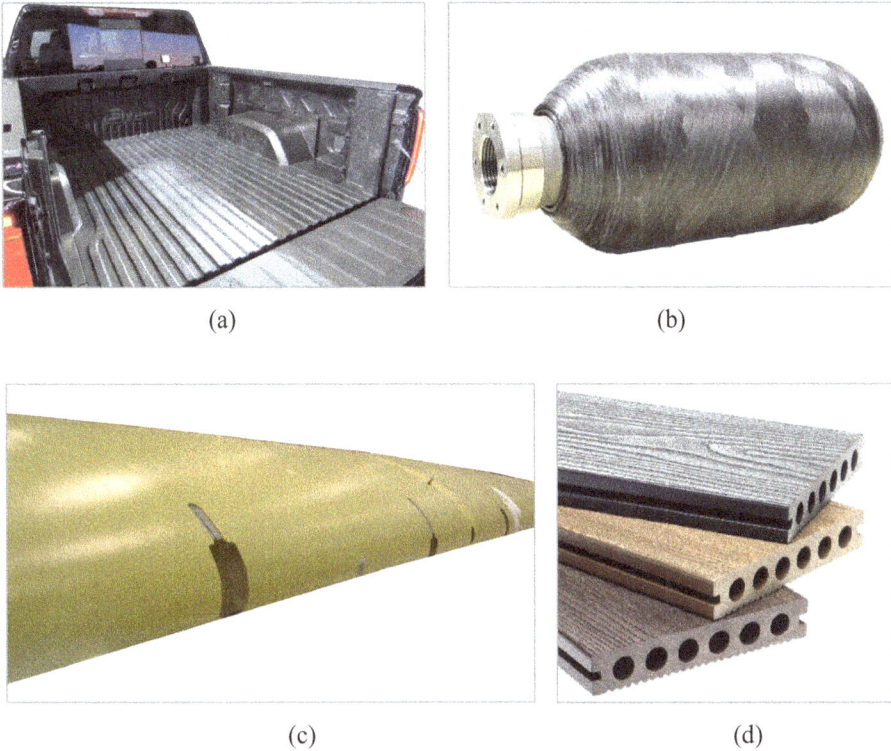

Fig. 1.3: Applications of thermoplastic composites: (a) bed liner in GMC truck (automotive application; adapted from Reference [2]); (b) pressure vessel (energy application; adapted from Reference [3]); (c) leading edge of Airbus A380 airplane (aviation application; reprinted from Reference [4] with permission); and (d) deck boards (construction application).

This book will comprehensively describe numerous aspects of thermoplastic composites, including their constituent materials, processing, additive manufacturing, characterization, micro- and macro-mechanics, and recycling.

References

[1]   https://www.marketsandmarkets.com/Market-Reports/thermoplastic-composite-market
      -111944669.html. Accessed: July 2020.
[2]   Halvorson B. Fiber supplement: in depth with the first ever carbon-fiber pickup bed. Car and
      Driver. 2018.
[3]   Malnati P. Thermoplastic composite pressure vessels for FCVs. Composites World. 2015.
[4]   Mathijsen D. Leading the way in thermoplastic composites. Reinforced Plastics. 2016;60(6):
      405–7.

Chapter 2
Constituents of thermoplastic composites

2.1 Introduction

A thermoplastic composite is composed of a thermoplastic polymer matrix and rein-
forcement(s) that are mixed together but remain physically distinguishable. Both
the matrix and the reinforcement have their unique characteristics. When combined
to form thermoplastic composites, the matrix and the reinforcement work synergistically to provide the composite high strength and modulus, lightweight, good wear
resistance, high impact performance, and so on.

The thermoplastic matrix and the reinforcement have different functions in the
thermoplastic composite. The thermoplastic matrix has significantly lower strength
and modulus than the reinforcement. It forms a continuous phase enclosing the
reinforcement and has several important functions in the composite. Firstly, the
thermoplastic matrix helps transfer loads to the reinforcement. This is the primary
function of the matrix. It is preferable that the reinforcement, instead of the matrix, bears a substantial amount of the load during use because of its much higher
strength. When the thermoplastic composite is under loading, the thermoplastic
matrix is loaded first because the reinforcement is enclosed inside the matrix. The
load is transferred to the reinforcement through the matrix/reinforcement interface. Since the reinforcement has much high modulus than the matrix, it bears
significantly more load than the matrix when both have the same strain. The percentage of the load shared by the reinforcement can be calculated by Eq. (7.20).
The bonding strength between the reinforcement and the matrix affects the efficiency
of the loading transfer. This will be illustrated more in Section 3.3.2. Secondly, the
thermoplastic matrix provides protection to the reinforcement from wear and abrasion that can adversely affect the performance of the composite. Generally, the performance of the reinforcement is highly sensitive to surface defects. The defects from
the abrasion, such as scratches, can easily propagate in the reinforcement during
loading due to stress concentration and cause premature failure. The thermoplastic
matrix protects the reinforcement from wear and abrasion by surrounding the reinforcement and minimizes or prevents such defects. Thirdly, the thermoplastic matrix
normally has a high toughness and good ductility. Its high strain to failure can reduce the stress concentration and diminish crack initiation and propagation in its
composites. As a result, matrix-dominated properties of thermoplastic composites
can be enhanced. In addition, the thermoplastic matrix provides various surface fin-
ishes, including roughness, texture, and color, for its composite. The matrix also
forms the bulk of the composite, keeps the reinforcement in place, and maintains
the shape of the composite. Furthermore, thermoplastics generally have a very low

https://doi.org/10.1515/9781501519055-002

thermal conductivity and electrical conductivity and, therefore, can provide electrical and thermal insulation to the composite.

Overall, the functions of the thermoplastic matrix are summarized as follow:
1. Transfers load to the reinforcement
2. Protects the reinforcement from abrasion and environment attack
3. Reduces propagation of cracks in the composite
4. Determines the surface quality and appearance (texture and color) of the composite
5. Keeps the reinforcement in place
6. Carries interlaminar shear stress
7. Provides certain functionality such as electrical insulation and thermal insulation

The reinforcement, on the other hand, is a discontinuous constituent enclosed in the continuous matrix in the composite. It has significantly higher modulus and strength and lower strain to failure than the matrix. Therefore, it determines the strength, modulus, and material orthotropy of the composite. There are several types of reinforcements used in the thermoplastic composite and these reinforcements vary significantly in material form, cost, and performance. Typical reinforcements used in the thermoplastic composite include fibers, particulates, and nanomaterials.

Among all of the reinforcements, fibers are dominantly used. The fibers used in the thermoplastic composite include both organic and inorganic fibers. The inorganic fiber includes carbon fiber, glass fiber, and basalt fiber. There are other inorganic fibers such as ceramic fiber, boron fiber, and metal fiber; however, those fibers are not commonly used in thermoplastic composites. The organic fiber includes aramid fiber, cellulose fiber, polypropylene (PP) fiber, and polyethylene (PE) fiber. Each fiber type will be discussed in Section 2.3.

A typical thermoplastic composite consists of one type of reinforcement and a thermoplastic matrix; however, some thermoplastic composites have multiple reinforcements. For example, both fibers and nanomaterials are used to reinforce thermoplastics at the same time. Different types of fibers can also be present in a thermoplastic composite. For example, glass fibers are added to carbon fiber-reinforced thermoplastic composites to enhance their electrical insulation; glass fibers are blended into cellulose fiber-reinforced thermoplastic composites to improve the mechanical properties of the composite.

There are other constituents existing in the thermoplastic composite besides reinforcement and matrix. Those constituents are the interface between the reinforcement and the matrix, that is, fiber/matrix interface, and voids, both of which are not as noticeable as the reinforcement and the matrix. However, they can considerably affect the performance of the thermoplastic composite. Those constituents are described in Sections 2.4 and 2.5, respectively.

2.2 Matrices in thermoplastic composites

There are a wide range of thermoplastics with versatile characteristics available as the matrix for composites. Some thermoplastics are amorphous whereas the others are semicrystalline. Some thermoplastics have linear molecular chains only while some others have branched chains. The wide variety of thermoplastics make the thermoplastic composite a highly tailorable engineering material for applications ranging from commodity to aerospace. The distinctly different molecules and function groups in thermoplastics provide significantly different mechanical properties, thermal properties, and tribological properties to the thermoplastic. In addition, the variety of fibers in thermoplastic composites including synthetic fibers (carbon, glass, and organic fibers) and natural fibers (cellulose fibers) with various architectures and fiber lengths further provide a wide range of options for end users to select for specific applications.

Typical thermoplastics used in thermoplastic composites are PP, PE, polyamide (PA) or nylon, polyphenylene sulfide (PPS), polyether ether ketone (PEEK), and so on. These thermoplastics have different backbones, function groups, as well as molecular chains with different lengths and arrangements, and thus, their properties, including physical property, chemical property, and mechanical property, can vary significantly. The difference in the property of thermoplastics considerably contributes to the property difference in their composites.

Several attributes of thermoplastics that can affect the properties of their composites are listed and elucidated as follow:

1. **Backbone of thermoplastic polymer chains.** The backbone, or main chain, of a thermoplastic is a long molecular chain composed of covalently bonded atoms. A common backbone is made of carbon–carbon bonds, as seen in PP, PE, and polystyrene (PS). Other atoms can also be present in the backbone. For example, PPS has sulfur in its backbone while nylon has nitrogen in its backbone. The bonding strength between the atoms in the backbone affects the behavior of the thermoplastic. Stronger bonds among the atoms in the backbone result in high strength and rigidity (less flexibility) in the thermoplastic, especially at its glassy state (below its glass transition temperature T_g).

   Phenyl group, or aromatic group, exists in the backbone of several thermoplastics. The phenyl group has a formula of $-C_6H_5$, a benzene ring with one hydrogen atom substituted by some other elements or compounds. Because of its ring structure, the phenyl group is sometimes called phenyl ring, sometimes simplified as Ph or Φ. Since the phenyl group is restricted from rotation at its glassy state (below glass transition temperature T_g) and has limited rotation at its rubbery stage (above T_g but below melting temperature T_m), the thermoplastic with the phenyl group has improved strength and modulus, enhanced glass transition temperature, higher melting temperature, and better chemical resistance compared to other thermoplastics without the phenyl group in their backbone. Typical thermoplastics with the phenyl group in their backbones include aromatic nylons (see Section 2.2.6), PPS (See Section 2.2.8), PEEK (See Section 2.2.9), and polyetherimide (PEI) (see Section 2.2.10).

2. **Average molecular weight of thermoplastics**. The average molecular weight of a thermoplastic matrix also affects the performance of its composites. It is calculated by dividing total weight of the thermoplastic polymer by the total number of its molecules. The average molecular weight is determined by the degree of polymerization (DOP). DOP is the number of repeat units in an average polymer chain and can be calculated by dividing the molecular mass of a polymer with the molecular weight of its monomer. When there is a higher DOP in a thermoplastic, its polymer chains are longer and the average molecular weight of the thermoplastic is higher. The thermoplastic with a higher DOP possesses higher mechanical properties and better heat resistance. Accordingly, the composite with a thermoplastic matrix that has a higher DOP possesses better mechanical properties and higher heat resistance. A typical example is the difference in mechanical and thermal properties among PE matrix composites that have different molecular weights or DOPs. High-density polyethylene (HDPE) matrix composites possess better mechanical properties and higher melting temperature than low density polyethylene (LDPE) matrix composites when other variables, such as the type and amount of the reinforcement, remain the same. The difference in the average molecular weight of each PE is detailed in Section 2.2.5. The average molecular weight of a thermoplastic also affects the processing of its composites. For example, PP with a low molecular weight results in faster impregnation with fibers because of its lower viscosity when melted at a certain temperature. Conversely, PP with a high molecular weight results in slower impregnation with fibers at the same temperature.

3. **Side chain of thermoplastics**. The side chain, or pendant chain, is the chemical group attached to the main chain. A side chain with a larger size and higher rigidity leads to higher rigidity of the polymer chain in the thermoplastic. As a result, the glass transition temperature of the thermoplastic increases. The side chain also affects the polarity of the thermoplastic, which in turn determines its surface energy and bonding with reinforcements. This will be described in Section 2.4.

4. **Arrangement of thermoplastic polymer chains**. Polymer chains in thermoplastics can have different arrangements that result in polymer chains with or without order. Section 2.2.1 will describe different arrangements of polymer chains and their effect on the properties of thermoplastics and thermoplastic composites.

2.2.1 Crystallinity of thermoplastics

2.2.1.1 Semicrystalline and amorphous thermoplastics

Thermoplastics can be classified into semicrystalline thermoplastics and amorphous thermoplastics based on the arrangement of their polymer chains. The semicrystalline thermoplastic is consisted of both ordered and disordered polymer chains. These polymer chains with ordered arrangement are called crystallites and the regions with ordered polymer chains are called crystalline regions. The regions without any ordered polymer chains are called amorphous regions. Figure 2.1a shows the polymer chains in a semicrystalline thermoplastic consisting of both crystalline and amorphous regions. Typical semicrystalline thermoplastics include PP, nylons, PEEK, and PPS. On the other hand, thermoplastics that do not have any ordered polymer chains are called amorphous thermoplastics. Figure 2.1b shows the disordered polymer chains in the amorphous thermoplastic. Amorphous thermoplastics include polycarbonate (PC), polystyrene (PS), polymethyl methacrylate (PMMA), and acrylonitrile butadiene styrene (ABS).

The arrangement of polymer chains in a thermoplastic plays an important role in its processing and properties. The amorphous thermoplastic has better geometry stability during processing than semicrystalline thermoplastics because there is no transformation between ordered and disordered polymer chains and less shrinkage is resulted. The amorphous thermoplastic also has better impact resistance than the semicrystalline thermoplastic. However, the polymer chains without any ordered arrangement are subjected more to chemical attack and, therefore, the amorphous thermoplastic has poor chemical resistance. The amorphous thermoplastic also has poor fatigue resistance.

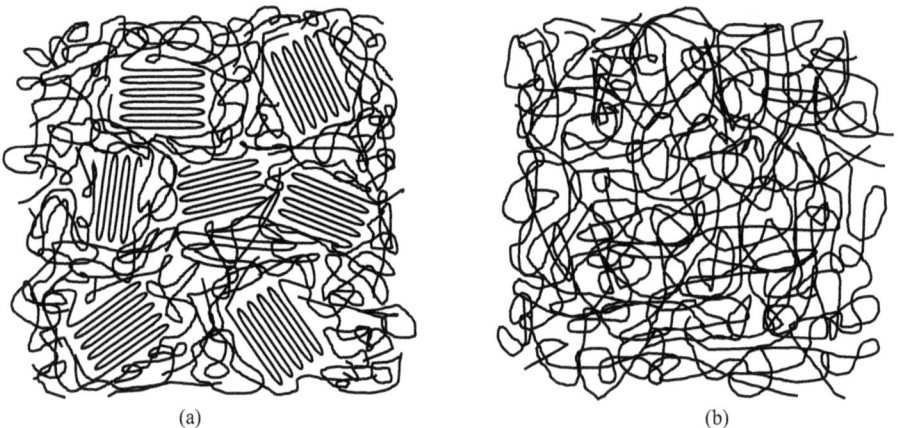

(a)   (b)

Fig. 2.1: The molecular chain arrangement in (a) a semicrystalline thermoplastic and (b) an amorphous thermoplastic. Note that the semicrystalline thermoplastic has both crystalline and amorphous regions while the amorphous thermoplastic does not have any crystalline regions.

2.2.1.2 Degree of crystallinity

The volume fraction of the crystalline regions over the total thermoplastic volume in a semicrystalline thermoplastic is defined as degree of crystallinity (DOC), or crystallinity. It is used specifically to describe the volume fraction of the crystalline regions in the semicrystalline thermoplastic. The thermoplastic with a higher DOC normally possesses enhanced strength, improved wear resistance, and better chemical stability and its composites benefit from those enhanced properties.

Crystalline regions in semicrystalline thermoplastics or their composites are generally developed from processing. During processing of a semicrystalline thermoplastic or its composite, the thermoplastic is firstly melted and molded followed by a cooling process. If the cooling rate is sufficiently low, the thermoplastic undergoes a recrystallization process that rearrangement of polymer chains takes place and crystallites form and grow in the thermoplastic. Slow cooling of the thermoplastic promotes the formation of crystalline regions and increases its DOC to a certain extent. A longer dwell time above the recrystallization temperature of the semicrystalline thermoplastic can also improve the DOC.

The rearrangement of molecular chains into crystallites in the semicrystalline thermoplastic during recrystallization is more difficult than that of the atoms in metals. Metals are normally a fully crystalline material and have 100% DOC. However, most semicrystalline thermoplastics in a bulk form as a matrix material used in composites normally have a DOC of 25–80%. The molecular chains in thermoplastics are in a much larger scale than single atoms in metals. Entanglement of the molecular chains imposes considerable difficulty in the rearrangement of the molecular chains into crystallite. In addition, other attributes, such as nonuniform molecular weight and branched polymer chains, hinder the rearrangement of polymer chains. On the other hand, linear polymer chain structure, narrow molecular weight, and high molecular weight are favorable to form thermoplastics with a high DOC.

Stretching or drawing can be used to induce more crystalline regions and improve DOC in thermoplastics. However, this method is mainly for producing thermoplastic polymer fibers. A number of thermoplastic polymer fibers achieve a high DOC through the approach (See Section 2.3.1.3). Drawing the thermoplastic can significantly promote crystallization of the polymer chains by creating stress-induced crystallites. A DOC up to 98% can be achieved through drawing (Fig. 2.54). Other factors may affect DOC of thermoplastics as well. For example, when fibers are added to the thermoplastic, the fibers play an important role in the nucleation and growth of crystallites (see Section 6.7).

Crystallization of polymer chains in a semicrystalline thermoplastic affects its geometry stability during molding. The local arrangement of the polymer chains in order causes decrease in volume because of the closely packed molecular chains and resultant higher density. A higher DOC means a greater volume change because more polymer chains transition from a disordered structure to an order structure during cooling.

This disordered-to-ordered transformation at the molecular level affects the geometry stability of the thermoplastics because of the resultant shrinkage. Thermoplastics with a higher DOC experience greater shrinkage than those with a lower DOC or amorphous thermoplastics. Adding reinforcements, such as carbon fibers and glass fibers, to the thermoplastic can effectively reduce its shrinkage and enhance its geometry stability because of the excellent geometry stability of the reinforcement.

Because the volume of the crystalline regions cannot be directly measured, an indirect method is normally used to determine the DOC of a semicrystalline thermoplastic. When the semicrystalline thermoplastic is heated to its melting temperature, the amorphous region in the thermoplastic does not experience any phase change, while the molecular chains in the crystalline region are transitioned from an ordered state to a disordered state. The transition is endothermic and heat is continuously absorbed until the ordered-to-disordered transformation is completed. The DOC of a semicrystalline thermoplastic can be calculated as follows:

$$\text{DOC} = \frac{\Delta H_s}{\Delta H_f} \times 100\% \tag{2.1}$$

where ΔH_s is the specific enthalpy of the thermoplastic and ΔH_f is the specific enthalpy of the thermoplastic that is assumably 100% crystalline. The specific enthalpy of a thermoplastic (ΔH_s) can be measured using differential scanning calorimetry (DSC) by integrating the endothermic peak in the DSC curve. The value of ΔH_f for common thermoplastics can be found in Tab. 6.3. It is noted that this equation only applies to neat thermoplastics, namely, thermoplastics without any reinforcements, fillers, or additives.

The DSC method can be adopted to measure the DOC for the thermoplastic matrix in a thermoplastic composite. However, the fiber content of the composite and any absorbed or released heat in crystallization have to be considered. The equation for calculating the DOC of the matrix in the thermoplastic composite is derived in Section 6.7.

Some thermoplastics, including polyethylene terephthalate (PET) and polyether ketone ketone (PEKK), can be either semicrystalline or amorphous. When PET is amorphous, it is translucent and has a good toughness. Amorphous PET is the material that is often used to make commodity products such as packaging films and disposable water bottles. More description of both amorphous and semicrystalline PET is provided in Section 2.2.7. Some of the amorphous thermoplastics can be converted to a semicrystalline thermoplastic by an annealing process, a heat treatment process in which the amorphous thermoplastic is heated to above its recrystallization temperature and held at that temperature for a duration of time. The hold allows some polymer chains to rearrange and form ordered regions in the thermoplastic.

2.2.1.3 Difference in crystallinity, polymerization, and cross-linking

Degree of crystallinity (DOC) differs from degree of polymerization (DOP) and degree of cross-linking. DOP is the number of repeat units in an average polymer chain. A polymer with the same chemical formula can have a different DOP that is indicated by "n" placed as the subscript behind the repeat unit. A higher DOP or a larger "n" value indicates a longer molecular chain or a higher molecular weight. The "n" value determines the mechanical property of the polymer. A polymer with larger "n" values generally possesses higher mechanical properties. For example, ultra-high molecular weight polyethylene (UHMWPE) has a higher "n" value and greater molecular weight than HDPE and possesses higher elastic modulus and tensile strength than HDPE.

Cross-linking is the mechanism mainly present in thermoset or elastomer polymers that have cross-linked molecular chains. Degree of cross-linking is defined as the ratio of molecular chains that are cross-linked to the total number of molecular chains. It is normally used to describe the extent of cross-linking in a thermoset or an elastomer. Table 2.1 compares DOC, DOP, and degree of cross-linking. It is noted that there are special cases, such as thermoplastics that have cross-linked structures and thermosets that have crystalline regions.

Tab. 2.1: Comparison of degree of crystallization, degree of polymerization, and degree of cross-linking.

Variable	Definition	Applicable to Polymers
Degree of crystallization	The ratio of the volume of crystalline regions to the overall polymer volume	Thermoplastics
Degree of polymerization	The number of repeat units in an average polymer chain	All polymers
Degree of cross-linking	The ratio of the number of cross-linked molecular chains to the total number of molecular chains	Thermosets and elastomers

A silicone elastomer is used as an example to describe the difference between degree of cross-linking and DOP. As Fig. 1.1c indicates, elastomers have a partial cross-linking structure. Some of the polymer chains in the elastomer are cross-linked and the others are not. Figure 2.2a shows a silicone that has 27 repeat units of (O–Si) group and its "n" value is 27.

Figure 2.2b shows another silicone that has a cross-linking structure. There are totally 20 (CH_3–Si–CH_3) units and 8 of them are cross-linked. Therefore, its degree of cross-linking is calculated to be $8/20 = 40\%$.

$$
\begin{array}{c}
\quad\quad\quad CH_3 \quad\quad\quad CH_3 \quad\quad\quad CH_3 \quad\quad CH_3 \\
\quad\quad\quad | \quad\quad\quad\quad | \quad\quad\quad\quad | \quad\quad\quad | \\
R-Si-\!\!\left[O-Si\right]_4\!\!-O-\!\!\left[Si\right]_4\!\!-O-Si-R \\
\quad\quad\quad | \quad\quad\quad\quad | \quad\quad\quad\quad | \quad\quad\quad | \\
\quad\quad\quad CH_3 \quad\quad\quad CH_3 \quad\quad\quad CH_2 \quad\quad CH_3
\end{array}
$$

$$
\begin{array}{cc}
\begin{array}{c}
CH_3 \quad\quad CH_3 \quad\quad CH_3 \\
| \quad\quad\quad\quad | \quad\quad\quad\quad | \\
R-Si-\!\!\left[O-Si\right]_{27}\!\!-O-Si-R \\
| \quad\quad\quad\quad | \quad\quad\quad\quad | \\
CH_3 \quad\quad CH_3 \quad\quad CH_3
\end{array}
&
\begin{array}{c}
CH_3 \quad\quad CH_3 \quad\quad CH_2 \quad\quad CH_3 \\
| \quad\quad\quad\quad | \quad\quad\quad\quad | \quad\quad\quad | \\
R-Si-\!\!\left[O-Si\right]_4\!\!-O-\!\!\left[Si\right]_4\!\!-O-Si-R \\
| \quad\quad\quad\quad | \quad\quad\quad\quad | \quad\quad\quad | \\
CH_3 \quad\quad CH_3 \quad\quad CH_3 \quad\quad CH_3
\end{array}
\\
(a) & (b)
\end{array}
$$

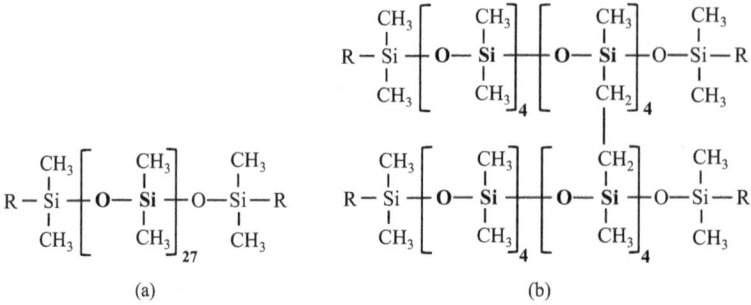

Fig. 2.2: (a) A silicone with a degree of polymerization of 27 and (b) another silicone with a degree of cross-linking of 40%.

2.2.2 Thermoplastics and their properties

Thermoplastics can be further categorized into commodity thermoplastics, engineering thermoplastics, and advanced engineering thermoplastics based on their performance. Figure 2.3 groups commonly used thermoplastics in a "thermoplastic tree" based on their performance, presence of crystallinity, and typical applications. Most of the thermoplastics in the "thermoplastic tree" are often used as the matrix in composites. The left half of the tree includes common amorphous thermoplastics such as polyvinyl chloride (PVC), PC, PEI. Common semicrystalline thermoplastics, including PP, PA, and PEEK, are listed on the right half of the tree. The performance of the thermoplastics increases from the bottom of the tree to the top. Their cost follows the same trend generally.

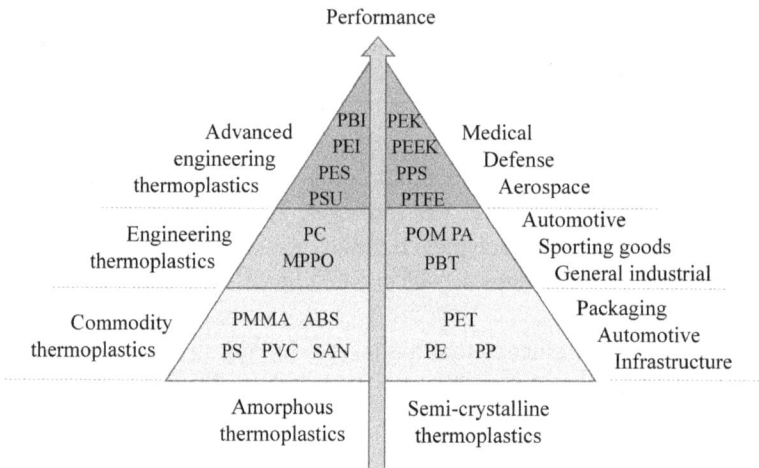

Fig. 2.3: Common amorphous and semicrystalline thermoplastics and their performance comparison in a "thermoplastic tree".

Commodity thermoplastics have low strength and their applications are limited to low load-bearing uses. They account for nearly 80% of all thermoplastics. Common commodity thermoplastics include PP, PE, PET, PS, and PVC. They are typically used in high volume and low cost applications where the structural integrity of the material is not critical, such as beverage container, grocery bag, packaging film, trash can.

Engineering thermoplastics including PC and nylon possess higher strength and toughness than the commodity thermoplastics and are widely used in applications that require the material to bear significant loads, including automotive, transportation, machinery, sports, etc.

Advanced engineering thermoplastics, or specialty thermoplastics, have the highest strength among all the thermoplastics. Their service temperatures are generally than the other thermoplastics. Those composites are used in high-end applications such as aerospace and aircraft. However, the synthesis of these advanced engineering thermoplastics is challenging and their cost is normally high.

The physical and mechanical properties of various thermoplastics are summarized in Tab. 2.2. The cost of each thermoplastic is listed for information only because of the price fluctuation in the market. A number of dollar signs is used to just specify the price range of each thermoplastic. Commonly used thermoplastics in composites are further described in Sections 2.2.4–2.2.12.

There is a remarkable difference in mechanical properties between the commodity thermoplastics and advanced engineering thermoplastics as shown in Tab. 2.2. The difference can be more than 400%, for example, between the elastic modulus of PP and PEEK. However, when fibers are added to the thermoplastics to form thermoplastic composites, the difference is diminished as the strength is largely stemmed from the fibers rather than the thermoplastic matrix. The properties of the thermoplastic composites with different thermoplastic matrix are compared in Section 3.2.1.

2.2.3 Temperature effect

The performance of thermoplastics is tremendously affected by temperature. With increasing temperature, its mechanical properties, such as tensile strength and elastic modulus, decrease. The following sections describe the change of different properties with temperature.

1. The **tensile strength** of thermoplastics is highly dependent on temperature. At elevated temperatures, the intermolecular bond between molecular chains and covalent bonding in the molecular chain of the thermoplastic are weakened. Therefore, their strength decreases and their performance deteriorates with increasing temperature. An analogy to the weakened covalent bonding at elevated temperatures is that heat

Tab. 2.2: Physical properties and mechanical properties of common thermoplastics used in composites.

Thermoplastics	Density (g/cm³)	Elastic modulus (GPa)	Tensile strength (MPa)	Elongation to failure (%)	T_g (°C)	T_m (°C)	Cost*
Acrylonitrile butadiene styrene (ABS)	1.0–1.2	1.1–2.9	27–55	1.5–100	88–128	N/A	$
Polyamide (PA)	1.0–1.34	0.9–4.7	40–135	2–59	44–135	170–322	$
Polyamide imide (PAI)	1.4–1.45	4.5–5.0	152	7.6–16	264–286	N/A	$$$$$
Polybutylene terephthalate (PBT)	1.3–1.38	1.9–3	56–60	50–300	22–43	220–267	$
Polycarbonate (PC)	1.1–1.2	2.0–2.4	60–72	70–150	142–158	N/A	$
Polyether ether ketone (PEEK)	1.3	3.8–4.0	70–103	30–150	143–157	322–346	$$$$$
Polyetherimide (PEI)	1.26–1.28	2.9–3.1	92–101	56–65	215–250	N/A	$$$
Polyether ketone (PEK)	1.29–1.31	3.6–4.2	88–120	44,124	152–163	340–373	$$$$$
Polyether ketone ketone (PEKK)	1.27–1.28	4.3–4.5	105–116	11–13	153–177	347–373	$$$$$
Polyethersulfone (PES or PESU)	1.37–1.46	2.4–2.8	68–95	30–90	220–230	N/A	$$$
Polyethylene (PE)	0.86–0.97	0.6–0.9	20–45	200–800	−25 to (−15)	105–138	$
Polyethylene terephthalate (PET)	1.3–1.4	2.8–3.0	48–72	20–300	59–86	212–250	$
Polylactic acid (PLA)	1.24–1.27	3.3–3.6	47–70	2.5–6	52–60	145–175	$

Tab. 2.2 (continued)

Thermoplastics	Density (g/cm³)	Elastic modulus (GPa)	Tensile strength (MPa)	Elongation to failure (%)	T_g (°C)	T_m (°C)	Cost*
Polymethyl methacrylate (PMMA)	1.17–1.2	2.2–3.2	48–72	2–5.5	101–109	N/A	$
Polyoxymethylene (POM)	1.39–1.43	2.6–3.2	60–90	15–75	−10 to (−8)	160–184	$
Polyphenylene sulfide (PPS)	1.34–1.36	3.2–3.8	48–86	1–6	81–97	280–295	$$
Polyphenylsulfone (PPSU)	1.29–1.3	2.3–2.7	66–84	60–120	210–230	N/A	$$$$
Polypropylene (PP)	0.89–0.92	0.9–1.5	27–42	100–600	−25 to (−15)	150–175	$
Polystyrene (PS)	1	1.2–2.6	36–56	1.2–3.6	74–110	N/A	$
Polysulfone (PSU)	1.23–1.25	2.6–2.7	70–104	40–100	186–192	N/A	$$$
Polytetrafluoroethylene (PTFE)	2.1–2.2	0.4–0.6	20–35	200–400	116–132	315–339	$$$
Polyvinyl chloride (PVC)	1.3–1.45	2.2–3.1	38–44	40–80	81–88	N/A	$
Polyvinylidene fluoride (PVDF)	1.77–1.78	1.4–2.5	24–50	12–200	−40 to (−27)	141–178	$$$
Styrene acrylonitrile (SAN)	1.06–1.08	3.3–3.9	69–82	2–3	102–110	N/A	$
Thermoplastic polyurethane (TPU)	1.12–1.24	1.3–2.1	31–62	60–540	−53 to (−28)	N/A	$

*The number of dollar sign is used to specify the range of price for each thermoplastic. $, less than 5 USD/kg; $$, 5–10 USD/kg; $$$, 10–20 USD/kg; $$$$, 20–40 USD/kg; $$$$$, more than 40 USD/kg. Those costs are provided for information only.

weakens the holding force (bonding) among the snakes (repeat units in molecular chains) and causes a lower strength and performance. Figure 2.4 shows the tensile strength decrease with increasing temperature for various thermoplastics [1]. Reinforcements, especially glass and carbon fibers, added to the thermoplastic can significantly improve its heat resistance. The thermoplastic with better heat resistance normally results in a thermoplastic composite with better mechanical properties at elevated temperatures.

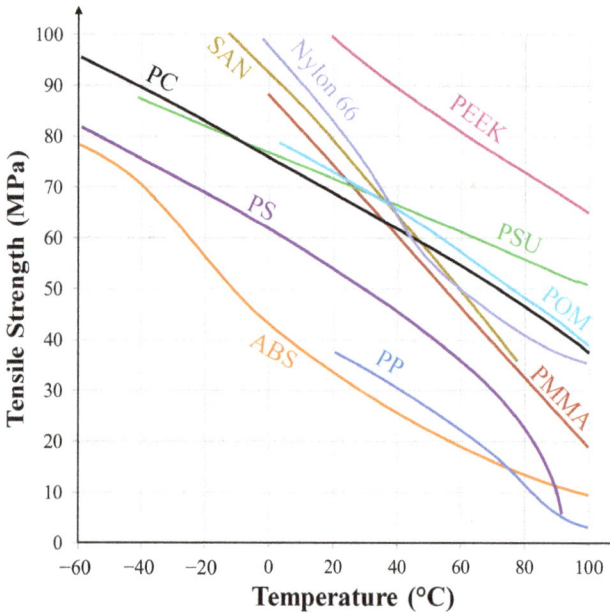

Fig. 2.4: Decrease in tensile strength with increasing temperature for several common thermoplastics used in composites (adapted from Reference [1]).

2. The **elastic modulus** of thermoplastics is also affected by temperature. Both amorphous and semicrystalline thermoplastics show a glassy state below their glass transition temperature; therefore, molecular chains are relatively immobile at a low temperature and the thermoplastics show high modulus. The rigid molecular chains are analogous to hibernating snakes that are less active and less mobile at low temperatures and therefore possess higher rigidity. When the thermoplastics reach their glass transition temperatures, the molecular chains in the amorphous regions have more tendency to move or slide across one another, resulting in drastically increased mobility. Glass transition temperature is defined as the temperature at which polymers undergo a reversible transition from glassy/brittle state to rubbery/flexible state. The drastic decrease of the modulus for both semicrystalline and amorphous thermoplastics at the glass transition temperature is illustrated in Fig. 2.5. Amorphous

thermoplastics have a larger modulus drop than semicrystalline thermoplastics because of the lack of crystalline regions.

Both the semicrystalline and amorphous thermoplastics have a glass transition temperature owing to the amorphous regions. In the semicrystalline thermoplastic, the amorphous regions contribute to the second-order phase transition, which is indicated by its glass transition temperature, T_g. The transition of crystalline regions from solid to liquid is a first-order transition and is indicated by the melting temperature, T_m. Figure 2.5 shows the modulus change with temperature for both semicrystalline and amorphous thermoplastics. The fully crystalline thermoplastic does not exist and is an ideal material. The semicrystalline thermoplastic shows a clear transition into the molten state, indicated by the dramatic decrease in property when the melting temperature, T_m, is reached. However, the amorphous thermoplastic does not have a distinct melting temperature due to the lack of crystalline regions.

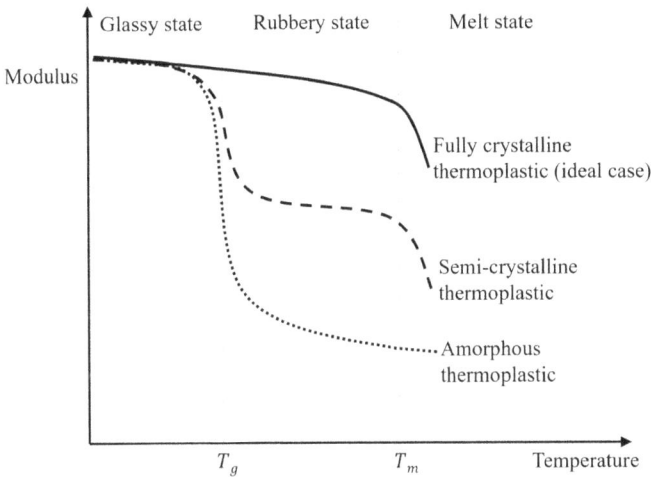

Fig. 2.5: The modulus change with temperature for amorphous, semicrystalline, and fully crystalline thermoplastics. It is noted that the fully crystalline thermoplastic is an ideal case and only used here to show its property change in comparison to the semicrystalline and amorphous thermoplastic.

3. The **specific volume** of the thermoplastic can change dramatically at its glass transition temperature. Specific volume is defined as the ratio of a material's volume to its mass. Materials, including thermoplastics, normally expand when heated because of increased distance among atoms or molecules. The increase rate of specific volume with temperature is determined by the coefficient of linear thermal expansion (CTE) of the material. Figure 2.6 shows the specific volume change with temperature for amorphous thermoplastics, semicrystalline thermoplastics, and fully crystalline thermoplastics (ideal case). The amorphous thermoplastic shows a rate change of the

specific volume with temperature at its glass transition temperature because of the increase of its CTE at that temperature. The semicrystalline thermoplastic also shows a similar rate increase at the glass transition temperature as well as a drastic specific volume change at its melting temperature. The CTE change of both amorphous and semicrystalline thermoplastics with temperature can be used to determine their glass transition temperatures (see Section 6.6.1). The drastic specific volume change in the semicrystalline thermoplastic at its melting temperature can considerably affect the geometry stability of the thermoplastic and its composites during processing, such as molding and additive manufacturing (see Section 5.8.5).

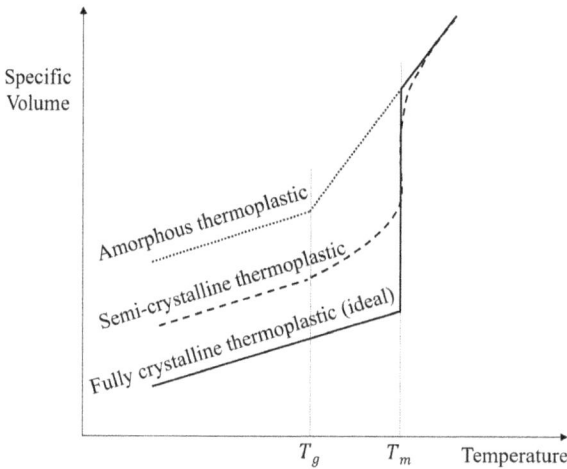

Fig. 2.6: The change of specific volume with temperature for amorphous, semicrystalline, and fully crystalline thermoplastics (an ideal case).

2.2.4 Polypropylene

PP is a versatile commodity thermoplastic that is used in a wide variety of applications, such as bottles, packaging film, food containers, toys, and automotive components. It offers a great combination of low density, low material cost, good impact resistance, great chemical resistance, excellent fatigue resistance, and great electrical properties. Its chemical formula is $(C_3H_6)_n$ as shown in Fig. 2.7. It belongs to the polyolefin family as it has a general formula of C_xH_{2x} and the ratio of the carbon atoms to hydrogen atoms is 1:2.

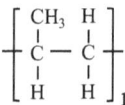

Fig. 2.7: The chemical formula of polypropylene, $(C_3H_6)_n$.

Most commercial PP is isotactic PP, a thermoplastic that was discovered in 1954 by chemist Giulio Natta. The isotactic PP has all its methyl groups, $-CH_3-$, located on one side of the C–C backbone. The other two types of PP are syndiotactic and atactic PP. The syndiotactic PP has the methyl group arranged alternatively along the carbon–carbon backbone while the atactic PP has a random arrangement of the methyl group along the carbon–carbon backbone. Figure 2.8 shows those PPs with different tacticity. The methyl group is highlighted in bold to indicate the difference in its arrangement among those PPs. The isotactic PP has crystalline regions while the atactic PP does not form any crystalline regions.

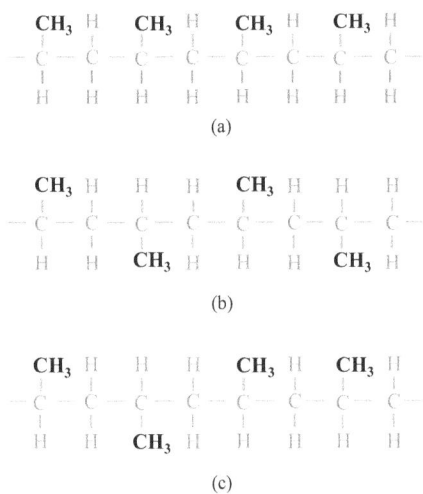

Fig. 2.8: Different types of polypropylene based on its tacticity: (a) isotactic polypropylene; (b) syndiotactic polypropylene; and (c) atactic polypropylene. Note the different arrangement of the methyl group for each polypropylene.

Because of the difference in tacticity, the arrangement of the polymer chains in those PPs is different and their mechanical properties vary. Most of the PPs used in composites are isotactic PP which has a semicrystalline structure. The isotactic PP has all the methyl groups oriented on one side of the backbone, which results in a higher DOC and enhanced strength and modulus. The syndiotactic PP is less crystalline while the atactic PP is amorphous. Unless otherwise noted, all PPs mentioned in this book are the isotactic PP. The isotactic PP is also the material for producing PP fiber because of its ability of forming a highly crystalline structure (see Section 2.3.1.3).

PP has a glass transition temperature below 0 °C and a melting temperature ranging from 150 to 170 °C. Because its glass transition temperature is below room temperature, it remains rubbery at room temperature. Therefore, it possesses great toughness and good impact resistance at room temperature. Its relatively low melting temperature and large melting range allow ease of processability, even with addition of reinforcements, such as glass and carbon fibers.

A typical DSC curve of PP is shown in Fig. 2.9. The curve shows its thermal behavior in a heat and cool cycle. The thermoplastic sample is heated to 200 °C at a rate of

20 °C/min and then cooled down to room temperature at the same rate. It can be seen that the melting of the thermoplastic starts at approximately 150 °C and ends around 170 °C. It is noted that those temperatures can vary with heating rate and molecular weight of PP. The cooling curve shows an obvious exothermal peak caused by recrystallization, namely, rearrangement of polymer chains and formation of crystallites. The recrystallization starts at 135 °C and ends at 124 °C approximately.

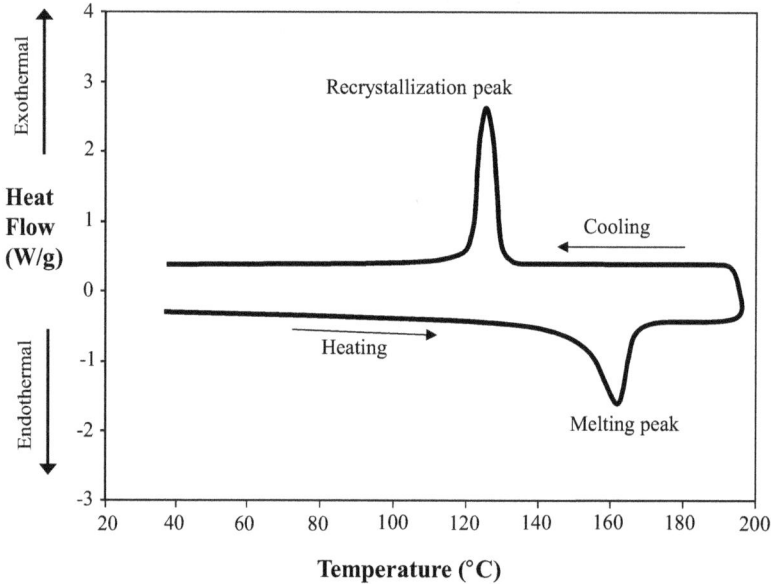

Fig. 2.9: A differential scanning calorimetry curve of polypropylene showing its melting and recrystallization peaks in a heat and cool cycle.

PP has a density of approximately 0.9 g/cm^3. Because of its low density, PP possesses relatively good specific modulus and strength, enabling it a material for lightweight applications. Specific modulus of a material is defined as the ratio of its modulus to density. In the same manner, its specific strength is the ratio of its strength to density. Both specific strength and modulus have been commonly used as a material selection criterion for applications where weight saving is a critical factor. PP is one of the most common matrix materials for thermoplastic composites for automotive applications. When fibers such as carbon fibers or glass fibers are added to PP to form a composite, the strength and modulus are significantly enhanced although the density is slightly increased because of higher densities of the fibers. As a result, the PP matrix composite possesses great specific properties along with the advantages stemmed from the PP matrix such as excellent chemical resistance, good impact resistance, and excellent processability.

PP has poor resistance to ultraviolet (UV) light. When exposed to sunlight, PP polymer chains are attacked by the UV radiation in the sunlight and undergo chain scissoring that cleaves the molecular chains and reduces the molecular weight of PP. This process is also called photodegradation. Discoloration, embrittlement, chalking (a phenomenon that powders form from degradation), and cracking can occur after the exposure, mainly on its surface as UV light can only penetrate a depth at the micron scale. The photo degradation eventually causes loss of mechanical properties especially after long-term exposure to sunlight.

UV stabilizers are normally added to PP to absorb the UV light and provide protection to PP. Carbon black is a typical UV stabilizer that offers great UV absorption ability. It is also common to add carbon black in PP matrix composites for the same reason, especially the ones for outdoor applications. A minimal amount of carbon black, such as 2% of PP mass, can provide effective protection to PP and its composites. The carbon black absorbs the UV light and converts it to heat that is dissipated through the composite. Other properties of PP including electrical and mechanical properties can also be affected. The PP possesses a higher electric conductivity with an increasing amount of carbon black as it is an electrically conductive material. The mechanical property of PP such as tensile strength and modulus increases while its elongation at break decreases with an increasing amount of added carbon black. The addition of the carbon black changes the original opaque color of PP to a black color. Long glass fiber-reinforced PP composite pellets in natural color and black color are shown in Fig. 4.32a,b, respectively. The black color of the composite is attributed to carbon black added to the composite (Fig. 4.32b).

In addition to its role as a matrix in composites, PP is also used for production of fibers which are increasingly used in composites as reinforcements. The details of how to process the PP fibers and their use as a reinforcement in self-reinforced composites will be discussed in Sections 2.3.1.3 and 3.2.5, respectively. PP has also been used to produce honeycomb cores for composite sandwich structures, which will be discussed in Section 3.4.2.

2.2.5 Polyethylene

PE is another polyolefin polymer besides PP. The formula of PE is shown in Fig. 2.10. It is another most commonly used thermoplastic because of its low cost, excellent toughness, great chemical resistance, low friction coefficient, low moisture absorption, and ease of processing. Over 100 million tons of PE is produced annually. Most of the PE is used for grocery bag, packaging film, and container bottle, while some of the PE is used as a matrix in PE matrix composites or a reinforcement (PE fiber) in thermoplastic composites.

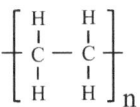

Fig. 2.10: The chemical formula of polyethylene, $(C_2H_4)_n$.

Based on the molecular weight, there are five types of PE, namely LDPE, linear low-density polyethylene (LLDPE), medium-density polyethylene (MDPE), HDPE, and UHMWPE. LDPE and HDPE are commonly used as the matrix in thermoplastic composites. UHMWPE, however, can be used as either matrix (in a bulk form) or reinforcement (in a fiber form) in a composite. UHMWPE fiber will be further discussed in Section 2.3.1.3. Table 2.3 summarizes the density, molecular weight, and melting temperature of each PE. Figure 2.11 compares typical DSC curves of LDPE, HDPE, and UHMWPE. UHMWPE shows the highest melting temperature while LDPE has the lowest melting temperature.

Tab. 2.3: Density, molecular weight, and melting temperature of different types of polyethylene (adapted from Reference [2]).

PE type	Density (g/cm³)	Molecular weight (g/mol)	Melting temperature (°C)
LDPE	0.91–0.93	<50,000	105–115
MDPE	0.92–0.94	200,000	120–130
LLDPE	0.91–0.95	50,000–200,000	120–130
HDPE	0.94–0.97	50,000–250,000	125–137
UHMWPE	0.92–0.97	3 million–7.5 million	132–140

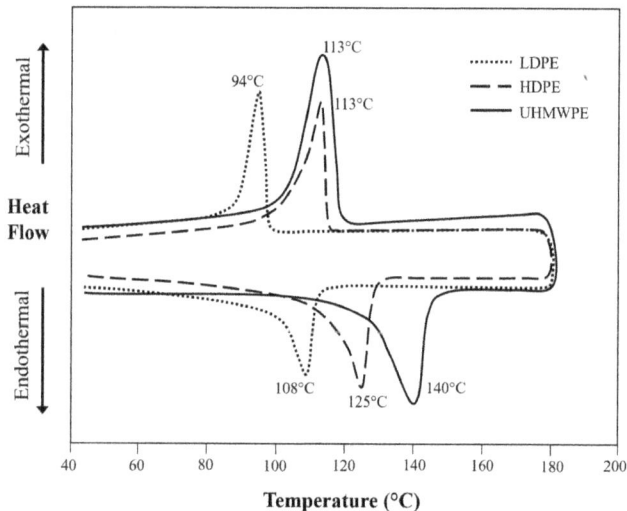

Fig. 2.11: DSC curves of different types of polyethylene showing their melting peaks (endothermal peaks) and recrystallization peaks (exothermal peaks) (adapted from Reference [3]).

1. **Low-density polyethylene (LDPE)** is a highly branched polymer and, thus, has a less compact structure. Figure 2.12a shows a schematic of branched molecular

chains of LDPE. It has the lowest density (around 0.92 g/cm^3) and molecular weight (less than 50,000 g/mol) among all types of PE. It is mainly used in flexible packaging and electrical insulation applications but also as the matrix in composites, for example, cellulose fiber-reinforced LDPE composites. LDPE has a low melting temperature around 110 °C that favors processing of cellulose fiber-reinforced LDPE composites by minimizing the thermal degradation of cellulose fibers.

2. **Medium-density polyethylene (MDPE)** has a density range of 0.92–0.94 g/cm^3 and a molecular weight of around 200,000 g/mol. It has a melting temperature of about 125 °C, slightly higher than LDPE. It is mainly used in pipes, bags, packaging film, etc. It has been also used as a matrix in glass fiber or nanomaterial-reinforced composites.

3. **Linear low-density polyethylene (LLDPE)** is another polyethylene. Its branched molecular chains are shorter than the branched chains in LDPE. The shorter branched molecular chains induce a more compact arrangement of polymer chains, therefore, its density (up to 0.95 g/cm^3) is higher than the density of LDPE. LLDPE also has higher toughness and larger elongation at break than LDPE because of the ease of its molecular chain sliding by one another.

4. **High-density polyethylene (HDPE)** has long linear molecular chains with minimal branching. Figure 2.12 shows the difference of the polymer chain structure between HDPE and other PEs such as LDPE and LLDPE. The linear chain structure in HDPE allows the molecular chains to arrange tightly and form a more compact structure, both of which result in a higher density of HDPE (up to 0.97 g/cm^3). The long linear chain structure with minimal branching also promotes the formation of crystalline regions. Therefore, HDPE has a higher DOC and higher density than both LDPE and LLDPE. The crystallinity of bulk HDPE can reach up to 80%. Because of its more compact polymer chains, higher crystallinity and molecular weight, HDPE has better strength, modulus, and chemical resistance than LDPE and MDPE. HDPE is used in a broad range of applications such as bottles, grocery bags, containers, pipes, and toys. HDPE is also the most common PE used in composites. Fibers, including glass fibers, carbon fibers, and cellulose fibers, are combined with HDPE to form high-performance engineered thermoplastic composites that find their use in various applications including automotive and construction.

5. **Ultra-high molecular weight polyethylene (UHMWPE)** is a PE with an extremely high molecular weight that ranges from 3 million to 7.5 million g/mol. It has excellent mechanical properties, great chemical resistance, low friction coefficient, and exceptional abrasion resistance. Different reinforcements, including carbon fibers and alumina particles, have been added to UHMWPE to further enhance its properties. Besides enhanced strength and modulus from the added reinforcement, the UHMWPE matrix composite also offers characteristics such as biocompatibility, chemical stability, and wear resistance that are derived from the UHMWPE matrix.

Furthermore, UHMWPE can be produced into a fiber form that is increasingly used in various applications, including high-strength ropes and lightweight high-performance reinforcements for composites. Detailed description of processing of UHMWPE fibers is provided in Section 2.3.1.3.

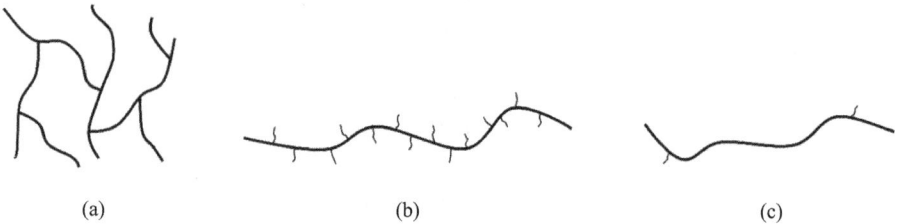

(a) (b) (c)

Fig. 2.12: The molecular chain structure for various types of polyethylene: (a) LDPE with long branched polymer chains; (b) LLDPE with a large number of short branched chains; and (c) HDPE with linear polymer chains with minimal branching.

2.2.6 Nylon

Nylon is a group of engineering thermoplastic polymers widely used as a matrix in composites. It has a combination of good mechanical properties and reasonable material cost. Nylon has a signature amide group (–NH–CO–) and is often called polyamide (PA) for that reason. A variety of other chemical groups can be present in the nylon and the number of those groups can vary, both of which make the nylon a thermoplastic with a wide variety of types.

Firstly, there are homopolymer nylons and co-polymer nylons. The homopolymer nylon can be further classified into monadic nylon and dyadic nylon. The monadic nylon is derived from amine and carboxylic acid monomers. The monadic nylon has a general formula as shown in Fig. 2.13. Its nomenclature is specified as "nylon $(x+1)$" or "PA $(x+1)$" because there are totally $(x+1)$ carbon atoms in its repeat unit. Typical monadic nylons used in composites include nylon 6, nylon 11, and nylon 12, among which nylon 6 is the most common type used in composites.

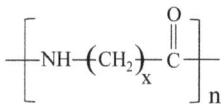

Fig. 2.13: The general formula of monadic nylons, including nylon 6, nylon 11, and nylon 12.

The dyadic nylon is derived from diamine and dicarboxylic acid monomers. It has a general formula shown in Fig. 2.14. Its nomenclature is "nylon x, $(y+2)$" or "PA x, $(y+2)$." The x and $(y+2)$ values correspond to the numbers of carbon atoms in -NH-(CH$_2$)$_x$-NH- and -CO-(CH$_2$)$_y$-CO-, respectively. When $x = 6$ and $y = 4$, the nylon type is nylon 6,6, another commonly used nylon in composites besides nylon 6. Other dyadic nylons include (nylon 5,10), (nylon 6,10), (nylon 6,11), (nylon 6,12), and (nylon

10,12). The comma between the numbers is normally omitted, and (nylon 6,6), (nylon 5,10), (nylon 6,10), (nylon 6,12), (nylon 6,11), (nylon 10,12) are often written as nylon 66, nylon 510, nylon 610, nylon 612, nylon 611, and nylon 1012, respectively.

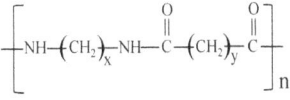

Fig. 2.14: The general chemical formula for dyadic nylons, including nylon 66 (when $x = 6$ and $y = 4$).

Nylon can also have aromatic groups instead of -CH$_2$- groups. Based on the presence of aromatic ring in its polymer chains, nylon can be categorized into aliphatic nylon, aromatic nylon, and semiaromatic nylon. The nylons mentioned above, such as, nylon 6 and nylon 66, belong to the aliphatic nylon family as they do not have any aromatic rings in their structures. The general chemical formula for the aromatic nylon or semi-aromatic nylon is shown in Fig. 2.15. The groups, R and R′, in the backbone determine the type and the characteristic of the aromatic nylon or semicrystalline nylon. R and R′ can be the same or different.

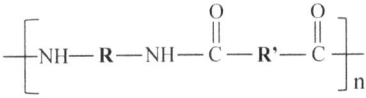

Fig. 2.15: The general chemical formula for aromatic or semiaromatic nylons.

When there are aromatic rings in R or R′, the nylon becomes an aliphatic aromatic nylon, or a semiaromatic nylon. These nylons include polyphthalamide (PPA) and polyarylamide (PAA or PARA). PPA is a nylon synthesized from condensation of an aliphatic diamine such as hexametylene diamine with terephthalic acid and/or iso-phthalic acid. PAA, often called as PA-MXD6 or nylon-MXD6, is synthesized from m-xylylenediamine and adipic acid. Both the semiaromatic nylons have been used as matrices in composites.

As mentioned in Section 2.2, the types of atoms/groups in the polymer backbone significantly affect the properties of the polymer. The aliphatic nylon has -CH$_2$- groups in R and R′ and the fully aromatic nylon has aromatic rings in both R and R′. The semiaromatic nylon has -CH$_2$- group or aromatic ring in R and R′. Since the aromatic ring provides enhanced rigidity, the semiaromatic nylon has a high modulus and high melting temperature than the aliphatic nylon. The fully aromatic nylon has better properties than the semiaromatic nylon.

When at least 85% of amide groups in a nylon are directly connected to aromatic rings, the nylon is classified as aramid. The aramid has great mechanical properties because of the high percentage of aromatic rings. It is used for producing high-performance fibers. For example, Kevlar® and Twaron® are two brands of aramid fibers that are fully aromatic nylons polymerized with the addition of terephthalic acid.

Nomex® is another aramid fiber but polymerized with the addition of isophthalic acid. Nomex® has been used as fire-resistant protective clothing in the space suit for the astronauts who went to International Space Station through Falcon 9 and Dragon spacecraft in 2020. These aramid fibers are also commonly used as reinforcements in thermoplastic composites for various applications, especially impact-resistant applications. More details of the aramid fiber will be provided in Section 2.3.1.3.

Table 2.4 summarizes the properties of different types of nylons that are used as the matrices in composites. The nylon with aromatic rings generally possesses greater modulus and strength and higher melting temperature (Fig. 2.16). The moisture absorption is relatively high for nylons, especially the ones with more carbonyl oxygen sites. The moisture absorption values listed in the table are obtained from a saturated condition (immersion in water) at room temperature. The moisture content can reach more than 10% of the nylon mass. The moisture in nylons or nylon matrix composites can affect their processing and mechanical properties. Steam generated from the moisture during heating in extruders increases the internal pressure in the extruder barrel unnecessarily. More seriously, the moisture can result in voids and polymer chains with lower molecular weight. Nylons and nylon matrix composites are usually required to be pre-dried to remove the absorbed moisture before processing.

Tab. 2.4: The properties of common nylons used as matrices in composites.

Nylon type	Density (g/cm³)	Melting temp (°C)	Glass transition temp (°C)	Young's modulus (GPa)	Tensile strength (MPa)	Elongation to failure (%)	Moisture absorption (%)
Nylon 6	1.13–1.15	210–220	44–56	0.9–2.6	33–62	41–59	8.2–11.5
Nylon 66	1.13–1.15	255–265	54–66	1.3–3.1	58–77	13–29	7–9.3
Nylon 12	1.0–1.02	170–178	40–43	1.1–1.4	45–55	41–59	1.2–1.8
Nylon 612	1.05–1.07	195–219	40–52	1.4–1.8	60–72	35–50	2.4–3.1
Nylon-MXD6	1.12–1.22	229–245	75–92	4.4–4.7	83–99	2–2.3	5.8
PPA	1.13–1.15	299–322	119–135	1.6–2.8	57–80	24–35	8.3–11

As seen from Tab. 2.4, nylons have different thermal, physical, and mechanical properties. These properties significantly influence the properties of nylon matrix composites. The sections below detail some common nylons used as the matrix in composites.

1. **Nylon 6** is one of the most widely used thermoplastics. It was developed by Paul Schlack in 1938. It is also called polycaprolactam because it is synthesized by a ring-opening polymerization process from its monomer, ε-caprolactam. The ring-opening polymerization reaction is shown in Fig. 2.17. The number "6" in nylon 6 is

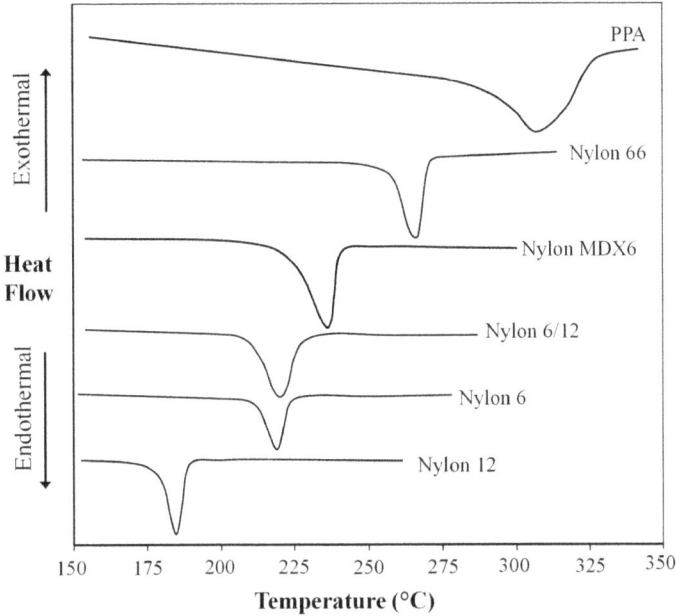

Fig. 2.16: DSC curves of common nylons used in composites.

used to describe the total number of carbon atoms (six carbon atoms totally) in its monomer, {NH–(CH$_2$)$_5$–CO}. Nylon 6 offers well balanced properties in modulus, strength, and chemical resistance. It is one of the most commonly used engineering thermoplastics in composites.

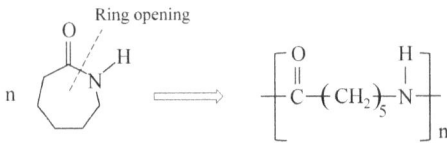

Fig. 2.17: The ring-opening polymerization process for producing nylon 6 from ε-caprolactam.

Nylon 6 is prone to absorbing moisture because of the amide groups in its repeat unit. The pendant carbonyl oxygen is negatively charged and attracts the positively charged hydrogen in water molecules. Most of the moisture absorption takes place at the carbonyl oxygen sites. In addition, the pendent hydrogen on the nitrogen atom can also draw moisture by attracting the negatively charged hydroxyl group in water molecules. The hydrogen attached to carbon atoms, however, is nonpolar and does not attract any water molecules. Figure 2.18 shows the hydrogen bonds at both the carbonyl oxygen and the pendent hydrogen on the nitrogen with water molecules.

Nylon 6 has more amide groups per unit of molecular length than the other nylons and therefore absorbs the most moisture. It can absorb moisture up to 11.5% of its mass in a saturated condition (see Tab. 2.4).

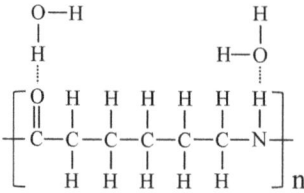

Fig. 2.18: Hydrogen bonds with water molecules at the carbonyl oxygen and the pendent hydrogen on the nitrogen in nylon 6, both of which contribute to its high moisture absorption.

A thermogravimetric analysis (TGA) curve of nylon 6 in Fig. 2.19 shows its mass change when heated in nitrogen. The mass loss at the initial heating stage (<100 °C) is caused by removal of the moisture absorbed in the nylon. The moisture absorbed by nylon 6 can adversely affect the processing of its composites. It is necessary to dry the material before processing. In addition, the absorbed moisture reduces the mechanical properties, including modulus and strength, of nylon 6 and its composites. The matrix-dominated properties in the composite, such as interlaminar shear strength and transverse tensile strength, drop with increasing moisture absorption.

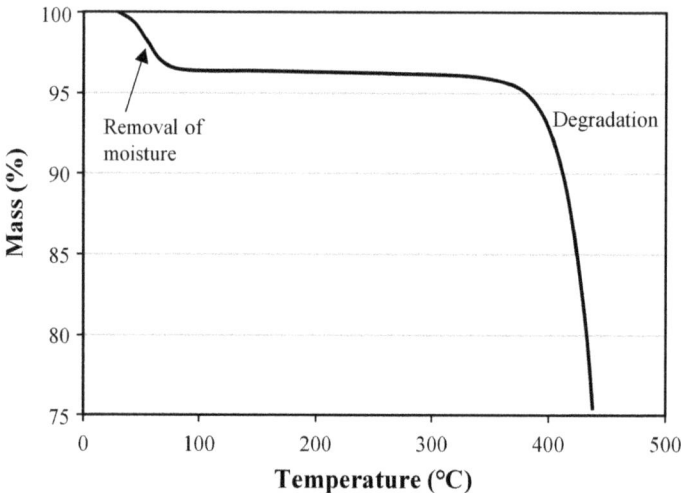

Fig. 2.19: TGA curve of nylon 6 in a nitrogen atmosphere. Note the mass loss at the initial heating stage because of removal of the moisture absorbed in the nylon.

2. **Nylon 66**, also known as nylon 6,6, was discovered by Wallace Carothers in 1935. It is the most used nylon system in composites because of its heat resistance and excellent mechanical properties. It has a higher glass transition temperature and melting

temperature than nylon 6. Its glass transition temperature reaches up to 66 °C and its melting temperature 265 °C.

Nylon 66 can be produced via a number of methods, including polycondensation of hexamethylene diamine and adipoyl chloride monomers, as well as hexamethylene diamine and adipic acid monomers. The latter method is more common in mass production of nylon 66 and the chemical reaction is shown in Fig. 2.20. Because the polymerized nylon has 6 carbon atoms in $-OC-(CH_2)_4-CO-$ group and 6 atoms in $-(CH_2)_6-NH-$, it is called nylon 66. Sometimes a slash sign "/" is placed between the numbers, namely, nylon 6/6. Nylon 66 matrix composites possess great strength and modulus and have been used in a wide range of applications across multiple market segments as a replacement for metals and thermoset composite materials.

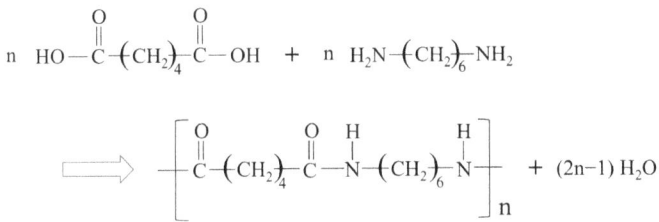

$$n \ \ HO-\overset{O}{\overset{\|}{C}}-(CH_2)_4-\overset{O}{\overset{\|}{C}}-OH \ + \ n \ H_2N-(CH_2)_6-NH_2$$

$$\Longrightarrow \ \left[-\overset{O}{\overset{\|}{C}}-(CH_2)_4-\overset{O}{\overset{\|}{C}}-\overset{H}{\overset{\|}{N}}-(CH_2)_6-\overset{H}{\overset{\|}{N}}-\right]_n \ + \ (2n-1) \ H_2O$$

Fig. 2.20: The condensation polymerization reaction for producing nylon 66 from adipic acid and hexamethylene diamine.

3. **Nylon 12** is another nylon used in thermoplastic composites. It has low density, low melting temperature, and low moisture absorption compared to other nylons. Nylon 12 has a large number of carbon atoms (12 atoms) separating the amine groups (Fig. 2.21). Therefore, it has a low percentage of the amide group number over the total group number in the repeated units of the polymer, or low amide group concentration. In fact, nylon 12 is a nylon with the lowest amide group concentration among commercial nylons. Its low amide group concentration results in a low moisture absorption. It absorbs 1.2–1.8% moisture in a saturated condition, much less than other nylons (Tab. 2.4).

Nylon 12 can be synthesized by condensation polymerization or ring-opening polymerization. The condensation polymerization involves condensing 12-aminolauric acid. The ring-opening polymerization is similar to the production of nylon 6 by opening the ring of laurolactam (Fig. 2.21). The ring-opening process is the preferred method for mass production of nylon 12.

$$n \ \ \langle \text{(ring)} \overset{O}{\underset{NH}{\overset{\|}{C}}} \ \text{-{-}-Ring opening} \ \Longrightarrow \ \left[-\overset{O}{\overset{\|}{C}}-(CH_2)_{11}-\overset{H}{\overset{\|}{N}}-\right]_n$$

Fig. 2.21: The ring-opening polymerization process for synthesizing nylon 12.

Nylon 12 has relatively low strength compared to other nylons. However, its strength can be significantly enhanced by adding fibers such as glass fibers and carbon fibers. Nylon 12 matrix composites have good processability because of the low melting temperature of nylon 12. Its low melting temperature facilitates the processing of composites with fibers that have a low degradation temperature, such as cellulose fibers (see Section 2.3.1.5). In addition, nylon 12 is one common material for selective laser sintering process because of its large processing window (see Section 5.4).

4. **Nylon 612** is a nylon that possesses mechanical properties between those of nylon 6 and nylon 12. It is produced through a polycondensation process from 1,6-hexamethylene diamine and 1,12-dodecanedioic acid (1,10-decane dicarboxylic acid). It has a chemical structure of $(NH-(CH_2)_6-NH-CO-(CH_2)_{10}-CO)_n$. Nylon 612 has a low moisture absorption because of its low amide group concentration. Its moisture absorption ranges from 2.4% to 3.1% in a saturation condition, which is slightly higher than that of nylon 12. However, it absorbs much less moisture than nylon 6. Nylon 612 matrix composites have been used in gears, cams, and structural and electrical components.

5. **Nylon-MXD6** is also called PA-MXD6 or polyarylamide (PAA or PARA). It is a nylon produced from m-xylylenediamine and adipic acid through polycondensation reaction. It is a semiaromatic nylon with an aromatic ring in its repeat unit as shown in Fig. 2.22. The aromatic ring in nylon-MXD6 contributes to its great strength and modulus, and high glass transition temperature. It is sometimes referred to as high-temperature nylon because of its high glass transition temperature and relatively high melting temperature. It also has low moisture absorption compared to PA6 and PA66 (Tab. 2.4).

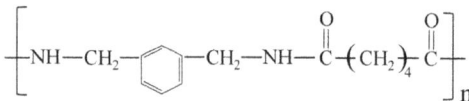

Fig. 2.22: The chemical formula of nylon-MXD6.

6. **Polyphthalamide (PPA)** is another semiaromatic nylon that has good mechanical properties and a high melting temperature besides nylon-MXD6. PPA offers great strength and modulus at elevated temperatures, excellent chemical resistance, good dimensional stability, and enhanced creep and fatigue resistance compared to the aliphatic nylons. Figure 2.23 shows the chemical formula of PPA. Its composites are used in various applications including automotive and electrical.

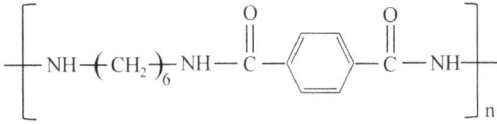

Fig. 2.23: The chemical formula of polyphthalamide (PPA), a semiaromatic nylon with great mechanical properties and good thermal stability.

2.2.7 Polyethylene terephthalate

PET or PETE is a thermoplastic polymer that is a widely used material for beverage bottles and fibers. It is synthesized by reacting terephthalic acid with ethylene glycol. The chemical formula of PET in Fig. 2.24 shows that PET has an aromatic ring in its backbone which provides stiffness to the thermoplastic. As a result, PET has relatively high elastic modulus and melting temperature (Tab. 2.2).

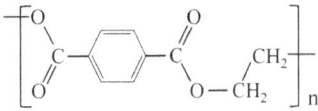

Fig. 2.24: The chemical formula of polyethylene terephthalate, consisting of aromatic ring in its backbone.

PET can exist as an amorphous or semicrystalline thermoplastic. For example, PET in disposable water bottles is amorphous, while PET in composites is generally semicrystalline. Cooling rate during the processing of PET usually determines if it is amorphous or semicrystalline. The blow molding process used in manufacturing the thin PET disposable water bottle induces a high cooling rate. PET-based composites are much thicker than the neat PET bottle thickness, the cooling rate during molding of the PET composite is significantly lower than the cooling rate in the blow molding process. Furthermore, the reinforcement such as carbon fibers or glass fibers in the composite promotes nucleation of PET crystallites during cooling, resulting in semicrystalline PET in composites. PET is categorized as a semicrystalline thermoplastic in the thermoplastic tree (Fig. 2.3) because it exists as a semicrystalline polymer in thermoplastic composites.

PET with different degrees of crystallinity, including amorphous PET that has 0% DOC, can be differentiated through DSC. The DSC curves (from bottom to top) in Fig. 2.25 demonstrate the thermal behavior of amorphous PET and semicrystalline PET polymers with increasing DOC. The curves are offset to better display the melting peaks, cold crystallization peaks, and glass transition temperatures. The amorphous PET shows cold crystallization that happens during heating the thermoplastic whose recrystallization was previously suppressed by fast cooling. The cold crystallization is a recrystallization process that happens because of an increase in the mobility of the molecular chains during heating and rearrangement of some molecular chains. When the enthalpy of cold crystallization is equal to the enthalpy of melting, it is indicated that the thermoplastic is amorphous. The cold crystallization becomes less obvious with the increase of DOC because of less available polymer chains for recrystallization during heating. The calculation of

DOC must consider the enthalpy change from the cold crystallization when the DSC method is used (Eq. (6.19)).

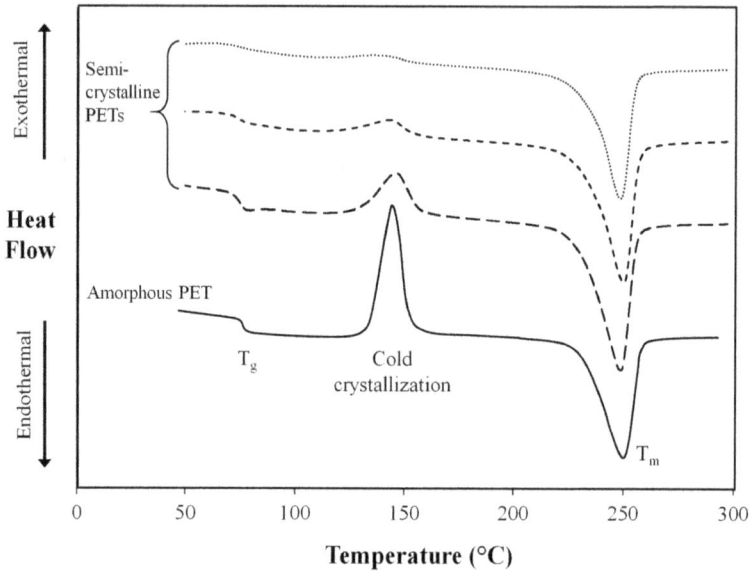

Fig. 2.25: DSC curves of amorphous PET and semicrystalline PET with different degrees of crystallinity (partially reprinted from Reference [4] with permission).

PET has also been made into fibers, commonly known as polyester fibers. The PET fiber has an elastic modulus of 11–14 GPa and tensile strength up to 1,250 MPa (Tab. 3.3). Those fibers are mainly used in textile and carpet applications. In addition, PET fibers have been used as a reinforcement in concrete to improve its fracture toughness and impact resistance. The fiber is also used as a reinforcement in thermoplastic composites, such as PET-, PP-, or PE-based composites. The composite consisting of PET fibers and PET matrix is one type of self-reinforced composite (see Section 3.2.5).

2.2.8 Polyphenylene sulfide

Polyphenylene sulfide (sometimes it is written as polyphenylene sulphide), or PPS, is one of the most widely used advanced engineering thermoplastics. This semicrystalline thermoplastic features aromatic rings linked by sulfides as shown in Fig. 2.26. The aromatic

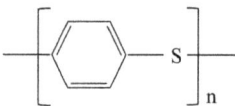

Fig. 2.26: The chemical formula of polyphenylene sulfide, consisting of aromatic rings linked by sulfides.

ring provides rigidity to the polymer chain, and therefore, PPS has a high elastic modulus (Tab. 2.2). In addition, PPS possesses high strength, high temperature resistance, excellent chemical resistance, good dimensional stability, low moisture absorption, and good flame retardancy. It is one of the popular advanced engineering thermoplastics used in composites as it offers a great performance to cost ratio.

PPS is a semicrystalline thermoplastic that has a glass transition temperature of 81–97 °C and a melting temperature ranging from 280 to 295 °C. The DSC curve of PPS in Fig. 2.27 shows its glass transition temperature and melting temperature. PPS recrystallizes during cooling and its recrystallization temperature, T_{rex}, is around 240 °C.

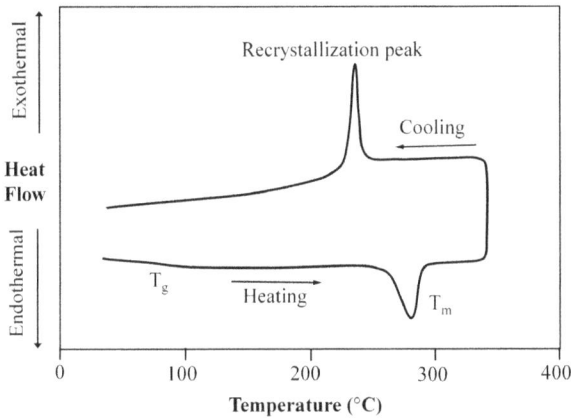

Fig. 2.27: A differential scanning calorimetry curve of polyphenylene sulfide showing its melting temperature at approximately 285 °C and recrystallization temperature at approximately 240 °C.

PPS has excellent chemical resistance. It does not dissolve in any common solvents below 200 °C, including concentrated acids and bases. PPS is also inherently flame retardant. The flammability rating of PPS is V-0/5 V according to UL 94 standard burn test (Standard for Safety of Flammability of Plastic Materials for Parts in Devices and Appliances testing). The burning of PPS can stop within several seconds on a vertical burning test setup after flame source is removed. The addition of reinforcements such as glass fiber can further improve its flame retardancy. The excellent chemical resistance, flame retardant characteristic, good thermal stability, and excellent mechanical properties of PPS matrix composites allow their use in harsh environments.

Fiber-reinforced PPS composites have been widely used in automotive applications. Typical automotive products made of PPS matrix composites include throttle body, inlet tank, water pump, crankshaft flange, fuel injection rail, diesel engine charge air duct, transmission, engine components, and coolant system. Aircraft industry is another sector that uses PPS composites extensively. Typical components include Fokker 50 undercarriage door, fixed wing leading edge for Airbus A340 and A380 (Fig. 1.3c).

2.2.9 Polyether ether ketone

PEEK belongs to the polyketone family. It is a semicrystalline thermoplastic that has superb mechanical properties, including strength, modulus, and impact resistance. It is composed of two ether groups $(R{-}O{-}R')$ and one ketone group $(R{-}\overset{O}{\underset{\|}{C}}{-}R')$ as shown in Fig. 2.28. The aromatic rings in the backbone of PEEK contribute to its high temperature stability and great mechanical properties (Tab. 2.2). Its glass transition temperature is 143 °C, however, it can be continuously used at temperatures up to 170 °C and the short-term operating temperature of PEEK can reach up to 260 °C because of the crystalline regions in this semicrystalline polymer. In spite of its excellent performance, the high price of PEEK has limited its use to certain high-end applications, such as aerospace, aircraft, dental, and other niche applications.

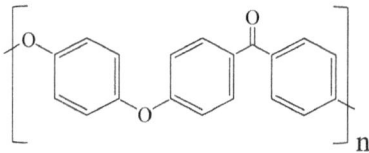

Fig. 2.28: The chemical formula of polyether ether ketone, consisting of ether groups, ketone groups, and aromatic rings.

Fibers are often added to PEEK to further improve its performance. The addition of the fiber can not only significantly enhance its performance, but also lower its unit price. The fiber, including carbon fibers, normally has a lower price than PEEK and resultant PEEK matrix composites have less unit price than neat PEEK. The PEEK matrix composite possesses great mechanical properties, chemical resistance, thermal

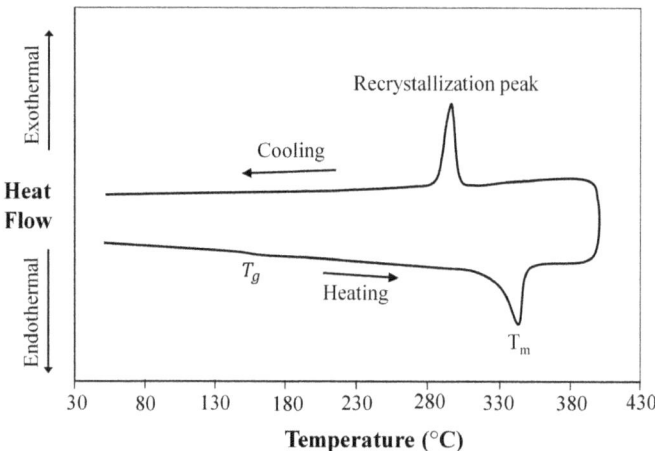

Fig. 2.29: The differential scanning calorimetry curve of polyether ether ketone showing its glass transition temperature, melting peak, and recrystallization peak.

stability, and wear resistance and have found its use in a wide variety of applications, including automotive, marine, nuclear, oil and gas, electronics, medical, and aerospace.

2.2.10 Polyetherimide

Polyimide (PI) is a group of high-performance polymers with an imide structure composed of two acyl groups (C=O) bonded to nitrogen (-N). Since its initial development in the 1950s, it has attracted tremendous interests for various applications owing to its excellent mechanical properties, thermal stability, flame retardancy, chemical resistance, and dielectric properties in addition to its low coefficient of thermal expansion.

Similar to nylons, PIs can also be classified into aliphatic (linear PI), semiaromatic, and aromatic PI based on the presence of aromatic rings in its repeat unit. A linear PI has the chemical structure shown in Fig. 2.30a. Figure 2.30b shows the general structure for aromatic PIs. The aromatic PI has great mechanical and thermal stability because of the presence of aromatic rings and it is commonly used as a matrix material for composites. Because of the polar characteristic of its imide group, PI is susceptible to polar solvents, which makes it suitable for solvent impregnation processes (see Section 4.3.5) besides melt-based processes.

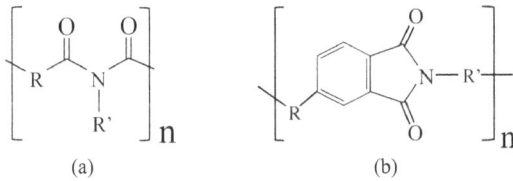

Fig. 2.30: The chemical formulas for (a) linear polyimides and (b) aromatic polyimides.

PIs can be thermoset or thermoplastic. Thermoset PI has its imide groups cross-linked into a three-dimensional network. Thermoplastic polyimide, or TPI, does not have cross-linking in its structure and exhibits thermoplastic characteristics. Commercial products of TPI include ULTEM ®, AURUM®, and EXTEM®. Those products are usually classified as polyetherimide (PEI) because of the presence of ether groups in those products. Table 2.5 lists general properties of those PEIs. PEI can be processed through injection molding or compression molding, which allows it to be a matrix material for composites. Additionally, as a derivative of PIs, PEI has great strength and excellent resistance to heat, chemical, and flame. PEI matrix composites are one of the most heat-resistant polymer matrix composites. However, the cost of PEI is relatively high and the PEI matrix composite has been mainly used for niche applications such as aerospace.

Tab. 2.5: The properties of common polyetherimides used in composites (reprinted from manufacturer's datasheets).

PEI products	Density (g/cm³)	T_g (°C)	T_m (°C)	Max continuous service temp (°C)	Elastic modulus (GPa)	Tensile strength (MPa)
ULTEM®	1.27–1.31	217–247	–	170	3.0–3.5	95–114
AURUM®	1.33	250	388	163	2.8	94
EXTEM®	1.31–1.37	260–311	–	230	3.3–3.8	90–120

One of the common PEIs used in composites is ULTEM®, an amorphous thermoplastic developed by General Electric Plastics Division. Figure 2.31 shows the chemical formula of ULTEM® 1000 series, a general-purpose PEI. It has two ether groups, two imide groups, and several aromatic rings in its repeat unit. ULTEM® is a common PEI used in composites, especially glass fiber-reinforced composites because of its great adhesion to glass fibers. The glass transition temperature of ULTEM® is around 217 °C, indicating its excellent thermal stability and great thermal resistance of PEI matrix composites when used as a matrix material.

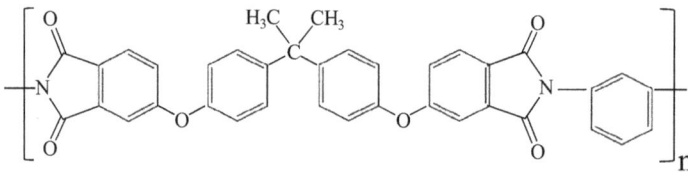

Fig. 2.31: The chemical formula of ULTEM® 1000 series, a general-purpose polyetherimide grade.

AURUM®, another PEI that has been used as the matrix material in composites, is developed by Mitsui Toatsu Chemical in Japan. Figure 2.32 shows its chemical formula, a similar structure to the ULTEM®. AURUM® has a glass transition temperature of 250 °C and a melting temperature of 388 °C. Because of its high melting temperature, the processing temperature for AURUM® is extremely high (at least 400 °C). AURUM® is normally amorphous after molding. However, a semicrystalline structure can be achieved via annealing above its glass transition temperature after the molding. The continuous service temperature of AURUM® can be increased from its amorphous state to semicrystalline state resulted from the annealing process.

EXTEM®, a PEI developed by SABIC, is one of the most heat-resistant polymers. One of the EXTEM® grades has the chemical formula shown in Fig. 2.33. It has a similar chemical structure as ULTEM® and AURUM® but with an addition of a sulfone group. It is sometimes called PEI sulfone. EXTEM® has a glass transition temperature up to 311 °C, indicating its excellent heat resistance.

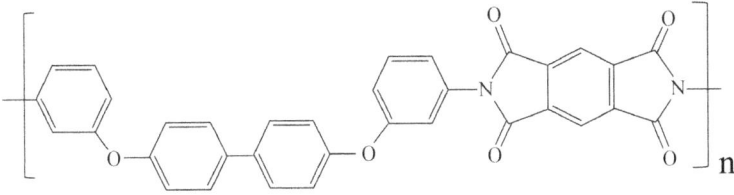

Fig. 2.32: The chemical structure of AURUM®, a polyetherimide with a glass transition temperature of 250 °C and melting temperature of 388 °C.

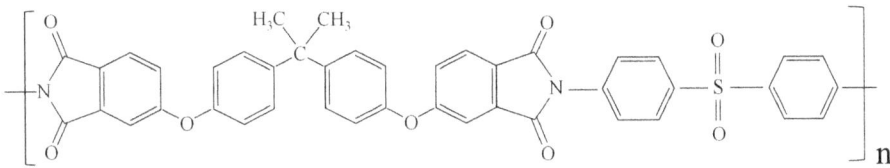

Fig. 2.33: The chemical formula of EXTEM®, a polyetherimide with an extremely high glass transition temperature.

Glass fiber and carbon fiber have been added to PEI to produce high-performance PEI matrix composites. Because of its high temperature resistance, PEI matrix composites exhibit great strength and modulus with the capability of continuous use at temperatures that are above the limit of other polymer matrix composites. A combination of superb mechanical properties, lightweight, and great thermal stability makes the PEI matrix composite a high-performance structural material for aerospace and other high-end applications.

2.2.11 Bio-based thermoplastics

Bio-based thermoplastics are polymers produced from biological resources, including starch, sugar, cellulose, and chitosan. These thermoplastics are developed mainly for their renewability, sustainability, and great potential for replacing their petroleum-based counterparts. The development of the bio-based thermoplastics using renewable and sustainable resources can tremendously benefit the environment by reducing the consumption of polymers derived from petroleum or natural gas. In addition to its renewability and sustainability, the bio-based thermoplastic has the following characteristics:
-   Recyclability
-   Low carbon footprint
-   Biodegradability of certain bio-based thermoplastics
-   Relatively high cost

- Poor durability
- Relatively low performance

Common bio-based thermoplastics include thermoplastic starch, polylactic acid (PLA), bio-based PE, bio-based PP, bio-based nylon, bio-based PET, bio-based PC, polyhydroxyalkanoate (PHA), bio-based polybutylene succinate (PBS), and bio-based polycaprolactone (PCL). These thermoplastics have found their use in applications such as packaging, automotive, agriculture, horticulture, construction, textile, electrical and electronics. A number of the thermoplastics have been used as the matrix in composites, typically reinforced by cellulose-based fillers or fibers, another renewable and sustainable material. In spite of its environmental benefits, the bio-based thermoplastic had only less than 1% of the overall global plastic market by 2015 mainly due to its relatively high price compared to their petroleum-based counterparts [5]. The bio-based thermoplastic is usually blended with petroleum-based thermoplastics to introduce a "green" component while still maintaining a reasonably low raw material cost. Nevertheless, the bio-based thermoplastic is having a steady increase in production and the plastic market share because of the development of more efficient production methods and resulted reduction in processing cost. Enhanced environmental awareness among the public and stricter government policies will also drive the market of bio-based thermoplastics and their composites.

Several common bio-based thermoplastics are described below:

1. **Thermoplastic starch** is the most used bio-based thermoplastics because of its abundance and biodegradability, occupying about half of the bio-based plastic market. Thermoplastic starch can be obtained from plant seeds and roots, including corn, rice, wheat, maize, and potato, that contain natural starch. Since starch is one of the most abundant biomass materials on earth, the cost of thermoplastic starch is relatively low. Thermoplastic starch can be produced by adding plasticizer such as water and glycerol to the natural starch. Thermoplastic starch products can be consequently manufactured by common processing methods for regular thermoplastics, such as injection molding, compression molding, film casting, and extrusion.

   The mechanical property of the thermoplastic starch is relatively low. Its tensile strength ranges from 0.4 to 38.0 MPa. Adding fillers or fibers can significantly increase its strength. For example, adding 5 wt% cellulose fiber to a thermoplastic starch can double its tensile strength and elastic modulus [6]. Also nanomaterials have been integrated into the thermoplastic starch to enhance its performance.

2. **Polylactic acid** (PLA) is the second most used bio-based thermoplastic. Microbial fermentation of feedstock materials, such as starch crops (corn, wheat, and so on) or sugar-containing agricultural products (sugarcane, beet, and so on), is the process used to produce the bio-based PLA. Lactic acid is generated from the fermentation process and then synthesized to PLA via ring-opening polymerization or direct polycondensation. PLA provides advantages such as biodegradability, good

strength, and biocompatibility. PLA has a tensile strength ranging from 14 to 70 MPa and an elongation at break of 1–8%. It exhibits low ductility but relatively good strength. PLA can undergo hydrolytic degradation with the presence of moisture. During its degradation, lactic acid, the monomer of PLA, is generated. Higher content of moisture and elevated temperature can lead to faster degradation. Amorphous regions in PLA are more susceptible to degradation than crystalline regions. Microorganisms in the environment can also degrade PLA through cleavage of its ester bonds and converting it into carbon dioxide and water.

3. **Bio-based polyethylene** is also produced from carbohydrate materials, such as sugarcane and corn. The carbohydrate material is microbially fermented and bio-based ethanol is produced. The ethanol is subsequently converted to ethylene that is finally synthesized to PE. The bio-based PE has similar physical, chemical, and mechanical properties to its petroleum-based counterparts, allowing them to be an alternative thermoplastic matrix to unrenewable petroleum-based PE. The bio-based PE can also be made into high-performance fibers. For example, UHMWPE fibers made of bio-based PE have been developed by DSM. Those fibers have great specific strength, cut and abrasion resistance, and high resistance to chemicals and UV light.

Although the biological resource of the bio-based thermoplastic is biodegradable, not all of the bio-based thermoplastics are biodegradable. Biodegradability is the capability of a material being broken down chemically by microorganisms to base materials such as carbon dioxide and water. The biodegradable bio-based thermoplastics are thermoplastic starch, PLA, PHA, and so on. Other bio-based thermoplastics, including bio-based PE, bio-based PP, and bio-based PET, are not biodegradable.

Bio-based thermoplastic composites have been continuously developed for their environmental friendliness and enhanced mechanical properties (compared to the neat bio-based thermoplastic). Cellulose-based fillers or fibers are normally the reinforcement used in the bio-based thermoplastic composite. Since the cellulose-based fillers or fibers are renewable and sustainable, the composite is 100% bio-based, renewable, and sustainable. The bio-based composite has a great potential to be used for applications such as automotive, packaging, agriculture, and horticulture.

Bio-based reinforcements can also be produced from biomass. Typical examples are cellulose fibers, cellulose nanowhiskers, chitosan nanowhiskers, and chitin nanowhiskers. The bio-based reinforcement will be detailed in Section 2.3.1.5 and Section 2.3.3.2.

2.2.12 Liquid crystal polymer

Liquid crystal polymer (LCP), or liquid crystalline polymer, is a unique group of thermoplastics that possess liquid crystal properties. A liquid crystal, or a liquid crystal state, is a state of matter that lies between solid and liquid. Liquid crystals change shapes like liquid but remain a long range order that normally exists in solid. LCP is a polymer that, when it is in a molten or dissolved state, contains ordered molecular chains that are normally seen in solid semicrystalline thermoplastic. LCP differs from conventional thermoplastics in structure and, therefore, in thermal behaviors, flow characteristics, and mechanical properties.

Overall, LCPs have the following characteristics:

- Excellent thermal stability
- Exceptionally dimensional stability (low coefficients of thermal shrinkage)
- High elastic modulus and tensile strength
- Excellent processability in thin features
- Excellent chemical resistance
- Flame retardancy
- Good weatherability
- Highly anisotropic property

The liquid crystal phase in LCP is called *mesophase,* a phase that is between a solid and liquid phase. The polymer chain group that induces the ordered molecular structure, or liquid crystals, in the mesophase is called *mesogen.* The mesogen results in a special molecular structure consisting of rigid and rod-like macromolecules that are ordered in the molten or dissolved state to form liquid crystals. The axis along which the ordered molecules in the liquid crystal phase align is called director. The liquid crystals have a degree of orientational order along the director but do not necessarily display any positional order, whereas the crystallites in the solid have both orientational and positional order. In addition, the liquid crystals are not as ordered as the crystallites in the solid. Figure 2.34 shows phase changes in LCP. LCP undergoes from a solid phase to a liquid crystal phase and finally to a liquid amorphous phase, and vice versa, through temperature change (thermotropic LCPs) or concentration change (lyotropic LCPs). The conversion between those different phases is reversible.

LCPs can be classified into thermotropic and lyotropic LCPs based on the method of inducing formation of the liquid crystal phase. In thermotropic LCPs, the formation of the liquid crystal phase is induced via heating or cooling. Since a thermal process is involved and the LCP has anisotropic properties because of the orientation preference of the liquid crystals, this type of LCP is called "thermotropic," a word that is created from both "thermally" and "anisotropic." The thermotropic LCP has a molecular structure consisting of a central rigid group (often aromatic) and flexible groups (often aliphatic) at both ends. The thermotropic LCP has been used as a matrix material in composites. It can be processed similarly as conventional thermoplastics through heating. It can also be remelted and reshaped. However, it exhibits a series of liquid crystal properties that cannot be found in the conventional thermoplastic, such as excellent thermal stability and excellent processability in thin features. Typical brands of thermotropic LCPs include Xydar® and Vectran®. Both

LCP brands are aromatic polyesters consisting of ester linkages directly attached with aromatics rings.

Lyotropic LCPs consist of liquid crystals that are formed through a solution method ("Lyo" is from Greek and means dispersion and dissolution). The LCP is dissolved into a solvent, e.g., concentrated sulfuric acid, and the formation of the liquid crystal phase is mainly dependent on the concentration of the solution. A solid lyotropic LCP is dissolved in the solvent to form liquid crystals. With the decreasing solution concentration, the liquid crystals decrease in number and finally disappear. The lyotropic LCP can be made into fibers or films. The fibers made from lyotropic LCPs can be used as a reinforcement in composites for their great mechanical properties. Kevlar®, a high-performance fiber with the chemical name of poly(p-phenylene terephthalamide) or poly-paraphenylene terephthalamide, is a classic example of the lyotropic LCP used in both thermoset and thermoplastic composites. It will be further discussed in Section 2.3.1.3.

- Solid phase
- 3D long range order
- Orientational order
- Positional order

- Liquid crystal phase
- 2D or 1D long range order
- Orientational order

- Liquid amorphous phase
- 0D no long range order
- No orientation order
- No positional order

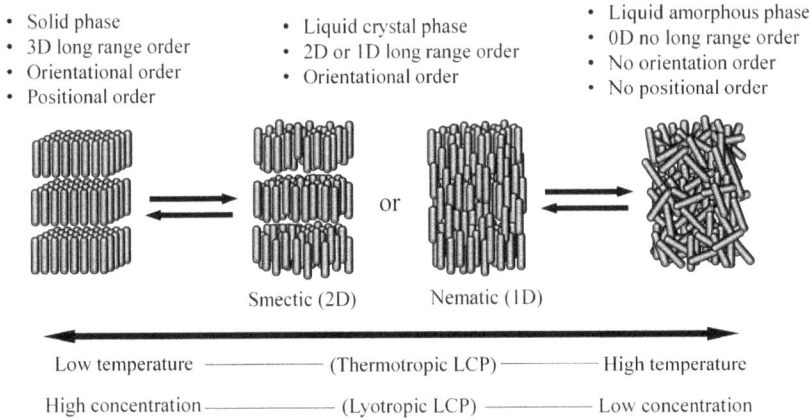

or

Smectic (2D)   Nematic (1D)

Low temperature ———— (Thermotropic LCP) ———— High temperature

High concentration ———— (Lyotropic LCP) ———— Low concentration

Fig. 2.34: The change of liquid crystal polymer from a solid phase to a liquid crystal phase and finally a liquid amorphous phase when it is heated (thermotropic LCP) or when it is dissolved (lyotropic LCP).

The liquid crystal phase, or the mesophase, can be defined as smectic and nematic according to its architecture. When liquid crystals form two-dimensional architecture, the mesophase is called a smectic mesophase. If the mesophase is one-dimensional, it is nematic. There is molecular alignment in its director but no special regularity in the nematic mesophase. By taking advantage of this characteristic, high-performance fibers such as Kevlar® fiber can be produced by highly aligning the one-dimensional liquid crystals (nematic mesophase) via a spinning process (Section 2.3.1.3). Figure 2.34 shows both the smectic and nematic mesophases. Because of the presence of the molecular chain orientation, both the solid and the liquid crystal phase exhibit bulk anisotropic material properties. The liquid amorphous phase possesses isotropic material properties due to the lack of any molecular chain orientation.

The mesogen can exist in different locations in the repeat monomer unit. The location of the mesogen can be located in the main chain only, side chain only, or both main chain and side chain of the repeat monomer unit. The LCP with mesogens in the main chain, such as Kevlar®, is attractive for its excellent mechanical properties, while the LCP with mesogens in its side chain is attractive for electronic and optical applications. Figure 2.35 shows the mesogens in the main chain only, side chain only, and both main chain and side chain. The mesogen can also be in a different shape, such as rod shape and disc shape. Both Figs. 2.34 and 2.35 show the rod-shaped mesophase.

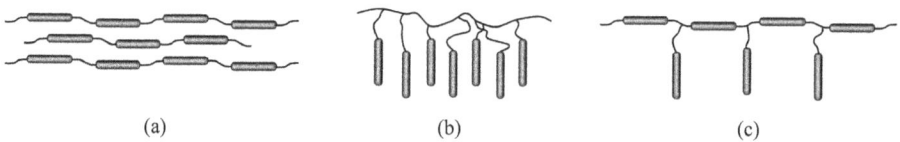

(a) (b) (c)

Fig. 2.35: Mesogens located in (a) main chain only; (b) side chain only; and (c) both main chain and side chain.

The repeated unit in lyotropic LCPs normally consists of three types of chemical groups that have distinct functions (Fig. 2.36). There is one molecular group at one end of the repeated unit that exhibits a hydrophobic characteristic and another group at the other end that exhibits a hydrophilic behavior. This type of molecule unit has both hydrophobic and hydrophilic properties and, therefore, it is called an amphiphilic molecule. The mesogen, normally a rigid rod-like molecular group, is located between the hydrophobic and hydrophilic groups. The rigid molecular group is consisted of aromatic rings, a group that provides the rigidity to the structure. The unique molecular structure of the lyotropic LCP helps to form liquid crystals by aggregating and self-aligning.

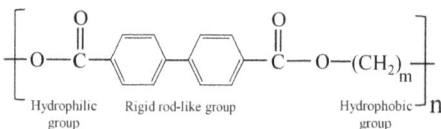

Fig. 2.36: The mesogen and the end groups in a polymer unit forming liquid crystals in a lyotropic LCP.

The thermal behavior of the LCP differs from that of the conventional thermoplastic. A typical DSC heating curve of the thermotropic LCP is shown in Fig. 2.37. It exhibits two endothermal peaks caused by phase transformation. The first endothermal peak is resulted from the transformation from its crystalline solid state to liquid crystal state at temperature $T_{SC\text{-}LC}$. The second endothermal peak represents the change from its liquid crystal state to amorphous liquid state at temperature $T_{LC\text{-}LA}$. More endothermal peaks can possibly appear between these two peaks if the LCP has polymorphs or mesophases with different degrees of order of the liquid crystal at different temperatures. For example, a smectic mesophase can have several polymorphs at different temperatures and their degrees of order and directors may vary. Peaks may display when there are transitions among the polymorphs, although the enthalpy change during the transition (indicated by the heat flow) in the DSC curve might be insignificant. The LCP normally has a high DOC, therefore, the glass transition is generally not noticeable from the DSC curve.

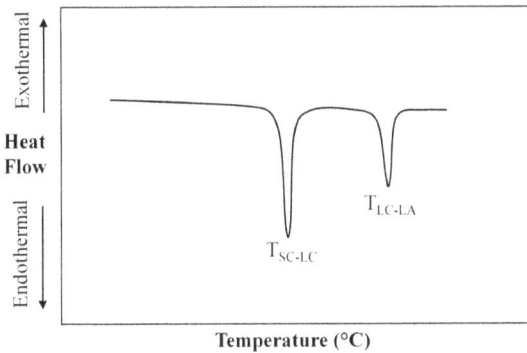

Fig. 2.37: A typical DSC curve of a liquid crystal polymer showing two endothermal peaks ($T_{SC\text{-}LC}$ and $T_{LC\text{-}LA}$) resulted from phase changes.

Fiber-reinforced LCP matrix composites are drawing more and more attentions because of their excellent mechanical properties, geometry stability, chemical resistance, thermal stability, good surface appearance, and so on. Typical fibers used in the LCP matrix composite include carbon fiber, glass fiber, and ceramic fibers. These LCP matrix composites can be processed using conventional thermoplastic processing methods, including the injection molding process.

LCPs can be used as reinforcements in a polymer form or a fiber form. LCPs in the polymer form have been blended with conventional thermoplastics, including PP, PC, nylon, and PPS, to improve their elastic modulus, tensile strength, processability, dimensional stability, and heat resistance. The blends can be injection molded or compression molded. The liquid crystals in the melt can be highly oriented along the flow direction during molding. Material properties such as tensile strength and elastic modulus are higher in the flow direction because of the alignment of the liquid crystals in that direction. In most of the blends, the liquid crystal phase remains immiscible with the conventional thermoplastic because of their incompatible nature. However, if the LCP and the thermoplastic have similar molecular groups, for example, a polyester-type LCP and a thermoplastic polyester, partial miscibility can be expected. Fibers made of LCPs are one important category of reinforcement that is used in both thermoset and thermoplastic composites. Those LCP fibers, such as Kevlar®, have been extensively used in ballistic armor applications because of their excellent impact performance.

2.3 Reinforcements in thermoplastic composites

Reinforcements are another critical constituent besides the matrix in thermoplastic composites. Reinforcements can be in different forms as mentioned before, such as fibers, particulates, and nanoscale materials. The following sections describe those different types of reinforcements used in thermoplastic composites.

2.3.1 Fibers

Fibers are the dominantly used reinforcement in thermoplastic composites for structural applications due to their high efficiency in load bearing. When there is loading applied to fiber-reinforced thermoplastic composites, the fiber is capable of bearing a majority of the loading. Depending on the fiber elastic modulus, fiber volume percentage, fiber orientation, and fiber aspect ratio in the thermoplastic composite, the fiber can carry more than 95% of the loading (see Eq. (7.20) and Example Question 7.1).

The fibers can be categorized into natural and synthetic fibers based on their material origin. Natural fibers are the fibers that can be found or made from natural resources such as minerals, animals, and plants. On the other hand, synthetic fibers are man-made and do not exist in nature. Those fibers, especially synthetic fibers, provide excellent strength and modulus to thermoplastic composites.

Figure 2.38 lists various types of natural fibers and synthetic fibers. Most of the fibers have been used in thermoplastic composites although some of the fibers, such as carbon fibers, glass fibers, basalt fibers, and cellulose fibers, are considerably more commonly used than the others.

The natural fiber includes mineral fibers (asbestos fiber and basalt fiber), animal fibers (wool, feather, silk, and hair), and cellulose fibers. These fibers, especially cellulose fibers, have attracted a tremendous amount of interest because of their environmentally friendliness, abundance, and renewability. Cellulose fibers are derived from plants and can be obtained from different parts of the plants, such as bast, leaf, seed, fruit, and stalk.

Synthetic fibers can be made of either organic or inorganic materials. Fibers made of organic materials include aramid fiber, PE fiber, and PP fiber. The fibers made of inorganic materials are glass fiber, carbon fiber, boron fiber, metal fibers (steel fiber), ceramic fibers (silica and alumina fibers), and so on. Table 2.6 lists the

Fiber

Natural Fiber

Synthetic Fiber

| Mineral Fiber | Animal Fiber | | | Cellulose Fiber | | | | | Organic Fiber | Inorganic Fiber |

| Asbestos Basalt | Wool Silk Feather Hair | Bast
Flax
Hemp
Jute
Kenaf
Ramie | Leaf
Abaca
Banana
Palf
Sisal | Seed
Cotton
Kapok
Luffa
Milk weed | Fruit
Coir
Oil Palm | Wood
Hard wood
Soft wood | Stalk
Barley
Maize
Oat
Rice
Rye
Wheat | Grass
Bagasse
Bamboo
Corn | Aramid
PE
PP
PET | Glass
Carbon
Boron
Silica
Alumina
Steel |

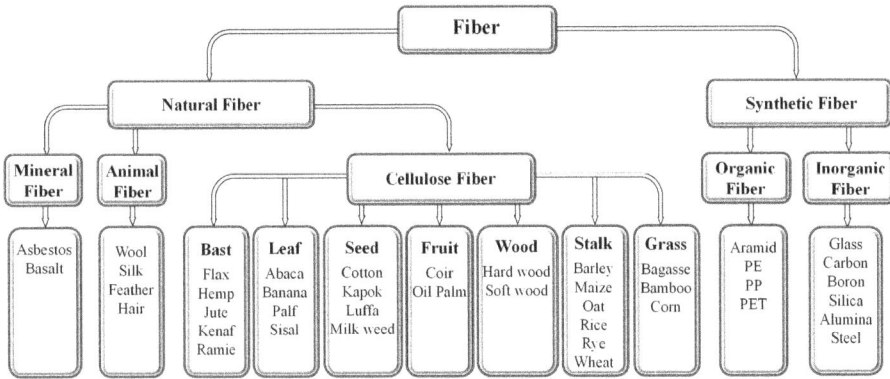

Fig. 2.38: Categories of natural fibers and synthetic fibers used in thermoplastic composites.

physical properties and mechanical properties of common synthetic fibers used in thermoplastic composites.

Tab. 2.6: Properties of common fibers used in thermoplastic composites.

Fibers	Density (g/cm^3)	Elastic modulus (GPa)	Tensile strength (MPa)	Elongation to failure (%)
E-glass	2.54–2.58	72.4	2,400–3,500	2.6–4.8
S-glass	2.46–2.49	85.5–86.9	4,300–4,890	2–5.7
Aramid	1.45	124–131	3,600	2.3–2.8
High-strength carbon fiber	1.8	253	4,500	1.1
High modulus carbon fiber	1.85–2.1	520–620	2,400	0.6
Basalt	2.6–2.8	71–110	3,000–4,800	3.1–3.3

Fibers possess excellent specific strength and modulus. Both specific strength and modulus are the main criteria besides cost during material selection in the applications that require lightweight, for example, automotive and aircraft applications that require high fuel efficiency. The specific strength and modulus of those common fibers are compared in Fig. 2.39. The high-modulus carbon fiber has the highest specific modulus. UHMWPE fibers and high strength carbon fibers are the fibers with the highest specific strength.

Fibers are known for their excellent modulus and strength. However, their strains to failure are generally limited, especially the fibers with high modulus. Figure 2.40

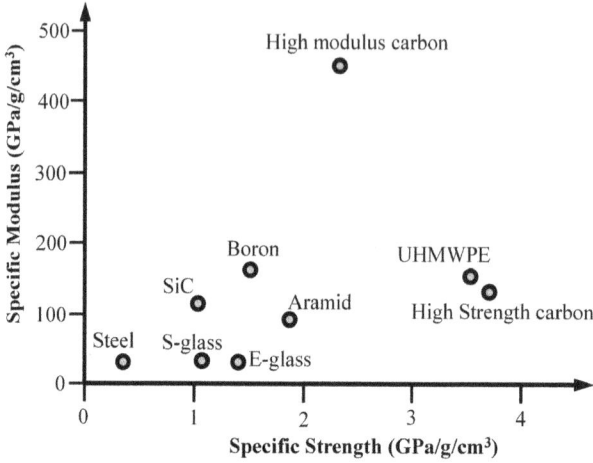

Fig. 2.39: Specific strength and modulus of different types of fibers.

shows the stress to strain curves for the fibers used in thermoplastic composites. Carbon fibers show the highest modulus while glass fibers show the highest strain to failure. The synthetic fibers made of inorganic materials, such as glass and carbon fibers, show a linearly elastic behavior when tested in tension as shown in Fig. 2.40. The fiber fractures when it reaches to peak load.

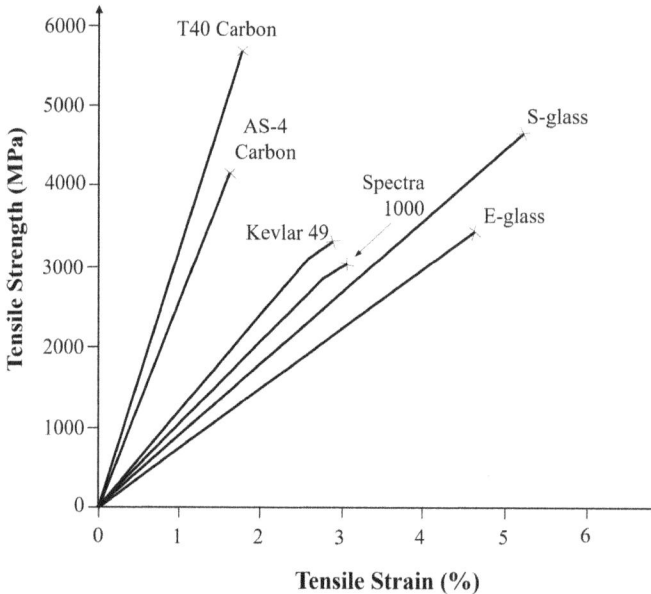

Fig. 2.40: Tensile stress and strain relationship of common synthetic fibers used in thermoplastic composites.

The sections below describe common fibers, including glass fiber, carbon fiber, synthetic polymer fibers (aramid fiber, UHMWPE fiber, and PP fiber), basalt fiber, and cellulose fiber, used to reinforce thermoplastics.

2.3.1.1 Glass fiber

Glass fibers are the most commonly used fiber in thermoplastic composites. The glass fiber is added to achieve high performance in strength and impact resistance in the thermoplastic composite at a relatively low material cost. Glass fiber-reinforced thermoplastic composites have a combination of great modulus and strength, excellent impact resistance, good chemical resistance, and lightweight, and have found their use in a wide variety of applications, including automotive, electrical and electronics, transportation, energy, construction, aircraft, marine, medical, military, and sports.

Glass fibers can be classified into E-glass, S-glass, R-glass, ECR-glass, C-glass, A-glass, D-glass, AR-glass, and T-glass based on their chemical compositions. Because of the difference in their composition, those glass fibers vary in their physical, chemical, and mechanical properties. Most of the glass fibers have been used in thermoplastic composites, including E-glass, S-glass, ECR-glass, A-glass, R-glass, and C-glass. Some other glass fibers, such as D-glass and AR-glass, are developed for special applications and rarely used in thermoplastic composites. For example, D-glass is developed for its great dielectric properties (D stands for dielectric) and used in special electrical applications; AR-glass is mainly used in cement for its excellent alkaline resistance (AR stands for alkaline resistant). Table 2.7 lists the properties and composition of the glass fibers commonly used in thermoplastic composites. It is noted that the composition of the same type of glass fibers from different manufacturers can vary. The composition can also vary in the same type of glass fiber manufactured by the same company at different times. The composition

Tab. 2.7: General properties and compositions of glass fibers used in thermoplastic composites.

	E-glass	S-glass	ECR-glass	A-glass	C-glass	R-glass
Density (g/cm³)	2.52–2.62	2.46–2.53	2.66–2.72	2.44	2.52–2.56	2.52
Tensile strength (MPa)	3,100–3,800	4,590–4,890	3,100–3,800	3,300	3,300	4,400
Elastic modulus (GPa)	72	87–91	80–81	69–72	69	86
Elongation at break (%)	4.5–4.9	5.2–5.8	4.5–4.9	4.8	4.8	4.8–5.1

Tab. 2.7 (continued)

		E-glass	S-glass	ECR-glass	A-glass	C-glass	R-glass
	SiO_2	52–56%	64–66%	54–62%	63–72%	64–68%	60–65%
	Al_2O_3	12–16%	24–26%	9–15%	0–6%	3–5%	17–24%
	B_2O_3	5–10%	–	–	0–8%	0–6%	–
	CaO	16–25%	–	17–25%	6–10%	11–15%	5–11%
	MgO	0–5%	8–12%	0–5%	0–4%	2–4%	6–12%
	ZnO	–	–	2.9%	–	–	–
Composition	Na_2O	0–1%	0–0.1%	1.0%	14–16%	7–10%	0–2%
	K_2O	trace	–	0.2%			0–2%
	TiO_2	0.2–0.5%	–	2.5%	0–0.6%	–	–
	Zr_2O_3	–	0–1%	–	–	–	–
	Li_2O	–	–	–	–	–	–
	Fe_2O_3	0.2–0.4%	0–0.1%	0.1%	0–0.5%	0–0.8%	–
	F_2	0.2–0.7%	–	–	0–0.4%	–	–

variation in glass fibers with the same designation might result in substantial difference in their properties.

The following sections describe each type of glass fiber used in thermoplastic composites.

1. **E-glass** fiber belongs to the alumina–calcium–borosilicate glass family and it is a general-purpose glass fiber. "E" in E-glass fiber stands for Electrical because of its low electrical conductivity and original use for electrical insulation applications. It accounts for more than 80% of the glass fiber products. The main compositions in E-glass fibers are SiO_2, Al_2O_3, CaO, B_2O_3, and MgO. B_2O_3 helps reduce the viscosity of molten glass during processing and lower the processing temperature.

    E-glass fibers offer a combination of low material cost and great mechanical properties, however, the fiber is susceptible to corrosion stress in an acidic environment due to the presence of boron. The boron present in the E-glass fiber is remarkably detrimental to the durability of the E-glass fiber and its composites. When exposed to acids, the E-glass fiber experiences an ion exchange process that the metallic ions in the fiber are replaced by hydrogen ions from the acid. The size difference between the metal ions and hydrogen ions results in a volume change in the fiber and causes stresses on the fiber surface. It is much easier for boron to form complexes with acids and leach out than other metallic ions present in the E-glass fiber. Therefore, the boron-containing glass fiber surfers much more severe acid attack, which shortens the lifespan of the E-glass fiber and its composites significantly. That prompts the development of glass fibers equivalent to E-glass fibers but without any boron, such as Advantex®. The boron-free glass fiber has a similar cost to E-glass fibers but same corrosion resistance as ECR-glass fibers.

2. **S-glass** fiber is a high-strength glass fiber and "S" stands for **S**trength. It is a magnesium alumino-silicate glass. Its strength can reach 4,890 MPa, much higher than the strength of E-glass fibers. It has a higher elastic modulus than the E-glass fibers. In addition, it has good fatigue resistance and high temperature resistance. However, the high manufacturing cost makes it more expensive than the other glass fibers and its use is limited to some niche applications, such as ballistic protection, aeronautics, aerospace, and sports.

3. **ECR glass fiber.** "ECR" stands for **E**lectrical and **C**orrosion **R**esistant in ECR-glass. It is a calcium alumino-silicate glass. The ECR glass fiber has excellent corrosion resistance, especially to acids. It is not susceptible to the ion exchange process, which minimizes the corrosion stress and prolongs its lifespan even in some highly corrosive environment. Compared to E-glass fibers, it possesses higher mechanical properties and better thermal resistance. Meanwhile, the ECR glass fiber offers good electrical insulation.

4. **A-glass fiber.** The letter "A" in A-glass fiber stands for **A**lkaline. It is the first type of glass fiber developed. The A-glass fiber is a calcium borosilicate glass. It is similar to the E-glass fiber but it is used in less demanding applications where electrical resistivity is not critical.

5. **C-glass fiber.** "C" in C-glass fiber stands for **C**hemical resistance because it possesses great resistance to chemicals, such as alkalis or acids. It is a calcium borosilicate glass. C-glass fibers are commonly used as surface veils that protect the composite beneath.

6. **R-glass** fiber is a calcium alumino-silicate glass fiber. "R" stands for **R**einforcement. The R-glass fiber possesses high mechanical performance, good fatigue resistance, excellent acid corrosion resistance, high thermal resistance, and low moisture absorption. Its use is normally limited to specific applications such as aerospace and ballistic armor. Due to its low volume production, the R-glass fiber has a relatively high cost.

Those glass fibers have a wide range of composition and properties (Tab. 2.7) and their applications also vary significantly. However, glass fibers share the same general characteristics regardless of their type and composition. For example, all the glass fibers are amorphous; all the fibers do not have any orientation preference in its structure and, therefore, they have isotropic material properties. Those general characteristics are listed below:

– Silica (SiO_2) being the main content (more than 50 wt%)
– Possessing isotropic material properties
– Being abrasive
– Being brittle
– Being nonflammable

- Being thermally insulating
- Possessing high strength and moderate modulus

The glass fiber is produced through a melt spinning process. Raw materials, including silica sand, limestone, and soda ash, or cullet (waste glass), or glass marbles, are fed into a furnace and heated to more than 1,370 °C. After removal of gaseous inclusions and homogenization, the molten glass flows by gravity into a refiner section. The melt consequently flows to the forehearth section that is located directly above fiber-drawing stations. The glass melt then goes through a bushing with arrays of holes or tips to be drawn into fibers. The bushing is made of high temperature materials such as platinum and rhodium alloy. The viscosity of the glass melt is well controlled to draw the glass fiber without breakage or without forming droplets. An excessively high viscosity causes breakage while an excessively low viscosity results in droplets, both of which will prevent successful manufacturing of glass fibers. During drawing, the fibers are cooled rapidly to prevent any crystallization while exiting the bushing. Glass fibers with diameters ranging from 5 to 24 μm are produced using this process. Finally, the fibers are coated with sizing on their surface and wound up on bobbins for consequent composite manufacturing after the sizing is dried.

2.3.1.2 Carbon fiber

Carbon fiber was initially invented as light bulk filaments in the late 1800s. However, high-performance carbon fibers were only developed in around 1960. Since then, the carbon fiber has become one of the most commonly used engineering materials for lightweight high-performance structures. Its demand has been continuously increasing. Figure 2.41 shows the worldwide demand of carbon fibers since its

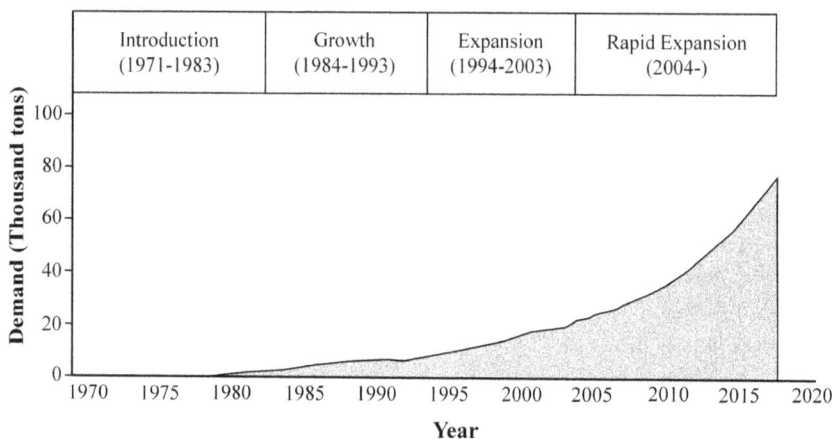

Fig. 2.41: Global demand for carbon fibers since its development (adapted from References [7–9]).

development. The global demand for carbon fibers in 2010, 2014, and 2017 were 33,000, 53,000, and 70,500 tons, respectively [7].

Carbon fibers have been increasingly used in thermoplastic composites and the market size of carbon fiber-reinforced thermoplastic composites has been steadily growing, especially in aerospace and automotive industries. The global market size of the carbon fiber thermoplastic composite was valued at USD 2.75 billion in 2015 and its compound annual growth rate was projected to grow at a 10.8% [10].

The pie chart in Fig. 2.42 shows the use of carbon fiber thermoplastic composites in the United States [10]. The applications include aerospace and defense, automotive, wind turbines, sports/leisure, construction, marine, and others. Aerospace and defense industries use most of the carbon fiber thermoplastic composites. Automotive industries are the second largest user of carbon fiber thermoplastic composites. With the emergence of low cost carbon fibers, such as lignin-based carbon fibers and recycled carbon fibers, the automotive industries can potentially become the largest consumer of the carbon fiber thermoplastic composite.

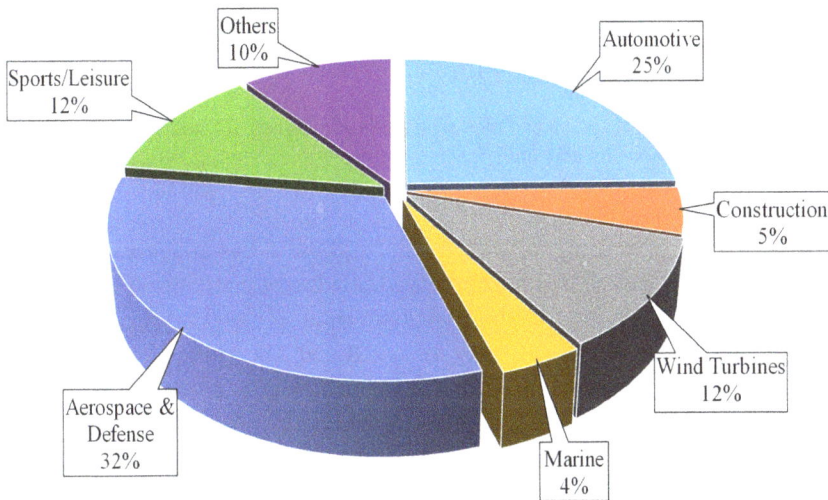

Fig. 2.42: The use of carbon fiber reinforced thermoplastic composites in the United States in 2015 (adapted from Reference [10]).

Carbon fiber is one of allotropes of carbon, structurally different forms of the same carbon element. Besides carbon fibers, there are several other allotropes of carbon, including graphite, graphene, diamond, pyrolytic carbon, carbon nanotube, fullerene, and amorphous carbon (carbon black). In those allotropes, carbon atoms have various arrangements and form different structures, which in turn can result in different materials with distinct properties. Figure 2.43 shows those allotropes of carbon.

(a) Graphite   (b) Carbon fiber

(c) Graphene   (d) Carbon nanotube   (e) Diamond

(f) Amorphous carbon   (g) Pyrolytic carbon

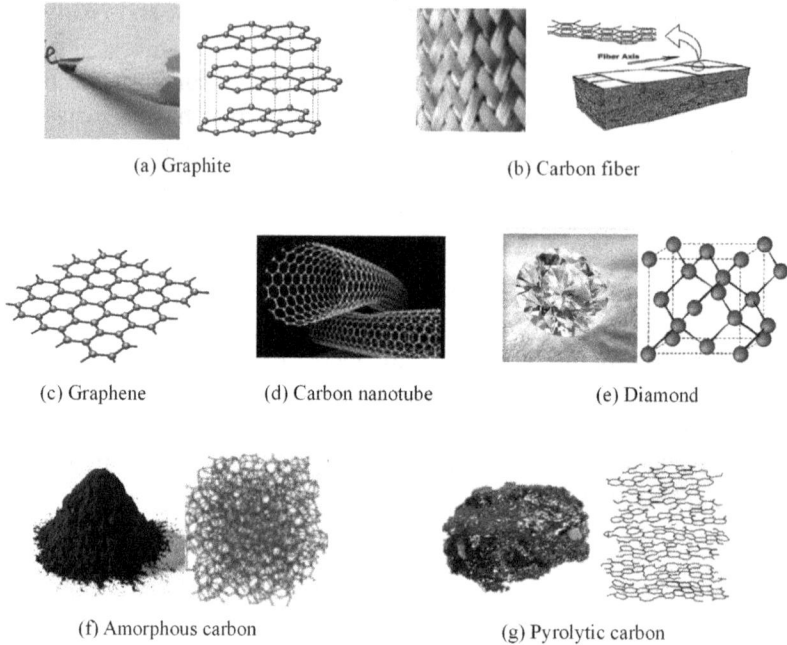

Fig. 2.43: Different allotropes of carbon and their structures: (a) graphite; (b) carbon fiber; (c) graphene; (d) carbon nanotube; (e) diamond; (f) amorphous carbon; and (g) pyrolytic carbon (partially adapted from Reference [11]).

Graphite is one of the common allotropes of carbon. It has a layered structure that consists of sheets of carbon atoms arranged in a hexagonal pattern. The plane of the sheet is called basal plane. The sheets are in an ordered arrangement and translational shift among the sheets may exist (AB and ABC stack sequence). The sheets are weakly bonded through van der Waals force, which allows the sheet slide across each other easily and makes graphite a soft material.

Each individual sheet in graphite is called graphene, another allotrope of carbon. It has a two-dimensional structure as its thickness (one carbon atom thick) is significantly smaller than the other two dimensions. It is often called nanographene because of its nanoscale in the thickness direction. It has been found that graphene has great mechanical, electrical, and thermal properties. Graphene is increasingly used to reinforced thermoplastics.

Carbon fibers consist of sheets that have carbon atoms with hexagonal arrangement, a similar atomic structure observed in graphite. Depending on the order of the carbon sheets, carbon fibers have a turbostratic or graphitic structure. The turbostratic structure has an order between amorphous carbon (disordered) and crystalline graphite (highly ordered). The carbon fiber with a turbostratic structure has carbon sheets that are not as ordered as seen in graphite and there is a random rotation (angle shift) and translation within the sheet planes. Crystallites with limited sizes normally exist in those carbon fibers with a turbostratic structure and vacancies may be present (Fig. 2.44a). The carbon sheets in the carbon fibers are curved, folded, and kinked, resulting in interconnection and interlock among the sheets (Fig. 2.44b) and, hence, excellent mechanical properties of the carbon fiber. Certain precursor materials such as polyacrylonitrile (PAN), cellulose, and lignin are used to produce carbon fibers with a turbostratic structure. Figure 2.44c, d show the reactions how PAN and cellulose are converted to the turbostratic carbon fiber structure (Fig. 2.44a), respectively.

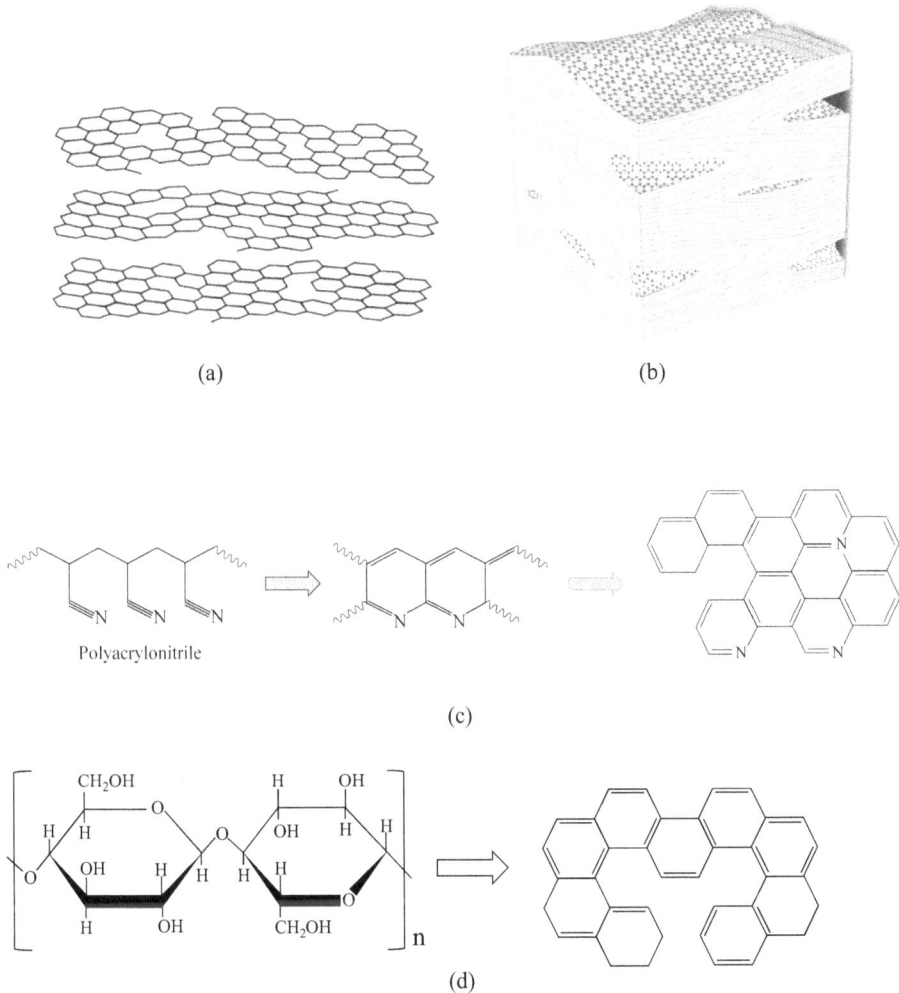

(a)

(b)

Polyacrylonitrile

(c)

(d)

Fig. 2.44: (a) A turbostratic carbon structure in a closer view showing vacancies; (b) a turbostratic carbon fiber structure (reprinted from Reference [12] with permission) showing curved, folded, and kinked carbon sheets; (c) polyacrylonitrile precursor material being converted to turbostratic carbon structure (Figure 2.44a); and (d) cellulose precursor material being converted to turbostratic carbon structure (Figure 2.44a).

Mesophase pitch can be used to produce carbon fibers with a highly ordered crystalline graphitic structure after a graphitization process above 2,400 °C. The heat treatment at high temperatures can induce the graphitic structure in the carbon fiber because it is a thermodynamically stable structure. The spacing between the sheets in the graphitic fiber structure is greater than that in the graphite and it decreases with the graphitization temperature. On the other hand, the crystallite size increases with the graphitization temperature. The carbon fiber with a graphitic structure has a highly ordered structure and possesses a high modulus as well as thermal and electric conductivities. In fact, high-modulus and ultra-high modulus carbon fibers are often referred to as graphite fibers. However, the production cost of the graphite fiber is high and its strength also deteriorates. Figure 2.45 shows the graphitization process converting mesophase pitch to carbon fibers with a graphitic structure. It is worth noting the difference in regularity of the carbon atom arrangement between the turbostratic carbon structure (Fig. 2.44a) and the graphitic structure (Fig. 2.45b).

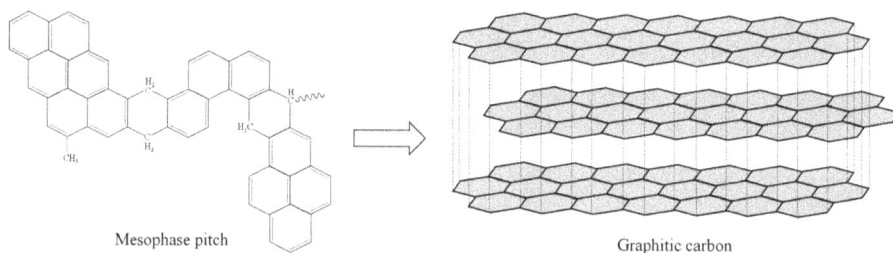

Mesophase pitch   Graphitic carbon

Fig. 2.45: The graphitization process converting mesophase pitch to carbon fiber with a graphitic structure.

Carbon fibers can have a wide range of tensile strength and modulus because of their large variation in structure, including spacing between adjacent carbon sheets, crystallinity, and crystallite size. Their tensile strength can range from 1,000 to 6,000 MPa and their tensile modulus from 100 to 700 GPa. Based on their modulus, carbon fibers are categorized into low-modulus carbon fibers that have tensile modulus less than 200 GPa, standard modulus carbon fibers that have tensile modulus of around 230 GPa, intermediate modulus carbon fibers that have tensile modulus of around 300 GPa, and high-modulus carbon fibers that have tensile modulus higher than 350 GPa. Carbon fibers with a tensile modulus greater than 600 GPa are often called ultra-high modulus carbon fibers. Figure 2.46 shows those carbon fibers with different ranges of tensile modulus and strength.

The raw material used in the manufacture of carbon fibers is called precursor. The cost of the precursor material normally makes up one third of the carbon fiber production cost. PAN and pitch are the most common precursors. In addition, cellulose and lignin are sometimes used. Even silk, chitosan, glycerol, lignocellulosic sugars,

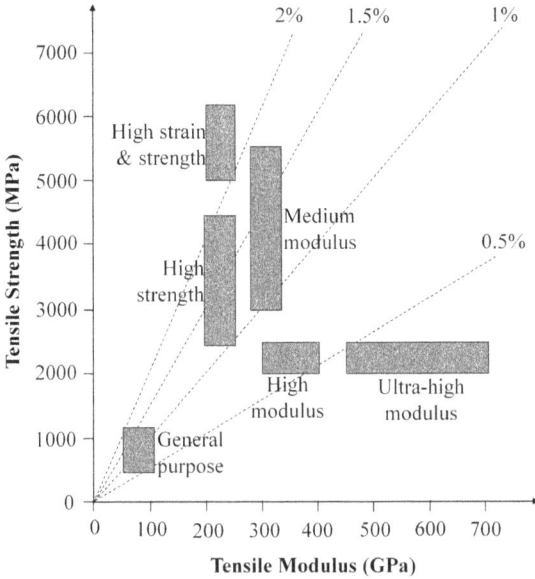

Fig. 2.46: Different categories of carbon fibers according to tensile modulus and strength.
The slanted dashed lines represent different strains.

and collagen are made into carbon fibers in rare cases. Their processing parameters,
such as carbonization temperatures and production time, may vary. The resultant
carbon fibers from those precursors also differ in their performance and cost. Below
are the descriptions of the common precursor materials and carbon fibers produced
from the precursor material.

1. **Polyacrylonitrile** or PAN is the most commonly used precursor material for pro-
    ducing carbon fibers. Nearly 90% of the carbon fibers in the market are made
    from PAN. The PAN-based carbon fiber has a turbostratic structure and possesses
    high strength. The manufacture starts with a PAN polymer solution that is used
    to produce PAN precursor fibers via a spinning process. The PAN precursor fiber
    then goes through a stabilization process that occurs at 200–400 °C under air at-
    mosphere with the presence of tension to maintain alignment of the molecular
    chains. Finally, carbonization is carried out at a high temperature (400–1,500 °C)
    while still maintaining tension on the fibers. PAN-based carbon fibers have a pro-
    duction yielding of 40–50%.

2. **Pitch**, a residual material from oil refining, is a precursor that can produce gen-
    eral-purpose and high-modulus carbon fibers. The pitch is firstly made into pre-
    cursor fibers through a melt-spinning process. The fibers are then stabilized,
    carbonized, and graphitized (Fig. 2.47). Pitch-based carbon fibers can be classi-
    fied into isotropic pitch-based and mesophase pitch-based carbon fibers based
    on the anisotropy of the atomic structure. Isotropic pitch-based carbon fibers

have relatively low tensile modulus and strength, around 100 GPa and 1,000 MPa, respectively, and are mainly for general-purpose use. Mesophase pitch can be used to produce a highly aligned molecular orientation from melt spinning because of its liquid crystalline characteristic. The pitch-based carbon fibers have a graphitic structure after the graphitization process up to 3,000 °C; therefore, the fibers have a high elastic modulus and excellent thermal conductivity in the fiber direction.

3. **Cellulose**. Rayon is the first precursor for carbon fiber production. It is mainly composed of cellulose from wood and agricultural products. It is converted into a soluble compound chemically and then made into fibers. The rayon-based precursor fiber then undergoes similar oxidization, carbonization, and graphitization processes. Rayon-based carbon fibers are first high-strength carbon fibers. However, the production of the rayon-based carbon fibers only accounts for less than 1% of all the carbon fiber production because of their low performance and productivity (yielding between 10% and 30%) compared to PAN-based and pitched-based carbon fibers. The rayon-based carbon fiber possesses a low heat conduction, making it a desired reinforcement material for light-weight and low thermal conductive carbon–carbon composites.

4. **Lignin** is a natural polymer that exists as one of the main components in the cell wall of plants. It is the second most abundant renewable material in nature (behind cellulose). Lignin has a carbon content up to 65%, which provides a carbon fiber yield of 30–50% [13]. Lignin has been used as a precursor material to produce low cost carbon fibers. The lignin carbon fiber has more than 35% less production cost than the PAN or pitch-based carbon fibers. Although its low price has raised interests, the lignin-based carbon fiber does not offer comparable mechanical properties achieved by either PAN-based or pitch-based carbon fibers. The main challenges lie in several areas, including high level of impurities in lignin, wide molecular weight distribution in lignin, difficulty in converting lignin into precursor fibers, and stabilizing the precursor fiber at an acceptable rate. Nevertheless, the combination of its low cost and relatively high modulus has enabled the lignin-based carbon fiber an alternative reinforcement to glass fibers for the thermoplastic composites used in automotive applications.

Production of carbon fibers involves a series of energy intensive processes. Overall, the processes are polymerization of the precursor, production of precursor fibers, stabilization, carbonization, and graphitization. The manufacture of carbon fibers from the precursor fiber is sketched in Fig. 2.47. The precursor fiber from spinning is firstly oxidized under stress at 200–400 °C. This process is also referred to as stabilization. The oxidization induces a cross-linked and nonmelting structure in the precursor fiber, which maximizes the carbon yield by avoiding the melting of the fiber and minimizing volatilization of elemental carbon in the following carbonization step. Maintaining tension of the precursor fiber during the stabilization process is critical to keep the polymer chain aligned in the fiber direction and produce high-performance carbon fibers. The stabilized fiber is consequently carbonized at higher temperatures. A nitrogen atmosphere is normally used to protect the fibers.The main purpose of this carbonization step is to remove hydrogen, nitrogen, oxygen and other noncarbon elements. A high carbon content up to 90% can be obtained from the carbonization step. The carbonized fiber undergoes a graphitization step to achieve a higher carbon content and degree of graphitization. Either high modulus or high strength, but not both, can be obtained from this process. The graphitization temperature ranges from 1,500 to 3,000 °C and argon gas is used for protection. Finally, the carbon fiber is treated with sizing chemicals before being taken up on bobbins for consequent composite processing. The carbon fiber normally has a diameter of 4–15 µm.

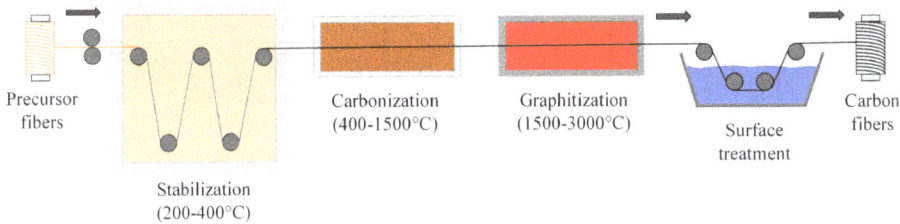

Fig. 2.47: A schematic of the processes involved in carbon fiber manufacturing from precursor fibers, including stabilization, carbonization, and graphitization.

Table 2.8 lists common commercially available carbon fibers and their mechanical, physical, thermal and electrical properties.

2.3.1.3 Synthetic polymer fiber

Both glass fibers and carbon fibers are inorganic and have been commonly used in thermoplastic composites. Another type of fibers, synthetic polymer fiber, has drawn increasing attention in reinforcing thermoplastics for their excellent mechanical properties, low density, and relatively low cost. Those fibers include aramid fiber, UHMWPE fiber, PP fiber, and PET fiber. The processing methods of those polymer fibers include melt spinning, gel spinning, solid-state extrusion, and melt blown process. The sections below describe the characteristic and processing of each of the synthetic polymer fibers.

1. **Aramid fiber** is one of the most widely used synthetic polymer fibers. It is made of aromatic polyamides consisting of long rigid crystalline molecular chains. The aromatic rings in the backbone of the polyamide mainly result in the rigid structure of the aramid fiber. The unique structure of the aramid fiber results in its excellent mechanical and physical properties. It has a high strength-to-weight ratio, outstanding dimensional stability, excellent flame retardancy, and high

Tab. 2.8: Mechanical, physical, thermal, and electrical properties of common commercially available carbon fibers (reprinted from manufacturer's data).

Carbon fibers		E (GPa)	σ_f (GPa)	Strain to failure (%)	ρ (g/cm³)	Diameter (μm)	Carbon%	CTE ((μm/m)/°C)	K (W/mK)	Electrical resistivity (Ω.m)
PAN based	HexTow AS4	231	4.4	1.8	1.79	7.1	94%	−0.63	6.83	1.7×10^{-5}
	HexTow IM7	276	5.5	1.9	1.78	5.2	95%	−0.64	5.4	1.5×10^{-5}
	HexTow IM10	310	6.9	2.0	1.79	4.4	95%	−0.70	6.14	1.3×10^{-5}
	Torayca T300	230	3.5	1.5	1.76	7	93%	−0.41	10.46	1.7×10^{-5}
	Torayca T700S	230	4.9	2.1	1.80	7	>93%	−0.38	9.6	–
	Torayca 1000G	294	6.3	2.2	1.80	5	>95%	−0.6	10.5	–
	Torayca M35J	343	4.5	1.3	1.75	5	>99%	−0.73	39.04	1.1×10^{-5}
	Tenex HTS40	240	4.4	1.8	1.77	7.0	–	–	10	1.6×10^{-5}
	Zoltek px35	242	4.1	–	1.81	7.2	95%	–	–	–

Tab. 2.8 (continued)

Carbon fibers		E (GPa)	σ_f (GPa)	Strain to failure (%)	ρ (g/cm³)	Diameter (μm)	Carbon%	CTE ((μm/m)/°C)	K (W/mK)	Electrical resistivity (Ω.m)
Pitch based	Thornel P-30	207	1.3	0.8	2.00	10	98%	–	62	1×10^{-5}
	Thornel P-55	414	1.3	0.5	2.00	10	99%	-1.3	120	8.5×10^{-6}
	Nippon YSH-50A	520	3.8	0.7	2.10	7	–	–	120	7×10^{-6}
	Nippon YSH-70A	720	3.6	0.5	2.14	7	–	–	–	–
	Nippon YSH-80A	785	3.6	0.5	2.15	7	–	-1.5	320	5×10^{-6}
	Nippon YS-90A	880	3.5	0.3	2.18	7	–	-1.5	500	3×10^{-6}
	Nippon YS-95A	920	3.5	0.3	2.19	7	–	-1.5	600	2.2×10^{-6}
	Dialead K63710	640	2.6	0.4	2.12	11	–	–	–	–
	Dialead K13 D2U	935	3.7	0.4	2.2	11	–	–	–	–
Cellulose based	Sohim	60	0.5–1.5	–	1.3–1.5	6–10	>90%	–	–	–
Lignin based	ORNL	82.7	1.07	2.0	–	–	–	–	–	–

melting temperature (more than 500 °C). The aramid fiber has been widely used in composites for demanding industries such as aerospace and defense.

Figure 2.48 shows typical chemical formulas of aramid fibers. The aramid fiber has a fully aromatic structure and hydrogen bonds between the aramid chains. The aromatic groups can be linked into the backbone chain of the aramid fiber through their 1 and 4 positions (para-aramid) or 1 and 3 positions (meta-aramid). Para-aramid fibers include brands such as Kevlar® (Du Pont), Technora® (Teijin), and Twaron® (Teijin). Meta-aramid fibers include Nomex® (Du Pont) and Teijinconex® (Teijin). Figure 2.49 shows the aromatic structure in the backbone chain and the hydrogen bonds between the backbone chains in Kevlar® fiber, a para-aramid fiber and the first aramid fiber developed.

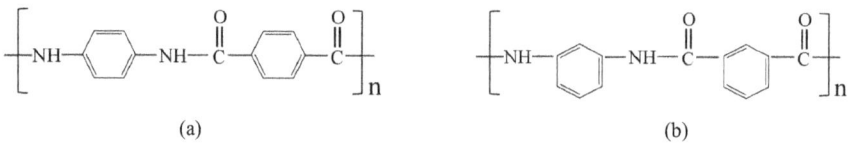

(a) (b)

Fig. 2.48: Typical chemical structures for aramid fibers: (a) para-aramid with aromatic groups linked through their 1 and 4 positions; (b) meta-aramid with aromatic groups linked through their 1 and 3 positions.

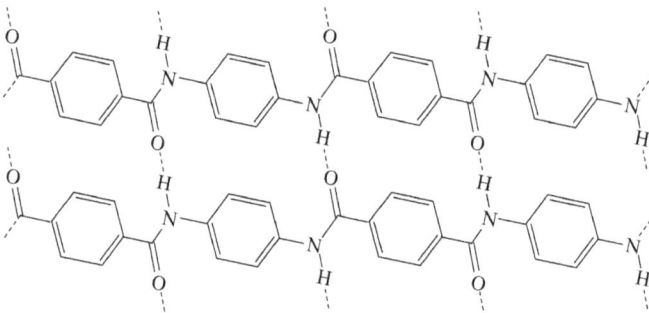

Fig. 2.49: The rigid aromatic backbone chains and the hydrogen bonds between the backbone chains in Kevlar® fiber, a para-aramid fiber and the first aramid fiber developed.

The aramid fiber is essentially made of a lyotropic LCP (see Section 2.2.12). The backbone of the aramid fiber consists of rigid aromatic groups that form mesogens in the liquid crystal. The hydrogen bonds between the carbonyl oxygen and the pendent hydrogen on the nitrogen in the backbone chains (Fig. 2.49) stabilize those chains.

Lyotropic LCPs normally decompose before melting when heated to elevated temperatures. But they can be dissolved in acids, such as sulfuric acid. Therefore, aramid fibers cannot be processed through melting methods but can be made via spinning methods. For example, Kevlar® fibers can be produced via a spinning process after p-phenylene terephthalamide (PPTA), a product of p-phenylene diamine and terephthaloyl dichloride, is dissolved into 100% anhydrous sulfuric acid at approximately 80 °C. The gel solution is consisted of liquid crystals that are randomly aligned. The gel is then extruded through spinning holes to form PPTA fibers that have aligned liquid crystals. The performance of the fibers can be further improved by a drawing process, which can induce an extremely high alignment of the liquid crystals. Figure 2.50 shows a schematic of PPTA solution with randomly aligned liquid crystals and the aramid fiber with highly aligned polymer chains after the spinning and drawing process.

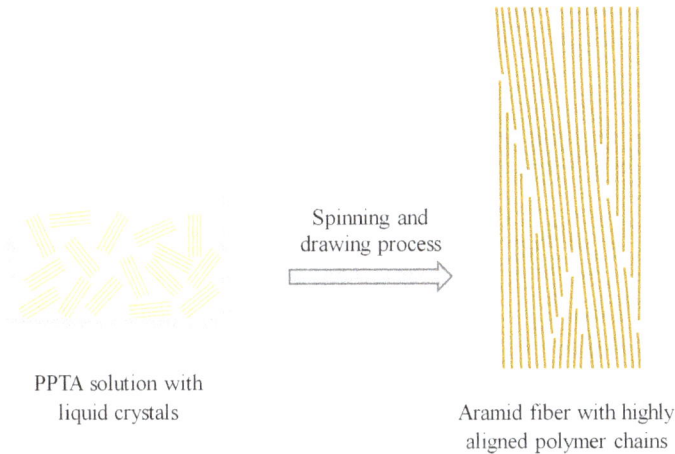

Spinning and
drawing process

PPTA solution with
liquid crystals

Aramid fiber with highly
aligned polymer chains

Fig. 2.50: Aramid fiber with highly aligned molecular chains produced from p-phenylene terephthalamide solution with randomly aligned liquid crystals through spinning and drawing processes.

Aramid fibers possess great mechanical properties. However, their properties, especially strength, can significantly suffer from exposure to elevated temperatures due to disruption of molecular bonds. Figure 2.51 shows the temperature effect on the tensile strength of Kevlar® fibers. Their tensile strength suffers a dramatic decrease when the exposure temperature is 200 °C or above. Therefore, when elevated temperatures are involved in processing, such as molding aramid fibers with thermoplastics and additively manufacturing aramid fiber thermoplastic composites, care should be taken to minimize the reduction in the tensile strength of the aramid fiber and its composite. Both lower processing temperatures and/or shorter exposure times can help minimize the detrimental effect. In addition, the use of aramid fiber-reinforced thermoplastic composites in the environment of chlorine, acids, bases, and UV radiation should be avoided as the aramid fiber is sensitive to those chemical compounds as well as ultraviolet radiation.

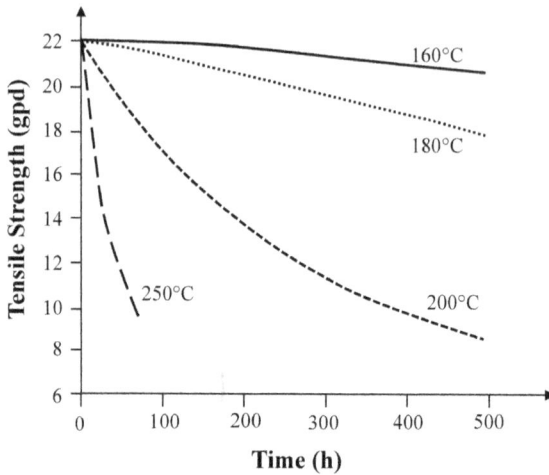

Fig. 2.51: Temperature and exposure time effect on the tensile strength of Kevlar® fiber (adapted from Reference [14]).

2. **UHMWPE fiber.** UHMWPE was initially invented and used as a bulk material for biomedical applications in the 1950s. UHMWPE fibers were only developed in the late 1970s. The UHMWPE fiber is made of UHMWPE whose molecular chains are highly crystalline and aligned in the fiber direction. The DOC and the level of alignment can reach up to 85% and 95%, respectively. The UHMWPE fiber has a low density, 0.97 g/cm³, because of the low density of its origin material, PE. Its unique characteristics such as low density, high strength, high strain to failure, low friction coefficient, nontoxicity, and near zero moisture absorption make the UHMWPE fiber a high-performance engineering material for niche applications such as bullet proof vest, cut proof gloves, maritime ropes, and advanced composites applications. Commercially available UHMWPE fiber products include Dyneema® (DSM, Netherlands), Spectra® (Honeywell, USA), and Tekmilon® (Mitsui Chemical Industries, Japan). Figure 2.52a,c show Dyneema® fabrics and Spectra® fabrics, respectively. The Dyneema® fabric is made of unidirectional fiber layers with a cross-ply layup (0° and 90° arrangement) (Fig. 2.52b) and the Spectra® fabric has a woven fiber architecture (Fig. 2.52d).

    UHMWPE fibers have an elastic modulus of 100 GPa and a tensile strength of 2–2.9 GPa in the fiber direction. The fiber has a relatively high strain to failure, 3–6.3%, which makes it an excellent impact-resistant material. Because of their low density and high impact performance, UHMWPE fibers have been increasingly used in various impact-related applications. The fibers have also been used to reinforce PE to form self-reinforced composites.

    Several processes are developed to produce UHMWPE fibers. Those processes include gel spinning process, melt spinning process, and solid state extrusion

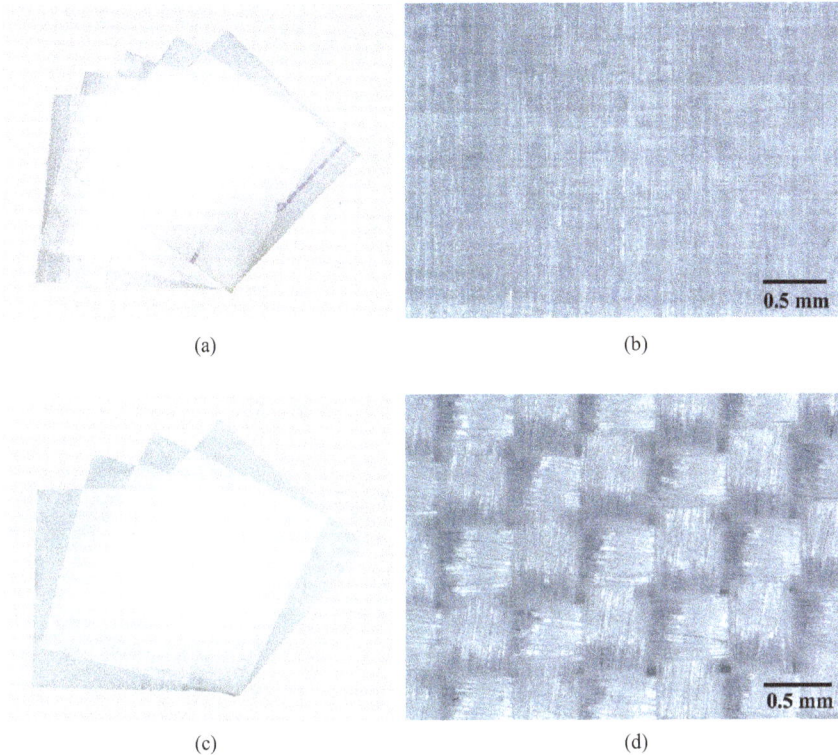

Fig. 2.52: Ultra-high molecular weight polyethylene fabrics. (a) Dyneema® fabrics (each fabric sheet has a size of 30 × 30 cm); (b) magnified view of the Dyneema® fabric showing its cross-ply layup; (c) Spectra® fabrics (each fabric sheet has a size of 30 × 30 cm); and (d) magnified view of the Spectra® fabric showing its woven fiber architecture.

process. Gel spinning is the most commonly used process. The schematic in Fig. 2.53 shows the gel spinning process of UHMWPE fibers. In the process, UHMWPE is firstly dissolved in a solvent such as paraffin, decalin, dodecane, p-xylene, 1,2,4-trichlorobenzene, or kerosene. The solution is then gel-spun through a spinneret and gel-like fibers form. The solvent in the fibers is then extracted by using another solvent such as diethyl ether, n-pentane, methylene chloride, tri-chlorotrifluoroethane, n-hexane, dioxane, or toluene. The fibers are consequently stretched with a drawing ratio of more than 30. Drawing ratio is the ratio of the resultant fiber length after a drawing process to its original length before the drawing process. A higher drawing ratio results in higher alignment of the polymer chains.

The drawing process induces extraordinary alignment of the long polymer chains in UHMWPE. The degree of alignment of the polymer chains can reach more than 95% in the fiber direction. The drawing process can also result in a

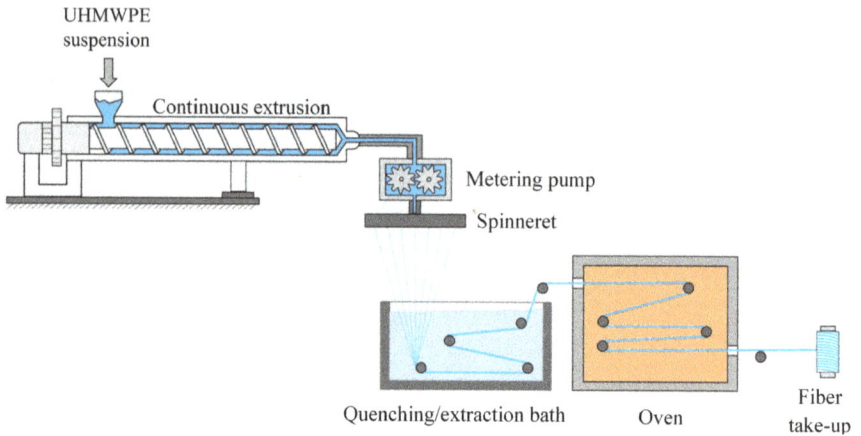

Fig. 2.53: A schematic showing the gel spinning process of ultra-high molecular weight polyethylene fibers (adapted from References [15, 16]).

DOC of more than 85%. Figure 2.54 schematically shows the highly aligned and crystalline structure in UHMWPE fibers and the typical semicrystalline structure of a regular PE in a bulk form for comparison purpose. The UHMWPE fiber with an increased DOC has a higher melting temperature. Figure 2.55 shows DSC curves of UHMWPE fibers with different drawing ratios. The UHMWPE fiber with a higher drawing ratio (up to about 60) possesses a higher melting temperature.

The UHMWPE fiber has anisotropic material properties. Long PE molecular chains are highly aligned in the fiber length direction and have a high load-bearing capacity in that direction. However, only weak van der Waals bond exists in the transverse direction between the polymer chains and hold the chains together. Therefore, the mechanical properties, including tensile modulus and tensile strength, in the fiber direction are significantly higher than those in the transverse direction. This property variation in different directions is indicated by the large difference between its tensile strength and compressive strength. The fiber tends to buckle and forms kink bands from compression loading, resulting in much lower compressive strength compared to tensile strength. This property anisotropy is typical for other fibers with similar structures such as PP fibers, aramid fibers, and carbon fibers.

UHMWPE fibers have been used as marine ropes, fishing lines and net, protective clothing and gloves. The fiber has also been used for producing ballistic helmets that offer a combination of great impact resistance and lightweight. Figure 2.56 shows helmets made of Dyneema® fabrics.

3. **PP fiber.** PP is another thermoplastic that can be used to produce high-performance fibers because of its linear molecular chains. The DOC of the PP fiber can achieve 80%. The PP fiber is resistant to wear, chemicals, and fungi.

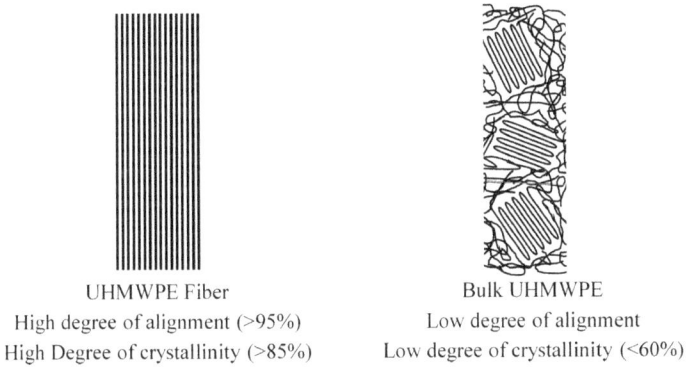

UHMWPE Fiber
High degree of alignment (>95%)
High Degree of crystallinity (>85%)

Bulk UHMWPE
Low degree of alignment
Low degree of crystallinity (<60%)

Fig. 2.54: Comparison of the molecular structures in ultra-high molecular weight polyethylene fibers and bulk ultra-high molecular weight polyethylene.

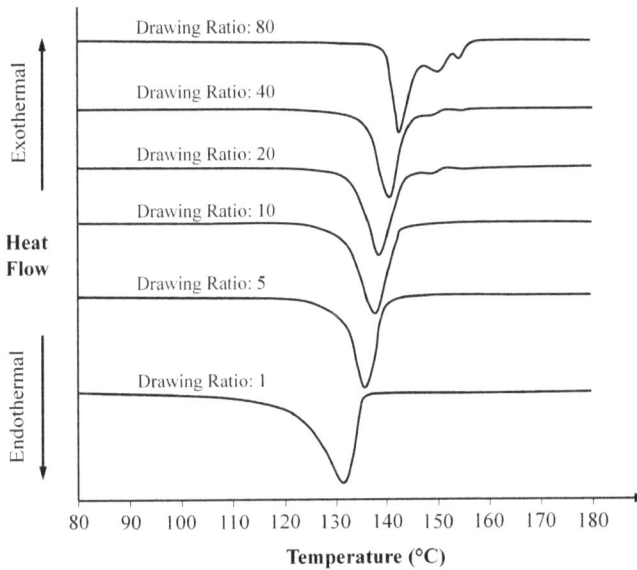

Fig. 2.55: DSC curves of UHMWPE fibers with different drawing ratios (adapted from Reference [17]).

It has a tensile strength up to 650 MPa and tensile modulus up to 19 GPa. Although the PP fiber has lower mechanical properties than UHMWPE fibers, it is more cost-effective compared to UHMWPE fibers. Figure 2.57 shows PP fiber rovings.

PP fibers are commonly manufactured using a melt-spinning process. PP is melted in an extruder and spun through a spinneret to form fibers. The fibers

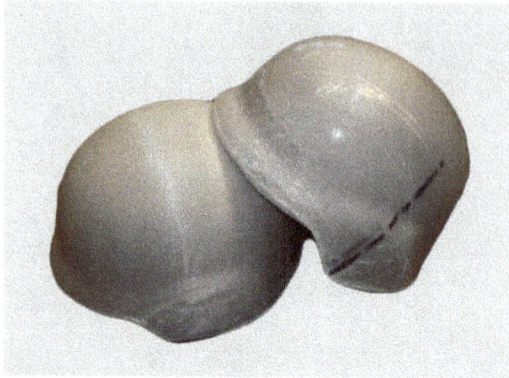

Fig. 2.56: Impact-resistant helmets made of ultra-high molecular weight polyethylene fabrics (Dyneema® fabrics) (reprinted from Reference [18] with permission).

go through multiple drawing stations to achieve maximum alignment of polymer chains and a high DOC.

The PP fiber has been used as a reinforcement in thermoplastic composites besides other applications such as ropes, twines, and so on. A typical use of PP fibers is self-reinforced composite. The self-reinforced composite is a composite that uses the same material for its reinforcement fiber and matrix. PP fiber-reinforced PP matrix composites are a self-reinforced composite with unique mechanical properties. The composite is described with more details in Section 3.2.5.

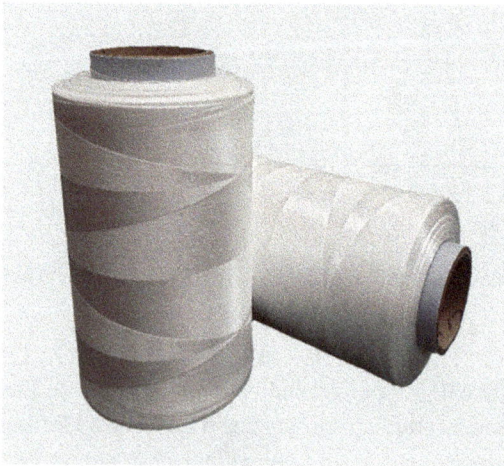

Fig. 2.57: Polypropylene fiber rovings.

2.3.1.4 Basalt fiber

Basalt fibers are produced from volcanic basalt rocks. The basalt rock is melted at approximately 1,500 °C and extruded into basalt fibers. The basalt fiber has excellent strength and modulus, great thermal stability, and excellent chemical resistance. The basalt fiber has been often compared to glass fibers because of their similarity in several aspects. The basalt fiber has a chemical composition similar to glass fibers. Table 2.9 lists the composition of the basalt fiber. Its main constituents are SiO_2, Al_2O_3, and CaO and their weight percentages are similar to the percentages of respective constituents in glass fibers. The basalt fiber possesses comparable tensile strength with S-glass fibers (Tab. 2.6) and its elastic modulus can reach as high as 110 GPa. Basalt fibers have a diameter ranging from 9 to 24 µm, which is similar to the glass fiber diameter. The price of basalt fibers is slightly higher than that of E-glass fibers but lower than that of S-glass fibers.

Basalt fiber-reinforced thermoplastic composites have been increasingly studied because of their large potential in competing with glass fiber-reinforced thermoplastic composites as well as carbon fiber-reinforced thermoplastic composites. There are several commercially available basalt fiber-reinforced thermoplastic composites, including VolcaLite®, a basalt mat PP composite developed by AZDEL Inc., and long basalt fiber composites with various thermoplastic matrices by PlastiComp, Inc. Those composites have found their use mainly as secondary structural materials in automotive applications.

Tab. 2.9: Composition of basalt fibers.

Constituents	SiO_2	Al_2O_3	$Fe_2O_3 + FeO$	CaO	MgO	Na_2O	K_2O	TiO_2	Others
Weight %	50–60	14–19	9–14	5–10	3–5	2.5–6.4	0.8–4.5	0.5–3.0	0.05–1.0

2.3.1.5 Cellulose fiber

Cellulose fiber belongs to the family of natural fibers as it can be directly obtained from plants. It is named after its main content, cellulose, which provides the structural integrity for the plant. The cellulose fiber is also called plant fiber at times. In addition, "natural fiber" is often used to refer to the cellulose fiber, even though "natural fiber" includes other types of fibers such as animal fiber in its original meaning. Cellulose fiber is one of the main natural materials used by mankind since the beginning of human civilization. For example, sticks or straws (consisting of cellulose fiber mainly) were used to reinforce mud bricks for buildings thousand years ago. The mixture of the stick (reinforcement) and the mud (matrix) have been considered the first composite material.

The cellulose fiber is one of the increasingly used fibers in composites, especially thermoplastic composites, because of its low density, relatively good properties, low cost, low abrasion, low carbon footprint, abundant availability, biodegradability, and renewability. In fact, cellulose is the most abundant renewable material on Earth. Although the mechanical properties of cellulose fibers are not as high as those of most man-made fibers, the great combination of their good specific properties, low cost, biodegradability, and renewability makes them an excellent material for secondary structural components that do not require significant load bearing capability. The main consumers for cellulose fiber thermoplastic composite include automotive and construction industries.

Cellulose is the main constituent in cellulose fibers regardless of the type of cellulose fibers. It is a linear polysaccharide of D-glucose units joined by beta-1,4-glycosidic bonds. Its repeat unit consists of two joined glucose units as shown in Fig. 2.58a. The formula is expressed in a Haworth projection to show the orientations of the side groups. The atoms connecting with thicker lines are closer to the observer. However, the ring form is not flat and the chemical formula in a 3D chair form (Fig. 2.58b) shows that certain atoms are out of plane. The linear cellulose molecules are aligned in parallel and interconnected by hydrogen bonds (Fig. 2.58c), which provides the tensile strength and rigidity to the cellulose fiber.

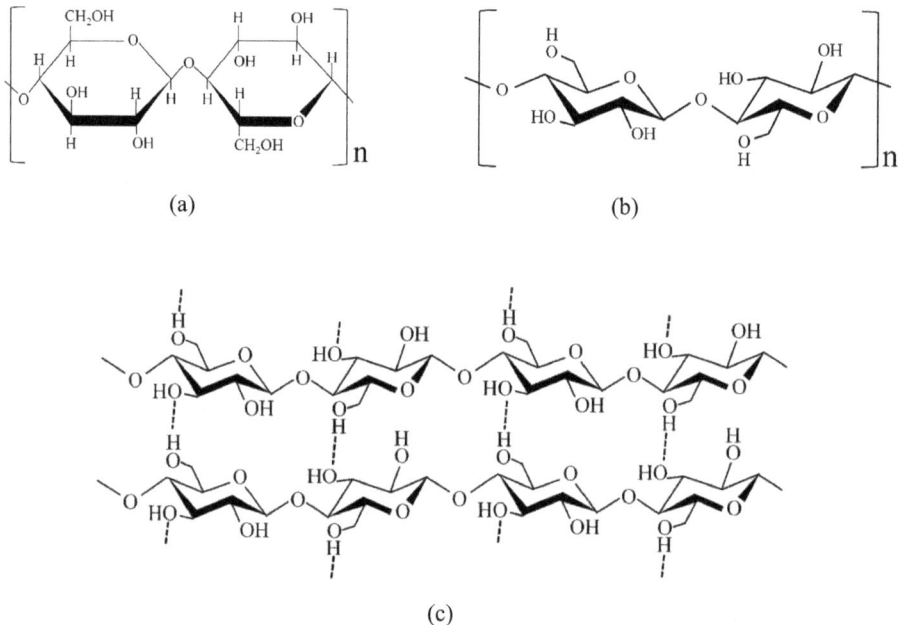

(a)   (b)

(c)

Fig. 2.58: (a) The chemical formula of the repeat unit for cellulose in a Haworth projection; (b) the chemical formula of the repeat unit for cellulose in a 3D chair form; (c) hydrogen bonds (in dashed lines) between linear cellulose chains that are aligned in parallel.

Besides cellulose, hemicellulose and lignin are other two main constituents in the cellulose fiber. Hemicellulose is composed of glucose and sugar monomers. It is amorphous because of a large number of side groups. Hemicellulose is least thermally stable and most hydrophilic compared to cellulose and lignin. Lignin is an amorphous polymer that is hydrophobic due to the presence of aromatic groups. Hemicellulose and lignin form the matrix for cellulose microfibrils. The contents of cellulose, hemicellulose, and lignin vary with the type of the cellulose fiber (Fig. 2.38). In addition, other constituents, such as pectin and wax, may exist in the cellulose fiber.

Cellulose fibers can be harvested from different parts of plants, such as bast, seed, leaf, and fruit. Accordingly, cellulose fibers are categorized into bast fiber (flax, hemp, jute, kenaf, and ramie), seed fiber (cotton, kapok, willow, and milk weed), leaf fiber (abaca, agave, banana, pineapple leaf fiber, and sisal), fruit fiber (coir, oil palm, and luffa), and so on. Those fibers and other cellulose fibers, including wood, stalk fibers, and grass fibers, are illustrated in Fig. 2.38.

Figure 2.59 shows the construction of hemp fiber, a typical bast cellulose fiber. The technical fiber is extracted from the bast of industrial hemp plants. It is commonly used in thermoplastic composites for automotive applications. The technical fiber is composed of a bundle of elementary fibers that are bonded together by pectin. Unlike synthetic fibers such as glass fiber and carbon fiber, each elementary fiber has a unique hollow structure intrinsically. The hollow space, or lumen, originally

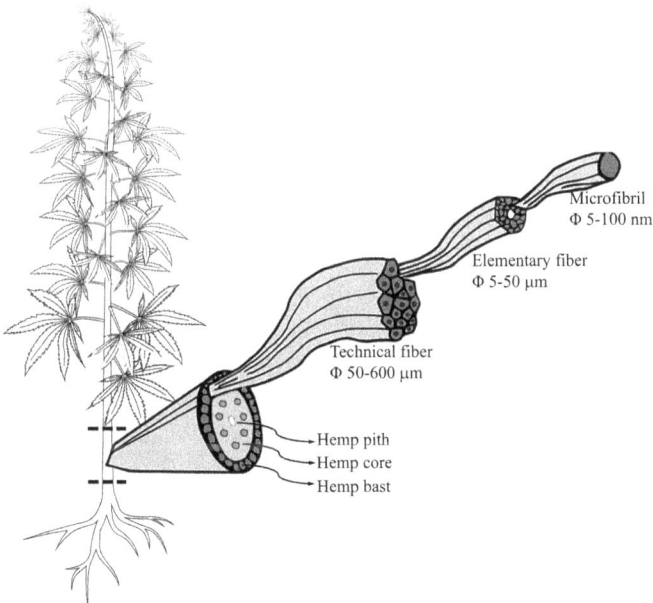

Fig. 2.59: The construction of hemp fiber, a typical bast cellulose fiber, showing technical fiber, elementary fiber, and microfibril (adapted from Reference [19]).

functions as the channel transporting water in living plants through capillary effect. After the plants are harvested and the cellulose fibers extracted from bast or other parts of the plants, the microsize lumens remain in the cellulose fibers. They can take up to 34% of the total volume in the cellulose fiber (except for some seed fibers that have lumens with a much higher volume percentage; however, those seed fibers are rarely used in thermoplastic composites).

Figure 2.60a shows a scanning electron microscopy (SEM) image of the cross section of a sisal technical fiber, a leaf fiber from agave plants. A magnified view of the cross section in Fig. 2.60b shows the elementary fibers bundled together and the contour of each elementary fiber. Each elementary fiber also has a lumen at its center and the lumen size and geometry vary among the elementary fibers. The lumen functions as a channel for transporting water and nutrients in the live plants through capillary effect. The lumen partially contributes to the excellent damping property of cellulose fiber-reinforced thermoplastic composites as it provides discontinuity in the composite. This is significant as the cellulose fiber-reinforced thermoplastic composite is mostly used in automotive application and the lumens in the composite can offer reduced noise, vibration, and harshness (NVH). The cell wall surrounding the lumen in the elementary fiber is composed of rigid and strong cellulose microfibrils embedded in the hemicellulose and lignin matrix (Fig. 2.59).

One of the main reasons that cellulose fibers are used in thermoplastic composites for secondary structures is their good mechanical properties. Table 2.10 lists the elastic modulus and tensile strength for the common cellulose fibers used in thermoplastic composites, including flax fiber, hemp fiber, jute fiber, kenaf fiber, ramie fiber, banana fiber, coir fiber, and sisal fiber. Their densities and compositions are also listed in the table. Those cellulose fibers are normally added as discontinuous

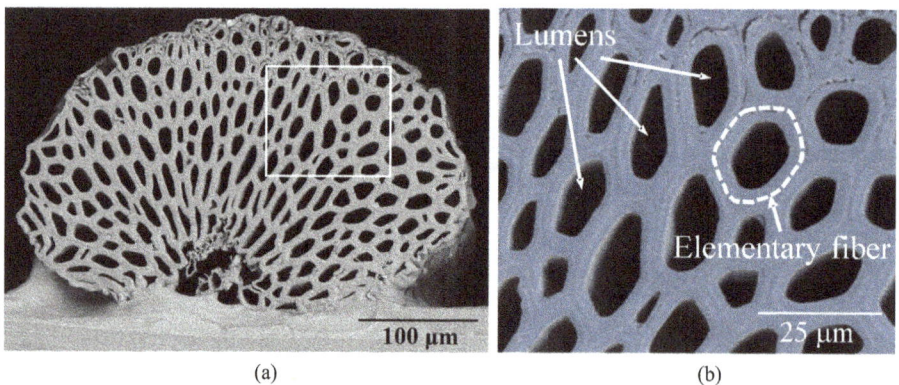

Fig. 2.60: (a) Cross section of a sisal technical fiber and (b) magnified view of the cross section showing elementary fibers bundled together and lumen in the elementary fiber.

fibers to thermoplastics to provide enhanced strength and modulus and reduced cost while still maintaining great processability.

Tab. 2.10: Typical cellulose fibers used in thermoplastic composites and their properties and compositions (reprinted from Reference [20] with permission).

Cellulose fiber	Density (g/cm³)	Elastic modulus (GPa)	Tensile strength (MPa)	Elongation at failure (%)	Cellulose content (wt%)	Hemicellulose content (wt%)	Lignin content (wt%)
Flax	1.4–1.5	27–60	350–1,040	1.3–3.2	62–72	18–21	2–5
Hemp	1.14–1.5	23–60	270–1,100	1–3.5	68–75	15–22	4–10
Jute	1.3–1.5	8–27	320–800	1–1.8	59–72	13–20	11–13
Kenaf	1.4–1.45	14–53	223–930	1.5–2.7	31–72	20–22	8–19
Ramie	1.0–1.5	24.5–44	400–1,000	1.2–4.0	68–85	13–17	0.5–0.7
Banana	1.35	12	500	1.5–9	63–68	10–19	5
Coir	1.15–1.46	2.8–6	95–230	15–51	32–44	0.2–20	40–45
Sisal	1.33–1.5	9–38	363–700	2–7	60–78	10–14	10

The cellulose fibers listed in Tab. 2.10 are commonly used in thermoplastic composites. Figure 2.61 shows images of some of the cellulose fibers, i.e., hemp fibers (Fig. 2.61a), sisal fibers (Fig. 2.61b), banana fibers (Fig. 2.61c), and kenaf fibers (Fig. 2.61d). The following sections describe typical cellulose fibers used in thermoplastic composites.

1. **Hemp fiber** is harvested from the bast section of industrial hemp plants. Industrial hemp differs from the medical-use hemp because of its lower tetrahydrocannabinol (THC) content and higher cannabidiol (CBD) concentration. The industrial hemp plant grows at a fast rate and produces long and strong hemp fibers. Besides its use in pulp and paper industries, the hemp fiber is one of the commonly used cellulose fibers in thermoplastic composites. The main applications of the hemp fiber thermoplastic composite are automotive and construction.
2. **Flax fiber,** another type of bast fiber, is derived from flax plants (Linum usitatissimum). The flax fiber has been traditionally made into textiles, often referred as linen. However, more and more attentions have been drawn by its good mechanical properties and low cost as a reinforcement for thermoplastic composites. The flax fiber-reinforced thermoplastic composite generally uses PP or PE as the matrix. Some typical applications of the flax fiber-reinforced

Fig. 2.61: (a) Hemp fibers; (b) sisal fibers; (c) banana fibers; and (d) kenaf fibers.

thermoplastic composite include automotive interior components, furniture, and commodity products.

3. **Kenaf fiber** is another bast fiber with high strength. The fiber is from kenaf plant, or Hibiscus cannabinus. The kenaf plant has a high rate of photosynthesis (23 μmol $CO_2/m^2/s$) [21], which results in a higher accumulation of CO_2 compared to other plants and makes its fiber a highly eco-friendly material. Like the other cellulose fiber-reinforced thermoplastic composites, kenaf fiber thermoplastic composites are mainly used in automotive application. For example, 50% kenaf fiber-reinforced PP composite has been used to mass produce the door bolster for 2013 Ford Escape.

4. **Sisal fiber** is extracted from the leaves of agave plants that normally grow in southwestern United States, Africa, South America, and Mexico. The fiber has great strength and resistance to corrosive environment. It is traditionally used in ropes and twines because of their high strength and readily availability. Sisal fiber has been gaining attraction as a reinforcement in thermoplastic composites for automotive application.

Cellulose fiber has a relatively low degradation temperature. Figure 2.62 shows the TGA curve of kenaf fibers in a nitrogen environment at a heating rate of 20 °C/min. There is a continuous mass loss initially because of removal of the moisture absorbed in the fiber when the fiber is heated. The moisture mass is about 10% of the fiber mass. The fiber shows a gradual mass loss when heated to around 205 °C, indicating the onset of degradation. Therefore, the processing temperature for cellulose fiber-reinforced thermoplastic composites is maintained below 200 °C to avoid any degradation of the cellulose fiber. The fiber mass reduces at a significantly higher rate when the temperature reaches approximately 250 °C.

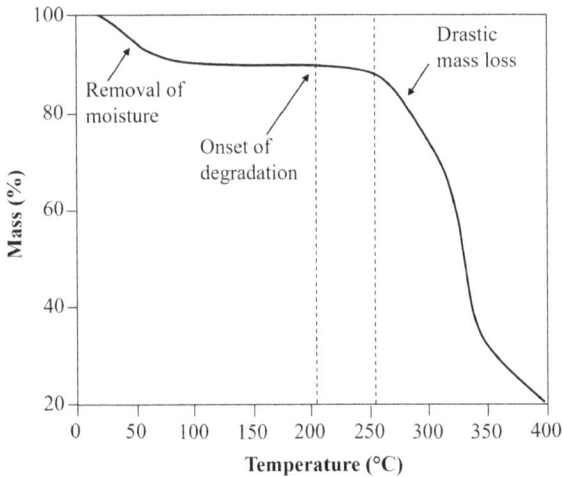

Fig. 2.62: TGA curve of kenaf fiber showing the mass changes caused by removal of moisture and onset of degradation.

Heating of both the thermoplastic matrix and reinforcement fibers is generally required to combine them during processing. The thermoplastic has to be heated above its melting temperature (for semicrystalline thermoplastics) or significantly higher than its glass transition temperature (for amorphous thermoplastics). However, the cellulose fiber can undergo degradation when heated above its degradation temperature (approximately 200 °C) during processing, which limits thermoplastic options to PP, PE, polyoxymethylene, PLA, and PCL, all of which have a processing temperature no more than 200 °C. Among all of the cellulose fiber-reinforced thermoplastic composites, PP and PE matrix composites are most studied and widely used in various applications.

Unlike synthetic fibers, cellulose fibers have uneven surfaces, varying sizes, irregular cross-sectional shapes, inhomogeneous structures, scattering of fiber properties, etc. Moreover, the same type of cellulose fibers does not necessarily provide similar mechanical properties. The properties of the fibers may vary from their different supply sources or even from the same source because of the variation in soil, sunlight, rainfall, nutrition, and so on, for each plant. Their inconsistent mechanical properties are indicated by the large range of elastic modulus and tensile strength for each cellulose fiber listed in Tab. 2.10. Those inherent drawbacks in the cellulose fibers and cellulose fiber-reinforced thermoplastic composites have to be considered during material selection and product design.

Inadequate adhesion between cellulose fibers and the matrix can occur in cellulose fiber-reinforced polyolefin matrix composites. Cellulose fibers have a polar and hydrophilic nature and polyolefins (PP and PE) are nonpolar and hydrophobic. Cellulose fibers are generally treated before being added into PP or other hydrophobic thermoplastics. The treatment of cellulose fibers can be found in Section 2.4.1.3. Additionally, coupling agents can be added to the composite to achieve adequate bonding between the cellulose fiber and the polyolefin matrix (See Section 2.4.2).

Cellulose fiber thermoplastic composites have affinity for moisture because of the highly hydrophilic nature of cellulose. The presence of the large amount of hydroxyl groups in the cellulose fiber attracts water molecules, resulting in high moisture absorption. The amount of moisture absorption increases with the fiber content of cellulose fiber-reinforced thermoplastic composites. The moisture in the composite can considerably deteriorate the mechanical properties of the composite with time. Approaches such as alkaline treatment on the cellulose fiber can reduce the moisture absorption.

The cellulose fibers are generally in a discontinuous form because the cellulose fiber length is limited to the length of the plant. The fiber length is also deliberately shortened to ensure adequate wetout and impregnation with the thermoplastic matrix and adequate flow during processing. The reduced fiber length results in lower fiber aspect ratios and decreased mechanical properties.

In spite of the aforementioned drawbacks, the cellulose fiber-reinforced thermoplastic composite is gaining significant attraction in secondary structural applications for their good specific properties, low cost, low abrasiveness, full recyclability, environmental friendliness, etc. The advancement in material development by improving variables, such as the cellulose fiber aspect ratio and compatibility between the cellulose fiber and the thermoplastic matrix, can further enhance the performance of the cellulose fiber-reinforced thermoplastic composite, which can promote its use in more fields and applications.

2.3.1.6 Other fibers

Other fibers have also been used in thermoplastic composites. Those fibers include boron fibers, metallic fibers, and ceramic fibers.

1. **Boron fiber** is a fiber produced via a chemical vapor deposition (CVD) process. Boron is coated on a core that is in the form of wire or fiber through the following chemical reaction at a temperature of 1,000–1,300 °C. Tungsten wire is normally used because of its high temperature resistance and stability. Carbon and glass fiber have also been used as its core material. Boron fiber does not have 100% boron because of the core material. The boron fiber has a very high modulus and strength. Its tensile modulus can reach 400 GPa and its tensile strength ranges between 3.4 and 4.2 GPa. It also has an extremely low co-efficient of thermal expansion and high compression strength. However, the high processing cost of boron fiber has limited its use to only high-end applications such as aeronautics.

2. **Metallic fibers** are fibers made of metals and alloys, such as steel, aluminum, magnesium, copper, molybdenum, and tungsten. Among those, steel fibers are widely used because of its high strength, excellent toughness, and low cost. The steel fiber has been extensively used to reinforce concrete and rubbers for car tires and conveyor belts. It has been also used to reinforce thermoplastics. Because steel is electrically and thermally conductive, the steel fibers can enhance the electrical and thermal conductivity of the thermoplastic. Other benefits include electromagnetic shielding and electrostatic dissipative. However, steel has a high density and the steel fiber can add significant weight to its composite, especially when its content is high. In addition, it is challenging to achieve good bonding between the steel fiber and the thermoplastic matrix.

3. **Ceramic fibers** are a group of fibers that are made of ceramic materials, including oxides (Al_2O_3, SiO_2, B_2O_3, ZrO_2), carbides (WC, SiC), nitrides (BN, Si_3N_4), and so on. Those fibers have exceptional elastic modulus. The elastic modulus of some commercially available ceramic fibers, e.g., DuPont FP fiber (more than 99 wt% α-Al_2O_3) and DuPont FP 166 fiber (15–25 wt% Al_2O_3 and balance of Al_2O_3), can reach 385 GPa. The ceramic fiber also has excellent tensile strength, ranging from 1,000 to 3,200 MPa, and great compressive strength. However, the ceramic fiber has low strain to failure and relatively high density (2.3–4.2 g/cm^3). Its strength can also suffer from surface flaws such as scratches and cracks. The ceramic fiber is dominantly used in metal matrix composites and ceramic matrix composites for their high strength and modulus at elevated temperatures. It is occasionally used in polymer matrix composites, including thermoplastic composites.

2.3.2 Particulate

Particulate refers to the group of reinforcements/fillers/additives that have small aspect ratios. It includes microsphere, carbon black, calcium carbonate, talc, and so on. Among those, microspheres function as reinforcements in thermoplastics, especially in enhancing compressive strength of the thermoplastic while lowering its density; carbon black is added to certain thermoplastics as protection from ultraviolet light attack; calcium carbonate, talc, and mica are added to thermoplastics, typically PP and PE, as a filler to lower their cost. The sections below describe those typical particulates used in thermoplastics.

1. **Microsphere** is also called microballoon, microbubble, or microbead. It is a hollow sphere with a thin wall. It offers unique properties such as high compressive strength, extremely low density, and thermal insulation. Microspheres can be made of different materials, including glass and carbon. Figure 2.63 shows SEM images of glass microspheres and the fracture surface of a glass microsphere-reinforced thermoplastic polyurethane (TPU) composite. Microspheres generally have thin walls and are brittle, which make them easy to break during processing. The breakage can be severe during compounding with thermoplastics because of high processing pressures involved. Care has to be taken to ensure processing parameters to be optimized to minimize the breakage of the microspheres. Low processing pressure, such as low shear during compounding and low forming pressure during molding, can minimize the breakage.

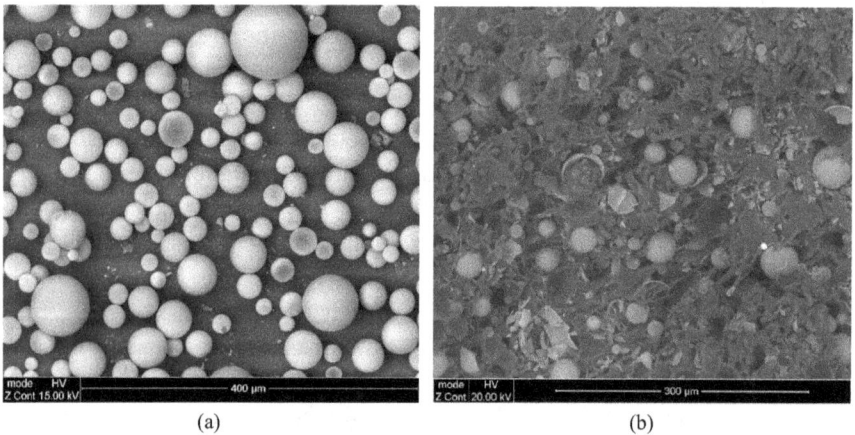

(a)   (b)

Fig. 2.63: (a) SEM image of glass microspheres and (b) fracture surface of a glass microsphere-reinforced thermoplastic polyurethane composite.

2. **Carbon black** is an amorphous carbon, an allotrope of the carbon element (see Section 2.3.1.2). It can be added to thermoplastics or thermoplastic composites (Fig. 4.32). Carbon black is found to offer several major benefits to the thermoplastic or the thermoplastic composite. Firstly, carbon black can improve the UV resistance of its composite for outdoor applications. Carbon black is known to protect the thermoplastic by absorbing the UV light and converting it into heat. This function makes the carbon black an excellent UV stabilizer for certain thermoplastics that are sensitive to UV attack, such as PP. Secondly, carbon black has self-lubricating effects. The thermoplastic or thermoplastic composite with carbon black added shows enhanced wear resistance. Thirdly, carbon black has excellent thermal and electrical conductivities. The thermoplastic or thermoplastic composite with carbon black exhibits good thermal and electrical conductivities. The conductivity increases with the content of carbon black. In addition, carbon black can improve the surface appearance of thermoplastics or thermoplastic composites as a pigment.

3. **Calcium carbonate**, or $CaCO_3$, is a common filler material added to thermoplastics, especially PP and PE. It is nontoxic, odorless, and stable over a wide temperature range. The main purpose of adding calcium carbonate is to lower the cost of thermoplastics and fiber-reinforced thermoplastic composites because of its abundance and low cost. Up to 40 wt% calcium carbonate is added to thermoplastics for certain commercial applications. In addition, it can provide several other benefits such as reduced shrinkage and warpage, enhanced modulus, flame retardancy, and improved impact resistance.

4. **Talc** is another common filler for thermoplastics, typically PP. It has a chemical formula of $Mg_3Si_4O_{10}(OH)_2$. Talc is the softest mineral and often used as a solid lubricant. It has a low shear strength because of the weak van der Waals force among the talc layers. The addition of talc in a thermoplastic can improve the modulus and creep resistance of the thermoplastic but decrease its elongation at failure and impact strength.

5. **Flame retardants**. Some thermoplastics, such as PPS and PEEK, are self-extinguishable, making them and their composites inherently fire retardant materials. However, other thermoplastics, including most engineering and commodity thermoplastics (Fig. 2.3), and their composites need fiber retardants for related applications. Typical flame retardants for the thermoplastic and thermoplastic composite include aluminum hydroxide, calcium carbonate, and ammonium polyphosphate, to name a few. Aluminum hydroxide, or $Al(OH)_3$, decomposes into Al_2O_3 and H_2O at about 200 °C. The decomposition is an endothermic reaction and water is released from the reaction, both of which lead to its fire retardancy. Ammonium polyphosphate, or $((NH_4)PO_3)_n(OH)_2$, is also used to protect thermoplastics or thermoplastic composites from fire through swelling and charring. Those

flame retardants help increase the ignition resistance and reduce flame-spreading speed and smoke generation for thermoplastics and thermoplastic composites.

2.3.3 Nanoscale reinforcements

Nanoscale materials, or nanomaterials, are the materials that have at least one of the dimensions in the nanometer range. Those materials have great aspect ratios and have been used to reinforce thermoplastics. Good dispersion of the nanomaterials in the thermoplastic matrix is critical. Poor dispersion causes agglomeration in the thermoplastic which can result in reduction in strength. The common nanomaterials used in thermoplastic composites are nanoclay, carbon nanotube, nanographene, nanofibers, and so on. Those nanomaterials are described below.

2.3.3.1 Nanoclay

Nanoclays are mainly made of layered aluminum silicates that form clay crystallite, with a variable amount of other elements including iron, calcium, and magnesium. Each individual layer in the nanoclay consists of octahedral and/or tetrahedral sheets. The octahedral sheet is made of aluminum-oxygen octahedrons with aluminum possibly substituted by magnesium. The tetrahedral sheet is made of silicon–oxygen tetrahedrons linked with the octahedrons. The arrangement of the octahedron and tetrahedron sheets and their compositions determine the type of nanoclays. Totally there are more than 30 types of nanoclays.

Nanoclays are commonly used in thermoplastic composite as a filler material because of their low cost and low environmental impact. Montmorillonite (MMT), with a 2:1 ratio of tetrahedral sheets and octahedral sheets, is the most common nanoclay used in thermoplastic composites. Besides its abundancy and low cost, it can enhance strength, modulus, thermal stability, and flame redundancy for thermoplastics and thermoplastic composites.

Nanoclays are commonly blended with thermoplastics via a melt-blending method in an extruder. It is important to induce intercalation or exfoliation during the processing. Intercalation is the insertion of polymer chains into naonclay sheets. The spacing among the sheets is expanded only to a limited extent. Exfoliation is a disruption of nanoclay sheets with spatially separation among the sheets. Exfoliation is highly desired as it provides effective reinforcement for the composite. Exfoliation can be achieved to an extent through a high shear force induced in extrusion processes, especially the twin screw extrusion process that can overcome the van der Waals bonds among the nanoclay sheets. Surfactants such as quaternary alkylammonium and alkylphosphonium compounds are often added to assist separation of the nanoclay sheets and attain a higher degree of exfoliation.

Based on the extent of intercalation, nanoclay-reinforced thermoplastic composites are categorized into microcomposite, intercalated nanocomposite, and exfoliated nanocomposite. Figure 2.64 shows the schematics of those thermoplastic composites. The exfoliated structure is the most desired structure for nanoclay-reinforced thermoplastic composites.

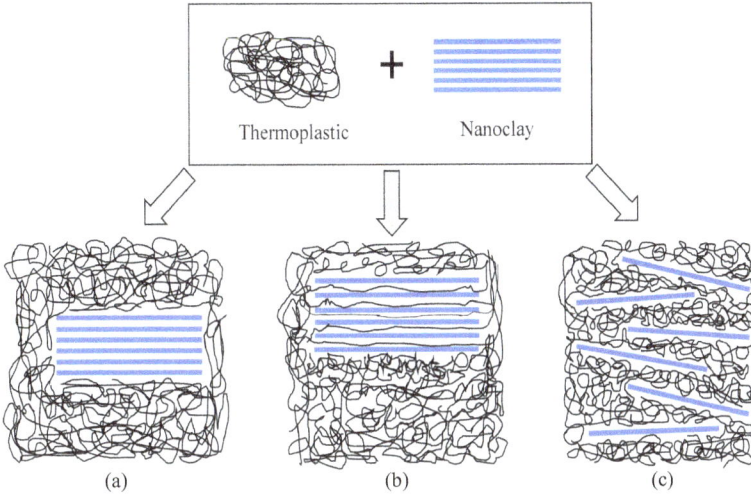

Fig. 2.64: Nanoclay-reinforced thermoplastic composites with different extents of intercalation; (a) microcomposite; (b) intercalated nanocomposite; and (c) exfoliated nanocomposite.

The addition of the nanoclay to thermoplastics can improve their modulus and strength. When the nanoclay is dispersed with exfoliation, its composite can exhibit enhanced strength and modulus even with a small quantity added. Figure 2.65

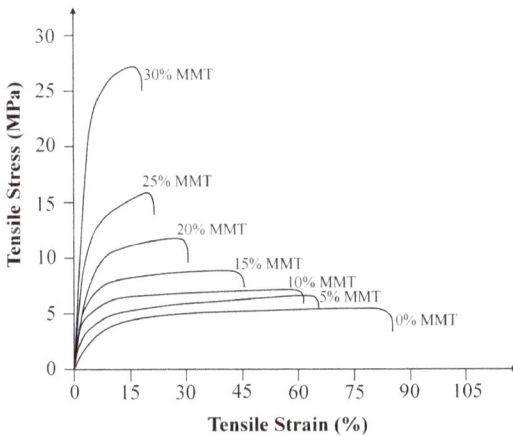

Fig. 2.65: Increase of tensile strength and modulus with increasing MMT nanoclay content in thermoplastic starch (reprinted from Reference [22] with permission).

shows the effect of MMT added to thermoplastic starch [22]. The tensile strength and modulus of the thermoplastic starch composite increase with the MMT content. In addition, the nanoclay contributes to increased heat distortion temperature and melting temperature of the thermoplastic starch. For the thermoplastic used in packaging film applications, nanoclays can help reduce its moisture permeability and oxygen permeability.

2.3.3.2 Other nanoscale reinforcements

Besides nanoclays, other nanoscale materials used to reinforce thermoplastics include nanotubes, nanographenes, nanofibers, and nanowhiskers. Although those materials are not commonly used in commercial products due to their high cost, research work has been extensively done on their effect on the structural performance of thermoplastics and thermoplastic composites as well as their thermal, electrical, and optical properties.

1. **Carbon nanotube (CNT)** is one allotrope of carbon (Fig. 2.43d). It is composed of sheet(s) of graphene formed into a seamless cylinder (single-walled carbon nanotubes) or multiple concentric cylinders (multiwalled carbon nanotubes). CNTs have superb mechanical properties. Their elastic modulus and tensile strength can reach 1,000 and 200 GPa [23], respectively. Additionally, CNTs possess unique electrical, optical, and thermal properties. Since its discovery in 1991, CNT has found its use in many areas, including the area of thermoplastic composites. CNT has been used as a reinforcement for thermoplastics to improve their structural performance. Figure 2.66a,b shows TEM images of multiwalled CNTs and a CNT-reinforced nylon 6 composite, respectively.

   CNTs have been added to various thermoplastics through different processing methods, including solution mixing, melt mixing, and extrusion. Overall, the CNT provides enhanced strength and modulus to the thermoplastic. Those properties increase with the CNT content to a certain level. However, the property enhancement from the addition of the CNT is very limited. The difficulty in dispersion and CNT alignment control as well as its discontinuity and inadequate bonding with the matrix are the main issues limiting its reinforcement effect. For those reasons, CNTs are not considered as an efficient and cost-effective reinforcement material for thermoplastic composites in spite of their superior mechanical properties individually. It is not practically viable to use the CNT as a mere structural reinforcement for thermoplastics due to its high cost and limited reinforcing function, unless other benefits such as electrical, optical, and thermal properties, are required in the meantime to produce multifunctional composites.

2. **Nanographene (NGP)** is another allotrope of carbon. It is a single sheet of carbon atoms with a two-dimensional honeycomb arrangement (Fig. 2.43c). NGP has tensile modulus of 1,000 GPa and tensile strength up to 130 GPa. Similar to

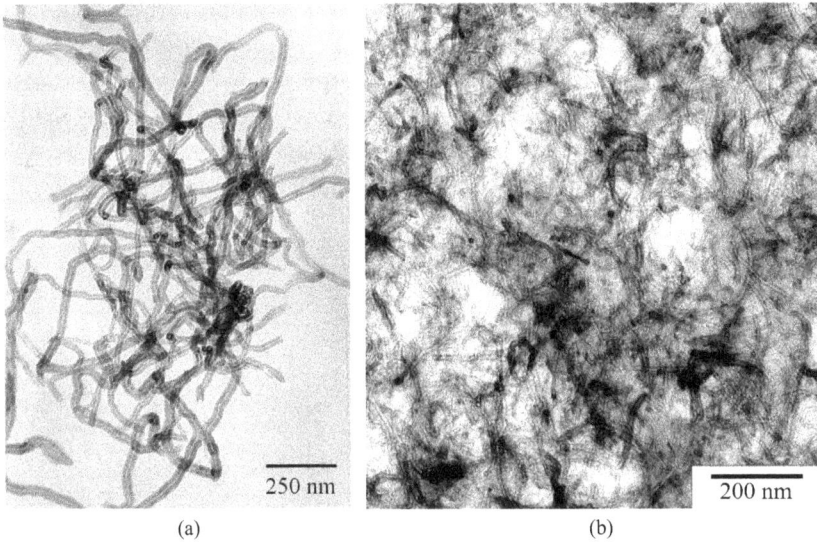

Fig. 2.66: (a) TEM images of multi-walled carbon nanotubes (reprinted from Reference [24] with permission) and (b) nylon 6 reinforced with 5 wt% carbon nanotubes (reprinted from Reference [25] with permission).

CNT, NGP possesses unique electrical, optical, and thermal properties. It has been added to thermoplastics to form multifunctional composite materials that provide not only enhanced strength but also other functions. For example, graphene-based thermoplastic matrix composite casings for electronics can provide great thermal conductivity to dissipate heat generated by the electronics besides enhanced strength and rigidity.

3. **Nanofibers (NFs)** are fibers with diameters in the nanometer scale. Typical materials used for producing NFs include carbon, cellulose, collagen, chitosan, and so on. Carbon nanofibers (CNFs) are a common nanofiber that has been added to thermoplastics in various research and development work. CNFs are normally produced through a CVD process, in which carbon from decomposed gas phase molecules at elevated temperatures is deposited onto a substrate and grows into CNFs in the presence of a metal catalyst. The tensile modulus and strength of the CNF can reach 600 and 7 GPa, respectively. However, its reinforcing function is limited when used as a reinforcement in thermoplastics and other polymers.

4. **Nanowhiskers** are highly crystalline needle-shaped particles with a nanometer scale size. The DOC of the nanowhisker is normally higher than 75%. Because of their high DOC, nanowhiskers are often called nanocrystals. Typical nanowhiskers include ceramic (TiO_2, Al_2O_3, SiC, etc.) nanowhiskers and polysaccharide (cellulose, chitin, chitosan, etc.) nanowhiskers. Nanowhiskers possess high

mechanical properties. For example, the tensile modulus and strength of cellulose nanowhiskers can reach 130 and 10 GPa, respectively.

Figure 2.67 shows a TEM image of chitosan nanowhiskers and a starch matrix composite reinforced with 1% chitin nanowhiskers. The same challenges mentioned in the nanotube section exist for nanowhiskers when used as reinforcements in composites albeit their great mechanical properties.

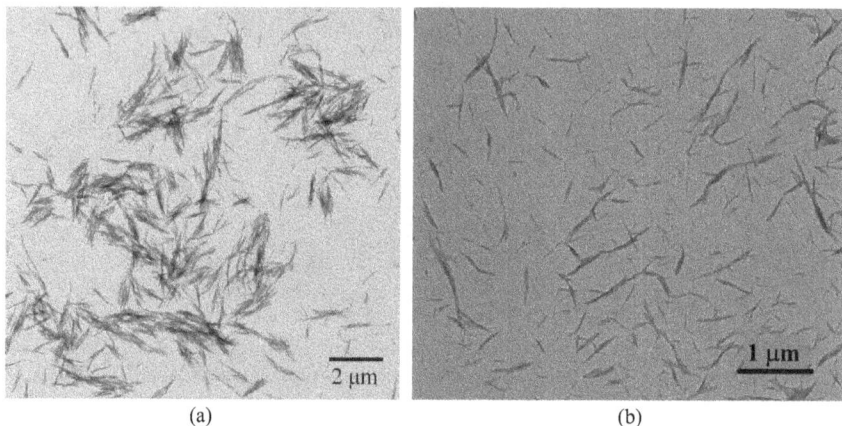

(a) (b)

Fig. 2.67: Transmission electron microscope image of (a) chitosan nanowhiskers and (b) a starch matrix composite reinforced with 1% chitin nanowhiskers (reprinted from Reference [26] with permission).

2.4 Fiber/matrix interface

The performance of a polymer matrix composite, including thermoplastic composite, is highly dependent on the properties of the fiber and the matrix. As the properties of the fiber and the matrix increase, the performance of their composites normally increases. However, there is another constituent that exists in the composite but is not as apparent as the fiber and the matrix. This constituent is the fiber/matrix interface, or the interface between the fiber and the matrix. It is a transition area that the material property changes drastically. The fiber/matrix interface plays a crucial role in determining the mechanical properties of the thermoplastic composite as well as its failure modes.

Load transfer from the matrix to fibers in any composites, including thermoplastic composites, relies on their interfacial bonding. When a composite structure is loaded, the load is applied onto the matrix initially as the matrix is surrounding the fibers. An efficient load-bearing composite structure is capable of transferring a majority of the load to the fibers (see Example Question 7.1). In contrast, if the load is not effectively transferred to the fibers, the composite may undergo premature failure such as fiber pullout and matrix fracture (see Fig. 3.14). Therefore, the interface between the fiber and the matrix is one major factor that determines the performance and the failure mode of the thermoplastic composite.

One of the factors affecting the fiber/matrix adhesion is the surface energy of the matrix. Surface energy is defined as the amount of external work required to create a new unit surface area on a material in vacuum. It is also referred to as interfacial free energy or surface free energy. Surface energy is calculated as energy over surface area and has a unit of mJ/m^2. It is often written as mN/m because J is equal to newton times meter (J=N.m). The surface energy of a material determines its adhesion force.

Thermoset composites such as epoxy matrix composites generally have a strong fiber/matrix adhesion. In addition to the sizing applied on the fiber tailored for thermosets and their low viscosity, both of which induce a great fiber/matrix adhesion, the high surface energy of thermosets also contributes to the adhesion. For example, epoxy has a surface energy of 42–50 mN/m and vinyl ester resin has a surface energy of up to 58 mN/m. On the other hand, thermoplastics can have a low surface energy. PTFE has the lowest surface energy (18–20 mN/m), followed by PVDF and polyolefins. Surface energy values of different thermoplastics used in composites are listed in Tab. 2.11. Polarity in thermoplastics is one main factor that determines their surface energy. For a thermoplastic that has side groups with polarity, it tends to have a high surface energy. PP has methyl groups that are not polar. Therefore, it has a relatively low surface energy (29–35 mN/m). PE is another polyolefin polymer and also has a low surface energy (30–36 mN/m). Both the polyolefin polymers exhibit intrinsically poor adhesion with fibers. On the other hand, a thermoplastic with a polar side group has a higher surface energy. For example, ABS with polar nitrile (–CN) groups has a high surface energy, 42 mN/m.

Other factors including surface characteristics of the fiber and compatibility between the fiber and the matrix also play an important role in the fiber/matrix adhesion. The adhesion between the fiber and the matrix is realized through processing when wetting of the fiber by molten thermoplastics and solidification of the thermoplastic take place under pressure. The viscosity of the molten thermoplastic and the pressure also significantly affect the fiber/matrix adhesion.

Different approaches have been developed to enhance the fiber/matrix adhesion in thermoplastic composites. Those approaches mainly involve the surface treatment on fibers. The surface treatment can induce both chemical and physical alterations on the fiber surface. Typical methods are coating of the fibers (or sizing), plasma treatment of the fibers, etching, chemical treatment, grafting, etc. In addition, coupling agents are also added to the thermoplastic matrix for providing compatibility and better adhesion with fibers. Figure 2.68 summarizes those typical treatment methods for enhancing the fiber/matrix adhesion.

Tab. 2.11: The surface energy of different thermoplastics used in composites in comparison with the surface energy of epoxy.

Polymers	Surface Energy (mN/m)
Polytetrafluoroethylene (PTFE)	18–20
Polyvinylidene fluoride (PVDF)	25–30
Polypropylene (PP)	29–35
Polyethylene (PE)	30–36
Polyoxymethylene (POM)	32–36
Polyphenylene sulfide (PPS)	38
Nylon 6	38–44
Polyethylene terephthalate (PET)	38–48
Polyvinyl chloride (PVC)	39–42
Polymethyl methacrylate (PMMA)	41
Polyether ether ketone (PEEK)	41–42
Nylon 12	41
Acrylnitrile butadiene styrene (ABS)	42
Nylon 66	43–47
Polyethersulfone (PES)	46
Epoxy	42–50

The failure mechanism in thermoplastic composites is an indication of the effectiveness of the fiber/matrix bonding. When the bonding is strong and the fiber length is adequately large (above the critical fiber length), fiber fracture mainly happens. However, when the bonding is not effective, the fiber/matrix interface is weak and separation of the fiber from the matrix can be easily resulted under loading. Thus, the fiber tends to be pulled out from the matrix as the main failure mechanism. These failure modes can be differentiated by the fracture surface under microscope, typically scanning electron microscope (see Section 6.9). Adhesion of the fiber/matrix interface can also be quantitatively measured using micromechanical testing methods described in Section 6.9.2.

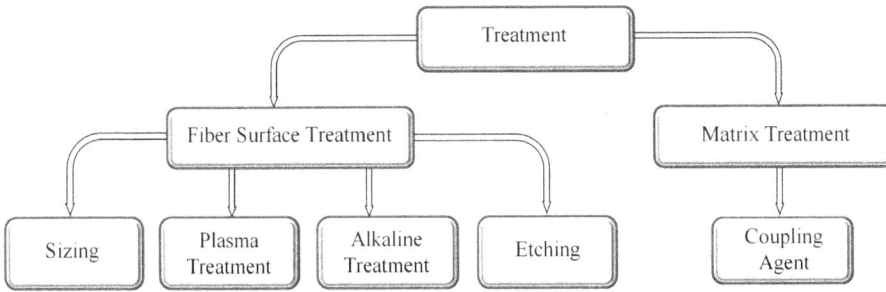

Fig. 2.68: Different treatment methods for enhancing the fiber/matrix adhesion in thermoplastic composites.

2.4.1 Fiber surface treatment

Several methods for treating fiber surface have been developed to enhance the adhesion between the matrix and the fiber. Those fiber surface treatment methods include application of sizing, plasma treatment, chemical treatment, and etching. The following sections detail those different methods.

2.4.1.1 Sizing

Sizing refers to a thin coating applied onto the fiber surface. It is also used to describe the process that a thin coating is applied onto fibers. Sizing is the most common method used to treat fibers because of its effectiveness in improving fiber/matrix adhesion as well as other functions, such as fiber protection and ease of handling. Application of sizing can also be achieved at a high rate, for example, the application rate can reach more than 40 meters per second for coating glass fibers. Sizing technology is one of the biggest commercial secrecy kept among fiber manufacturers. The chemical composition in the sizing, which is the core information of sizing technology, is normally kept as a secret by the fiber manufacturer because sizing is such an important factor affecting the performance of the composites and the sales of the fibers. This also reflects the important role of the fiber/matrix interface on the performance of the composite.

Sizing ingredients normally include coupling agents, film formers, modifiers, and water. Those ingredients are mixed to form emulsion and applied to the fiber surface during the sizing process. A small amount of sizing chemicals is normally applied on the surface of the fibers. Typically, 0.5–2.0 wt% sizing (relative to the fiber weight) is applied to the fibers. Although the sizing accounts for a small portion of the final thermoplastic composite product, it plays a critical role from handling of the fibers to processing of their composites and enhancing the performance of the final composite product. Overall, the sizing mainly has the following functions for better handling of the fibers and processing of thermoplastic composites.

1. **Enhance lubrication for protecting fibers from abrasion**. When the fiber is used for processing pre-impregnated thermoplastic composite preforms, it makes contact with other fibers, fiber guide plate, dies, impregnation pins, etc. The abrasion resulted from the contact can induce defects on the fiber surface and even breakage. When any pre-stress is involved during the processing, the fiber can suffer more damage from the contact. The defects can considerably diminish the strength of the fiber because of its high sensitivity to surface defects. One of the main functions of the sizing applied on the fiber is to protect the fiber surface from damage and minimize defects. The sizing coated on the fiber surface enhances the lubrication for the fiber during processing, reduces friction, and protects the fiber from abrasion and breakage.

2. **Facilitate easier handling and processing**. Fuzz is severe separation of filaments from the fiber strand/tow when there is excessive abrasion, tear of filaments in the tow, lack of pre-stress tension, etc. Once fuzz is initiated during handling or in a process, such as melt pre-impregnation process (see Section 4.3.1), the quality of the composite will deteriorate. In worst cases, processing needs to be stopped and the fuzz has to be cleared for any further processing. Sizing helps holding filaments together and maintaining the integrity of the strand/tow and, therefore, minimizes the occurrence of the fuzz during handling and processing.

3. **Promote adhesion between the filament and the matrix.** The fiber and the matrix in a thermoplastic composite are normally composed of different materials with distinct chemical structures. Without any fiber sizing, the adhesion between those different materials can be insufficient to achieve efficient load transfer in the composite. The sizing applied to the fiber surface can significantly promote the fiber/matrix adhesion and improve mechanical properties of the thermoplastic composite.

Sizing is very specific to the polymer type used in the composite. There is no sizing that can fit all polymer systems. The effort invested in the research and development of proper sizing technology, including chemical recipe, has gone through a long duration for thermoset composites. The sizing technology for thermoset composites is relatively mature compared to that involved in thermoplastic composites. The low viscosity and high surface energy of the thermoset resin systems, such as epoxy (see Section 2.1), also facilitate development of desired sizing for the thermoset composite. Most of the sizing agents for thermoset composites are epoxy-based and mainly for epoxy matrix composites.

Silane-based sizing is another common sizing for fibers used in composites. Silanes are a group of chemicals that are applied to the surface of fibers to achieve good adhesion with both thermosets and thermoplastics. Its typical formula is 80–90% film former, 5–10% silane coupling agent, and 5–15% size modifiers. Silane treatment, or chemical treatment of fibers using silanes, is normally used for fibers that have hydroxyl groups, including glass, basalt, and cellulose fibers. A generic chemical formula of silanes can be written as $R_{(4-n)} - Si - (R'X)_n$ (n=1,2), where R is alkoxy, X represents an organofunctional group, and R' is an alkyl bridge (or alkyl spacer) connecting the silicon atom and the organofunctional group [27]. The silane functions as an agent that couples the fiber and the thermoplastic matrix. The coupling effect is resulted from the function groups in the silane. One of the silane function groups reacts with the function group on the fiber, for example, hydroxyl groups, while the other function group reacts with a certain function group in the thermoplastic matrix. The most common types of silanes as the coupling agent for thermoplastic composites include γ-aminopropyl triethoxy silane (often abbrieviated as APTES or APS), γ-methacryloxypropyl trimethoxy silane, trichlorovinyl silane, and vinyl trimethoxy silane.

The sizing technology for the fibers used in thermoplastic composites has been continuously advanced. However, development of sizing technology in thermoplastic composite is much more challenging than that in thermoset composites. First of all, there are a wide variety of thermoplastics that have distinct characteristics such as polarity, function groups, and surface energy. The sizing developed for one thermoplastic cannot be simply adopted for another thermoplastic. The sizing chemical differs significantly with the polarity of the matrix. For nonpolar thermoplastics, such as PP and PE, the sizing with nonpolar surface groups is desirable. On the other hand, it is desirable to introduce carbonyl and hydroxyl groups to the fiber surface for polar thermoplastics such as PAs and PET. Therefore, sizing chemistry in thermoplastic composites could vary significantly. For example, Michelman Inc. has developed Hydrosize PP2-01 sizing for glass fiber-reinforced PP. It is a maleic anhydride grafted PP dispersion. Another sizing, HP PA845, is anionic PA dispersion and it is used for sizing carbon fibers that reinforce nylon matrices. These two types of sizing have distinctly different chemistry. Secondly, the high viscosity of thermoplastics poses a great challenge in the fiber wetout and impregnation during processing. The sizing has to function as an agent that can enhance spreading of the molten thermoplastic, fiber wetout, and impregnation besides other roles. Finally, thermoplastic composites are normally processed at elevated temperatures, which require the sizing to be thermally stable at those temperatures. If the sizing is not thermally stable, degradation will likely occur to the sizing and result in poor fiber/matrix interfacial bonding. Sizing developed for variable thermoplastic matrices has different chemical and thermal stability. Figure 2.69 compares the thermal stability among polyurethane (PU)-based sizing, PA-based sizing, and PI-based sizing [28]. The PU-based sizing is suited for PA and PBT matrix composites. The PA-based sizing has a high degradation temperature and is suited for high temperature PAs such as PPA. The PI-based sizing, a highly thermally stable sizing, has a degradation temperature of more than 500 °C and is suitable for high temperature thermoplastics such as PPS, PEEK, and PEI.

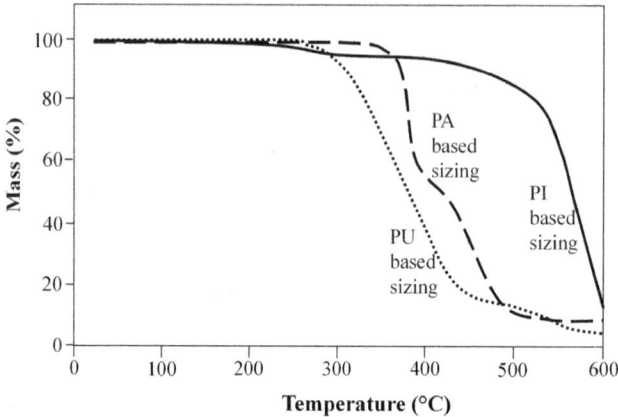

Fig. 2.69: Thermal behaviors of polyurethane-based sizing, polyamide-based sizing, and polyimide-based sizing (adapted from Reference [29]).

2.4.1.2 Plasma treatment

Plasma treatment can be applied on fiber surface to enhance the adhesion between the fiber and the matrix. Plasma, one of the four phases of matter, is made up of a mixture of charged ions, neutral atoms, and free electrons. It can be artificially generated by stripping electrons from gas atoms and creating positively charged ions via heating or applying a strong electromagnetic field. Plasma essentially is a fluid that can interact with electromagnetic fields. It has been used to treat fiber surfaces to induce chemical and physical alterations, both of which help promoting the adhesion fibers with thermoplastics. The plasma treatment introduces functional groups and therefore functionality to the fiber surface that can couple with the function group in the thermoplastic matrix. The function group introduced through the plasma treatment can increase the polarity and the surface energy of the fiber. It can also decrease the contact angle of the molten thermoplastic with the fiber surface, facilitating better fiber wetout with the molten thermoplastic and enhanced fiber/matrix adhesion. In addition, the plasma can alter the fiber surface morphology and roughness. The fiber surface is normally found to have increased roughness after plasma treatment. A larger surface roughness of the treated fiber can lead to better mechanical interlock and increased contact area with thermoplastic, both of which can improve the fiber/matrix adhesion.

Several types of gases, such as Ar, H_2, CO_2, O_2, NH_3, xylene, and helium, are used to generate plasmas for treating fibers. Those gases can introduce different function groups to the fiber surface. Typical function groups include $-OH$, $-C=O$, carboxyl, and amide. The function group can form strong bonding with the function groups in the thermoplastic matrix and therefore increase the performance of the thermoplastic composite.

Plasmas generated at different temperatures are used for treating different types of fibers. High temperature plasmas are desired to treat thermally stable fibers such as glass fibers and carbon fiber (under inert atmospheres). Low temperature plasma treatment is suited to treating organic fibers, such as PE fibers, PP fibers, and cellulose fibers to avoid degradation.

2.4.1.3 Alkaline treatment

Alkaline treatment involves applying alkaline solutions, such as NaOH solutions, to the fiber surface and modifying its chemical characteristics for better fiber/matrix adhesion. Alkaline treatment is normally used for cellulose fibers to improve their adhesion with the thermoplastic matrix in cellulose fiber reinforce PP or PE composites. Cellulose fibers have a large amount of hydroxyl groups (-OH) and exhibit hydrophilic characteristics. On the other hand, PP and PE are hydrophobic, which makes them incompatible with cellulose fibers. In addition, there is polarity difference between those thermoplastics and the cellulose fiber, resulting in a poor adhesion between the cellulose fiber and the polyolefin matrix.

The alkaline treatment is a process adopted from the pulp and paper industry. Cellulose fibers are soaked into a NaOH solution typically to reduce their hydrophilicity. Its adhesion with the thermoplastic is enhanced by mechanical interlocking because of the increase in surface roughness and the amount of exposed cellulose after removal of lignin, wax and oil that cover the external surface of the fiber cell wall. The solution generally has a NaOH concentration no more than 10%. A higher concentration may induce damage to the cellulose fiber that outweighs the benefits from the treatment. The soak time needs to be well controlled to maximize the effect of the alkaline treatment. Silane treatment after the alkaline treatment on the cellulose fiber can further enhance its adhesion with the thermoplastic matrix.

Physical alteration, such as splitting of the technical fiber and collapse of the lumen, can occur to the cellulose fiber after the alkaline treatment. The technical fiber is consisted of individual elementary fibers that are bonded together by pectin. With the removal of the pectin, a technical fiber bundle can split into fibers with smaller diameters. Those smaller fibers provide a higher length-to-aspect ratio because the length of the fiber is largely maintained during the alkaline treatment process. A higher aspect ratio of the fibers results in better mechanical properties of their composite. In addition, fibers with a smaller size offer more surface areas for bonding with the thermoplastic matrix. Another effect from this alkaline process is the collapse of the lumens. It is equivalent to reducing empty spaces, or voids, in the cellulose fiber-reinforced composite, which can contribute to the improvement of the properties for the composite.

2.4.1.4 Etching on fiber surface

Carbon fiber surfaces are sometimes etched before any sizing is applied. The process using chemicals to modify the surface morphology and surface energy is called etching. The fiber surface morphology is modified to have more surface area for bonding with the thermoplastic matrix. The etching can be done with different chemicals, typically acids. For example, nitric acid is a common chemical to etch the surface of carbon fibers and induce a rough surface and higher surface energy. Figure 2.70 shows the striations on a carbon fiber surface developed through etching in 68% nitric acid at 110 °C for 3 h [30]. Striations are created to increase the surface area for better adhesion. The etching also removes amorphous carbon and defective layers from the carbon fiber. The fiber diameter can be reduced due to the material removal. The mechanical property of the etched fiber can deteriorate if there is excessive etching. It is necessary to closely control etching parameters such as etchant type, etchant concentration, and etching time to achieve the maximum adhesion between the fiber and the thermoplastic matrix.

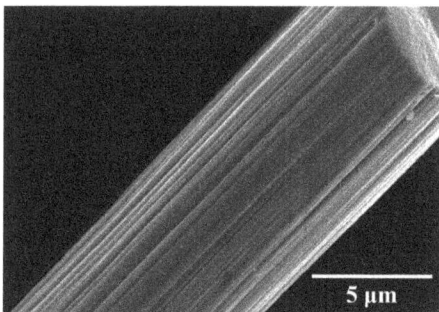

5 μm

Fig. 2.70: A carbon fiber after being etched in nitric acid for improved adhesion with thermoplastics (adapted from Reference [30]).

2.4.2 Coupling agent added to matrix

Coupling agents are compounds which provide chemical coupling between two dissimilar materials. The coupling agent has functional groups, each of which can react with one of the two dissimilar materials and form chemical bonds, thus coupling these two materials together. Other than the treatment applied onto the fibers to induce coupling effect between the fiber and the thermoplastic (such as silane treatment), another approach has been used to achieve the coupling function, e.g. adding coupling agents into thermoplastic matrices. A typical example is that a coupling agent is added to PP to enhance its adhesion with cellulose fibers. PP is one of the most used matrices in the thermoplastic composite because of its good performance to cost ratio. It is commonly combined with cellulose fibers to form a composite with low cost and good mechanical properties for secondary structure applications.

However, PP has a nonpolar and hydrophobic characteristic while the cellulose fiber is polar and hydrophilic because of its significant number of hydroxyl groups. The difference has posed challenges in achieving good adhesion between PP and the cellulose fiber. In addition to the treatments applied to the cellulose fiber as previously mentioned, namely, alkaline treatment and silane treatment, coupling agents, such as maleic anhydride, are added into the PP matrix to produce maleic anhydride grafted PP or maleated PP (MAPP). MAPP reacts with the hydroxyl group in the cellulose fiber and good adhesion is realized between the PP matrix and the cellulose fiber. Figure 2.71 shows that MAPP reacts with the hydroxyl group in a cellulose chain. In addition, MAPP is also added to glass fiber-reinforced PP matrix composites that have glass fibers coated with silane sizing. The MAPP acts as coupling agent between the silane and the PP matrix to achieve better fiber/matrix adhesion (see Fig. 6.13).

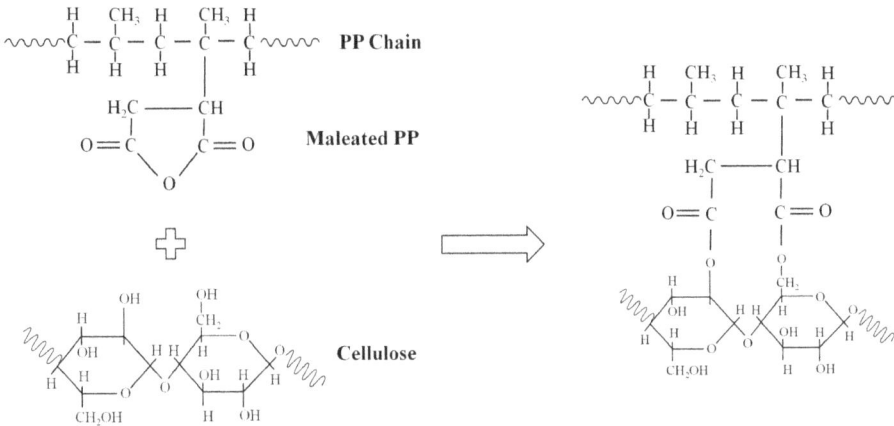

Fig. 2.71: Maleic anhydride polypropylene that is formed from polypropylene and maleic anhydride reacts with the hydroxyl group in cellulose fibers.

2.5 Voids

Voids in a composite are empty spaces that are not occupied with fibers or matrix. The void exists in all composites, including thermoplastic composites. It is another constituent in the thermoplastic composite and plays a critical role in determining the physical property, mechanical property, and thermal property of the composite. Compared to thermoset composites, the void content in thermoplastic composite is generally greater due to the high viscosity of thermoplastics.

Voids in thermoplastic composites are originated from several sources: (1) air entrapment inside the matrix or at the fiber/matrix interface during pre-impregnation; (2) absorbed moisture in the matrix that is not removed during pre-impregnation or consolidation; (3) volatiles arising from the matrix at elevated temperatures; and (4) air entrapment between preforms during consolidation.

Voids are usually undesired because of their detrimental effect on the mechanical properties of the composite. A larger void content results in a composite with inferior mechanical properties because of high stresses developed around the void. Certain mechanical properties are highly sensitive to voids. Those properties include fatigue resistance and matrix-dominated properties, such as interlaminar shear strength and transverse tensile strength.

The void occupies space in the composite but does not contribute to the mass of the composite. Therefore, the presence of voids in a composite reduces its density. Although most of the applications benefit from removal of the void, there are applications that require a low density of the composite and voids are intentionally introduce into the composite. A thermoplastic syntactic foam for subsea applications is a typical example in which voids are beneficial. Low density core materials used for thermoplastic composite sandwich structures are another example that a large number of voids are intentionally introduced to the material to achieve lightweight design (see Section 3.4.2).

Several methods have been developed to characterize the void content in thermoplastic composites. Typical methods are density method, X-ray method, and micro-CT method, and microscopy method. Those methods are described in Section 6.5.

2.6 Summary

- A thermoplastic composite is composed of a thermoplastic polymer matrix and a reinforcement. The matrix and the reinforcement work synergistically to provide superior specific strength and modulus, excellent corrosion resistance, wear resistance, high impact performance, etc., for the thermoplastic composite.
- Constituents in the thermoplastic composite are matrix, reinforcement, matrix/reinforcement interface, and void. All the constituents contribute to the properties of the thermoplastic composite.
- Common thermoplastics used in composites include PP, PE, nylons, PET, PPS, and PEEK.
- Thermoplastics can be categorized into commodity, engineering, and advanced engineering thermoplastics based on their performance.
- Reinforcements used in thermoplastic composites are fibers, particulates, and nanomaterials. The fiber is the main reinforcement used in the thermoplastic composite and it bears a majority of the applied load and provides strength to the composite.
- The aspect ratio of the fiber refers the ratio of its length to diameter. It is one of the main parameters that determine the mechanical properties of the thermoplastic composite.

- DOC is defined as the ratio of the volume of crystalline regions to the overall polymer volume. A thermoplastic with a higher DOC possesses enhanced strength, improved chemical resistance, and better thermal stability.
- Common reinforcements used in thermoplastic composites are glass fibers, carbon fibers, and cellulose fibers.
- A strong fiber/matrix interface facilitates efficient load transfer from matrix to fibers and results in improved strength for the thermoplastic composite.
- Sizing, plasma treatment, alkaline treatment, and etching are common methods applied to fibers to improve the adhesion between the fiber and the thermoplastic matrix.
- Coupling agents can be added to the thermoplastic matrix to prove improve the fiber/matrix adhesion.
- Sizing refers to a thin coating applied on the fiber surface. It is also used to describe the process that a thin coating is applied onto fibers. Sizing can enhance lubrication and protect fibers from abrasion, facilitate easier processing and handling, and, most importantly, promote adhesion between the fiber and the matrix.
- Voids are usually not desired in thermoplastic composites because of their detrimental effect on the mechanical properties of the composite. Properties including fatigue resistance and matrix-dominated properties are highly sensitive to voids.

References

[1]   Bierögel C, Grellmann W. Quasi-static tensile test–tensile properties of thermoplastics-data. Polymer Solids and Polymer Melts–Mechanical and Thermomechanical Properties of Polymers: Springer. 2014. 88–99.
[2]   Khanam PN, AlMaadeed MAA. Processing and characterization of polyethylene-based composites. Advanced Manufacturing: Polymer & Composites Science. 2015;1(2):63–79.
[3]   Li D, Zhou L, Wang X, He L, Yang X. Effect of crystallinity of polyethylene with different densities on breakdown strength and conductance property. Materials. 2019;12(11):1746.
[4]   Bailey NA, Hay JN, Price DM. A study of enthalpic relaxation of poly (ethylene terephthalate) by conventional and modulated temperature DSC. Thermochimica Acta. 2001;367:425–31.
[5]   Babu RP, O'connor K, Seeram R. Current progress on bio-based polymers and their future trends. Progress in Biomaterials. 2013;2(1):8.
[6]   Wattanakornsiri A, Tongnunui S. Sustainable green composites of thermoplastic starch and cellulose fibers. Songklanakarin Journal of Science & Technology. 2014;36(2).
[7]   Witten E, Mathes V, Sauer M, Kühnel M. Composites market report 2018: Market developments, trends, outlooks and challenges. AVK & Carbon Composites 2018.
[8]   Matsuhisa Y, Bunsell A. Tensile failure of carbon fibers. Handbook of Tensile Properties of Textile and Technical Fibres. Elsevier: 2009. 574–602.
[9]   Kraus T, Kühnel M, Witten E. Composites Market Report 2014 Market developments, trends, challenges and opportunities. Federation of Reinforced Plastics. Frankfurt, Germany: 2014.

[10] https://www.grandviewresearch.com/blog/carbon-thermoplastic-cfrtp-composites-market-size-share. Accessed in Jan 2021.

[11] Scarselli M, Castrucci P, De Crescenzi M. Electronic and optoelectronic nano-devices based on carbon nanotubes. Journal of Physics: Condensed Matter. 2012;24(31):313202.

[12] Wang J, Salim N, Fox B, Stanford N. Anisotropic compressive behaviour of turbostratic graphite in carbon fibre. Applied Materials Today. 2017;9:196–203.

[13] Souto F, Calado V, Pereira Jr N. Lignin-based carbon fiber: a current overview. Materials Research Express. 2018;5(7):072001.

[14] DuPont F. Kevlar® Aramid Fiber: Technical Guide. 2000.

[15] van der Werff H, Heisserer U. High-performance ballistic fibers: ultra-high molecular weight polyethylene (UHMWPE). Advanced Fibrous Composite Materials for Ballistic Protection. Elsevier: 2016. 71–107.

[16] Fukushima Y, Murase H, Ohta Y. Dyneema®: Super fiber produced by the gel spinning of a flexible polymer. High-Performance and Specialty Fibers. Springer: 2016. 109–32.

[17] Yeh J-T, Lin S-C, Tu C-W, Hsie K-H, Chang F-C. Investigation of the drawing mechanism of UHMWPE fibers. Journal of Materials Science. 2008;43(14):4892–900.

[18] Marissen R, Duurkoop D, Hoefnagels H, Bergsma O. Creep forming of high strength polyethylene fiber prepregs for the production of ballistic protection helmets. Composites Science and Technology. 2010;70(7):1184–8.

[19] Khalil HA, Tehrani M, Davoudpour Y, Bhat A, Jawaid M, Hassan A. Natural fiber reinforced poly (vinyl chloride) composites: A review. Journal of Reinforced Plastics and Composites. 2013;32 (5):330–56.

[20] Dittenber DB, GangaRao HV. Critical review of recent publications on use of natural composites in infrastructure. Composites Part A: Applied Science and Manufacturing. 2012;43(8):1419–29.

[21] Lam TBT, Hori K, Iiyama K. Structural characteristics of cell walls of kenaf (Hibiscus cannabinus L.) and fixation of carbon dioxide. Journal of Wood Science. 2003;49(3):255–61.

[22] Huang M-F, Yu J-G, Ma X-F. Studies on the properties of montmorillonite-reinforced thermoplastic starch composites. Polymer. 2004;45(20):7017–23.

[23] Takakura A, Beppu K, Nishihara T, Fukui A, Kozeki T, Namazu T, et al. Strength of carbon nanotubes depends on their chemical structures. Nature Communications. 2019;10(1):1–7.

[24] Jia Z, Wang Z, Xu C, Liang J, Wei B, Wu D, et al. Study on poly (methyl methacrylate)/carbon nanotube composites. Materials Science and Engineering: A. 1999; 271 (1-2): 395–400.

[25] Meincke O, Kaempfer D, Weickmann H, Friedrich C, Vathauer M, Warth H. Mechanical properties and electrical conductivity of carbon-nanotube filled polyamide-6 and its blends with acrylonitrile/butadiene/styrene. Polymer. 2004;45(3):739–48.

[26] Qin Y, Zhang S, Yu J, Yang J, Xiong L, Sun Q. Effects of chitin nano-whiskers on the antibacterial and physicochemical properties of maize starch films. Carbohydrate polymers. 2016;147:372–8.

[27] Xie Y, Hill CA, Xiao Z, Militz H, Mai C. Silane coupling agents used for natural fiber/polymer composites: A review. Composites Part A: Applied Science and Manufacturing. 2010;41(7): 806–19.

[28] https://iacmi.org/wp-content/uploads/2019/02/Michelman_FiberSizing.pdf. Accessed in Jan 2021.

[29] Michelman IACMI Winter MeetingJanuary 30, 2019; Indianapolis. Accessed in Jan 2021.

[30] Chukov D, Nematulloev S, Zadorozhnyy M, Tcherdyntsev V, Stepashkin A, Zherebtsov D. Structure, mechanical and thermal properties of polyphenylene sulfide and polysulfone impregnated carbon fiber composites. Polymers. 2019;11(4):684.

Chapter 3
Continuous and discontinuous fiber-reinforced thermoplastic composites

3.1 Introduction

Among different types of reinforcements, namely fibers, particulates, and nano-materials, fibers are dominantly used in thermoplastic composites for their excellent efficiency in load bearing. Based on the continuity of the fiber, there are continuous fiber-reinforced thermoplastic composites and discontinuous fiber-reinforced thermoplastic composites. The continuous fiber-reinforced thermoplastic composite has fibers that run from one end of the composite to the other end. On the other hand, the discontinuous fiber-reinforced thermoplastic composite is reinforced with discrete fibers with a length ranging from microns to centimeters. The continuous fibers have considerably better load-bearing capacity and are more effective in reinforcing the thermoplastic than the discontinuous fibers.

Fiber arrangements in continuous fiber-reinforced thermoplastic composites and their preforms vary significantly. Therefore, their mechanical properties can be distinctly different. The general fiber arrangements are unidirectional fiber (roving), woven fabric, braided fiber, noncrimp fabric (NCF), and knitted fibers. The fiber roving is the starting material for manufacturing continuous fiber-reinforced thermoplastic composites. It consists of a large number of filaments that are bundled into a tow. The tow size is defined by the number of filaments in one tow. Common tow sizes are 1K, 3K, 6K, 12K, 15K, and 24K. The letter "K" denotes 1,000 filaments. For example, a "12K" tow is made of 12,000 filaments. The fiber roving can be woven, braided, or stitched together to form woven fabrics, braided fabrics, and NCF, respectively. Composites that are reinforced with fiber rovings (Fig. 3.1a), woven fabrics (Fig. 3.1b), braided fabrics (Fig. 3.1c), and NCF are called unidirectional fiber-reinforced thermoplastic composites, woven fabric-reinforced thermoplastic composites, braided fiber-reinforced thermoplastic composites, and NCF-reinforced thermoplastic composites, respectively.

Discontinuous fiber-reinforced thermoplastic composites can be classified into random fiber mat thermoplastic composite, long fiber thermoplastic (LFT) composite, and short fiber thermoplastic (SFT) composite. Figure 3.1d–f show a random glass fiber mat, and long and short glass fibers for producing those discontinuous fiber-reinforced thermoplastic composites, respectively.

The classification of fiber-reinforced thermoplastic composites based on the continuity and arrangement of fibers is shown in Fig. 3.2.

Fiber length can significantly affect both the mechanical property and processability of thermoplastic composites. With increasing fiber length, the processability of the

https://doi.org/10.1515/9781501519055-003

Fig. 3.1: Different forms of glass fibers used in thermoplastic composites: (a) glass fiber roving; (b) woven glass fabric; (c) braided glass fabric; (d) glass fiber mat; (e) long glass fiber; and (f) short glass fiber.

composite decreases while its mechanical properties such as modulus, strength, and impact resistance increase. Figure 3.3 schematically shows the effect of the fiber length on the mechanical properties and the processability of the thermoplastic composite. The thermoplastic composite with a large fiber length, for example, continuous fiber-reinforced thermoplastic composites, possesses excellent tensile modulus and strength, flexural modulus and strength, and impact resistance. However, its processability is low and the production rate for the continuous fiber-reinforced thermoplastic composite is relatively low (see Section 4.3). For example, continuous fiber-reinforced thermoplastic composite products with excellent structural performance can be produced from processes such as autoclave molding process or compression molding process; however, the processing time required for molding the continuous fiber thermoplastic composites is long and the production rate is relatively low.

Discontinuous fiber-reinforced thermoplastic composites, on the other hand, can be molded at a much higher production rate. For instance, SFT has fibers that are normally less than 1 mm long and its processability is close to that of neat thermoplastics, thermoplastics without any reinforcements. The SFT composite can be readily

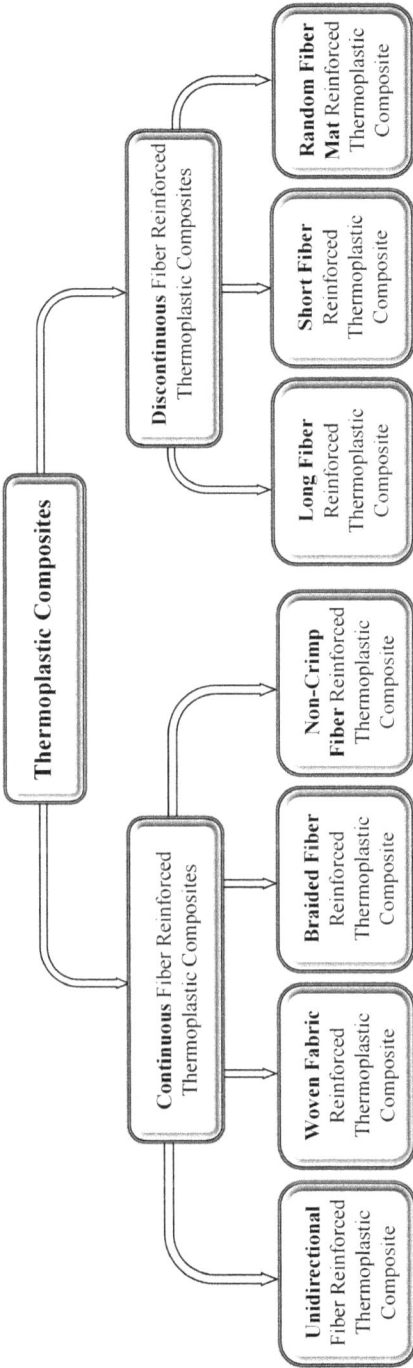

Fig. 3.2: Different categories of fiber-reinforced thermoplastic composites based on the continuity and arrangement of fibers in the composite.

manufactured using processes such as injection molding or compression molding at a high production rate. One molding cycle can be completed within seconds. However, the SFT composite has relatively low mechanical properties. Although the short fibers provide enhanced strength and modulus compared to the neat thermoplastic, the improvement is limited because of their low fiber aspect ratios. Processing of the continuous and discontinuous fiber-reinforced thermoplastic composite will be discussed in detail in Chapter 4.

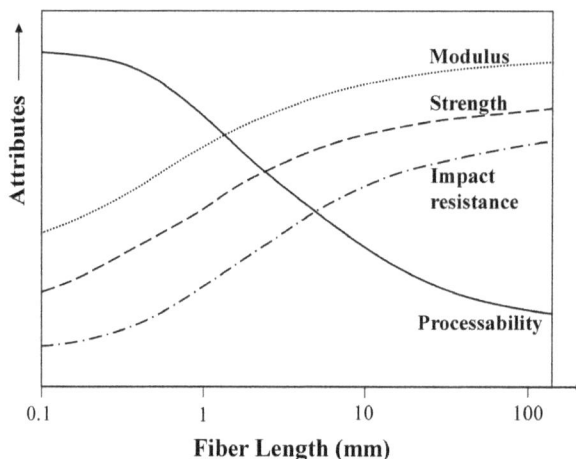

Fig. 3.3: The effect of fiber length on the mechanical properties and processability of fiber-reinforced thermoplastic composites. The processability of the thermoplastic composite decreases while its mechanical property increases with fiber length.

The diameter of the fiber used in thermoplastic composites is normally at a microscale and its length varies and determines the fiber aspect ratio. Fiber aspect ratio is defined as the ratio of fiber length (l) to fiber diameter (d), or l/d. Fiber length is the main factor that determines the fiber aspect ratio because the fiber diameter is normally predetermined from production for a specific type of fiber. The fiber aspect ratio is one of the most important parameters that affect the performance of fiber-reinforced thermoplastic composites. The effect of the fiber aspect ratio will be discussed in Section 3.3.2. The fiber aspect ratio is also frequently used as an input for models that predict the properties of discontinuous fiber-reinforced composites (see Section 7.2).

3.2 Continuous fiber-reinforced thermoplastic composite

The continuous fiber-reinforced thermoplastic composite is commonly used in primary structural components, and it bears most of the externally applied load during use. It possesses not only great strength and modulus but also low density. The combination of both superior properties and low density results in excellent specific properties. Therefore, the continuous fiber-reinforced thermoplastic composite has been increasingly used in aviation, energy, and other industry sectors in which high specific properties are required.

The fiber architectures used in continuous fiber-reinforced thermoplastic composite preforms include unidirectional fiber, woven fabric, braided fabric, noncrimp fiber, and knitted fiber. The following sections describe each of the fiber architecture and its role in the processing and properties of the thermoplastic composite. General terms, "fiber tow" or "fiber tows," are used in these sections to describe the construction of individual architecture. The "fiber tow" includes not only the reinforcement fiber tows, but also commingled fiber yarns and powder impregnated fiber tows.

3.2.1 Unidirectional fiber-reinforced thermoplastic composite

Unidirectional fiber-reinforced thermoplastic composites are one type of continuous fiber-reinforced thermoplastic composites and have all of the fibers aligned in one direction. That direction is normally called fiber direction, longitudinal direction, or 0° direction. The direction perpendicular to the fiber direction is called transverse direction or 90° direction. The unidirectional fiber-reinforced thermoplastic composite is usually produced as a tape, an intermediate composite preform that is further processed into composite structures. The unidirectional fiber-reinforced thermoplastic composite tape is also called UD tape, uni-tape, or uni. Manufacturing of the uni-tape is generally achieved through a melt impregnation process (see Section 4.3). Uni-tapes are consequently molded into unidirectional fiber-reinforced thermoplastic composites or other thermoplastic composites with uni-tapes oriented in desired orientations, for example, cross-ply thermoplastic composites or quasi-isotropic thermoplastic composites. The unidirectional fiber-reinforced thermoplastic composite offers excellent strength and modulus in the longitudinal direction because all of its fibers are aligned in that direction. However, the mechanical properties in the transverse direction are significantly lower as no fiber is aligned in that direction.

Thermoplastic composites reinforced with unidirectional glass and carbon fibers, especially carbon fibers, are one of the engineering materials with the highest specific property thus far. Figure 3.4 shows the specific strength and modulus of both glass fiber and carbon fiber-reinforced nylon 6 composites in the fiber direction. The fiber contents in both the composites are 60 wt% and the carbon fiber is an intermediate modulus carbon fiber. Other structural materials such as aluminum alloy (7068-T6511), steel (martensitic steel), and titanium alloy (Ti-10V-2Fe-3Al) are also compared in the figures. Those metallic materials have the highest strength in their individual alloy category. It should be pointed out that the variation of fiber type, fiber grade, fiber content, and fiber orientation can result in a large range of elastic modulus and tensile strength of the composite, and, therefore, a large range of specific modulus and strength. In addition, the strength and modulus in the unidirectional fiber thermoplastic composites are also much higher than the transverse direction while the metallic materials generally have same properties in all directions because of their isotropic nature.

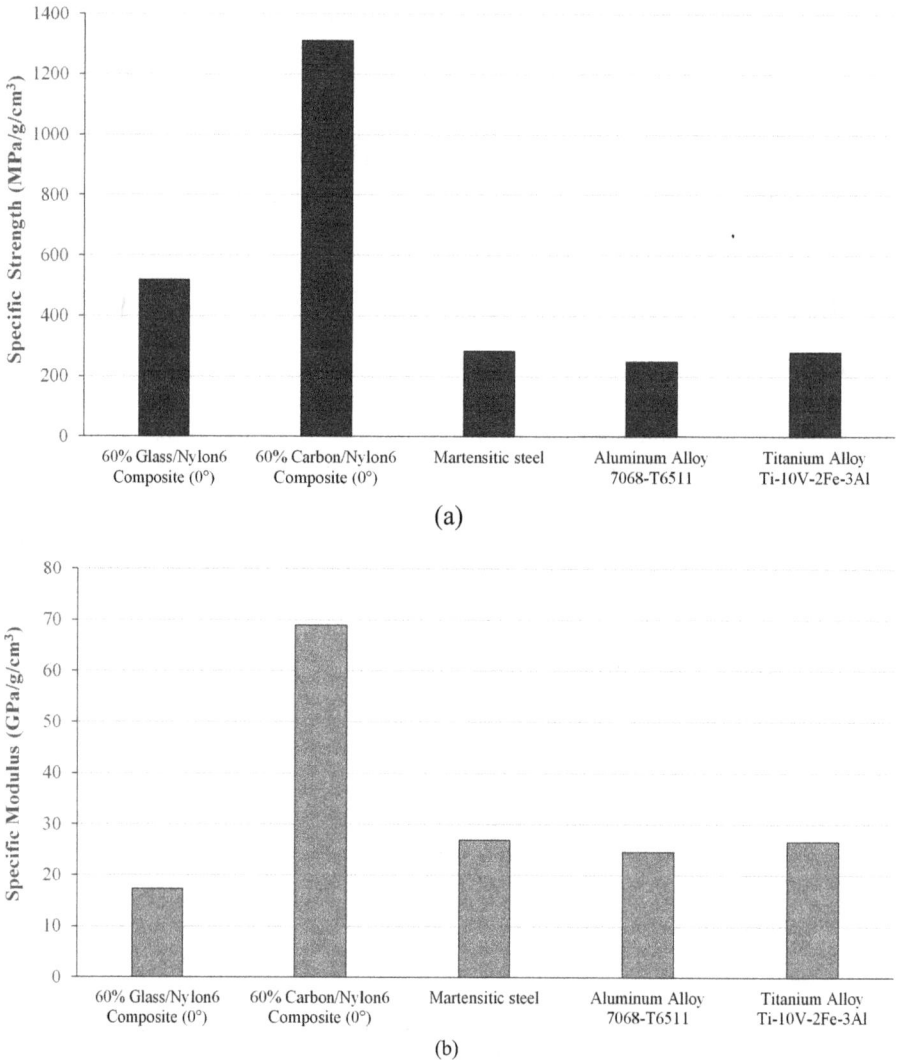

(a)

(b)

Fig. 3.4: (a) Specific strength and (b) specific modulus among different materials, including unidirectional glass and carbon fiber-reinforced nylon 6 composites in their longitudinal direction. The fiber contents in both the nylon matrix composites are 60 wt%.

Tables 3.1 and 3.2 list the tensile strength and modulus in the fiber direction (0°) for some commercially available glass fiber and carbon fiber uni-tapes for producing unidirectional fiber-reinforced thermoplastic composites, respectively. The fiber contents of some uni-tapes are provided in only volume percentages (highlighted with bold font) by the manufacturers. The volume percentage can be converted to weight percentage (wt%) if the fiber density and matrix density are known, or vice versa (see Eqs. (6.1) and (6.2), and Example question 7.1).

Tab. 3.1: Tensile strength and modulus of glass fiber uni-tapes in 0° direction (reprinted from manufacturer's datasheets).

Glass fiber uni-tapes	Matrix	Fiber content	Density (g/cm³)	Tensile strength (0°) (MPa)	Tensile modulus (0°) (GPa)
CELSTRAN® HDPE GF70-01	HDPE	70 wt%	1.71	864	35.3
CELSTRAN® PA6 GF60-03	Nylon 6	60 wt%	1.69	679	29.7
CELSTRAN® PA66 GF60-02	Nylon 66	60 wt%	1.73	759	33.4
CELSTRAN® PP GF60-13	PP	60 wt%	1.49	732	25.7
CELSTRAN® PP GF70-13	PP	70 wt%	1.66	931	33.9
CELSTRAN® PPS GF60-01	PPS	60 wt%	1.88	782	34.7
MaruHachi MCP1239	Nylon 6	40 **vol%**	–	730	38
MaruHachi MCP1228	PP	35 **vol%**	–	660	30
MaruHachi MCP1238	PP	52 **vol%**	–	890	40
Polystrand ThermoPro™ IE7010	PP	70 wt%	1.66	610	33.1
SABIC UDMAX™ GPP 45-70	PP	45 **vol%**	1.65	959	35
Solvay APC-2 (S2-glass)	PEEK	71 wt%	–	1,140	55
TOPOLO GPE900	HDPE	70 wt%	1.72	900	35
TOPOLO GPE800	HDPE	65 wt%	1.63	800	29
TOPOLO GPA66	Nylon 66	60 wt%	1.72	700	31
TOPOLO GPPS	PPS	48 wt%	1.75	550	26
Toray Cetex® TC910	Nylon 6	60 wt%	1.73	900	30
Toray Cetex® TC1200 (S2-glass)	PEEK	71 wt%	1.97	1,520	52
Toray Cetex® TC940	PET	60 wt%	1.89	960	32
Toray Cetex® TC960	PP	60 wt%	1.49	750	27.8

Tab. 3.2: Tensile strength and modulus of carbon fiber uni-tapes in 0° direction (reprinted from manufacturer's datasheets).

Carbon fiber uni-tapes	Matrix	Fiber content	Density (g/cm^3)	Tensile strength (0°) (MPa)	Tensile modulus (0°) (GPa)
BÜFA® Slittape PEKK CF 66	PEKK	66 wt%	–	2,439	144
CELSTRAN® PA6 CF60-03	Nylon 6	60 wt%	1.45	1,910	100
CELSTRAN® PA66 CF60-02	Nylon 66	60 wt%	1.44	1,930	93.7
CELSTRAN® TP PPS CF60-01	PPS	60 wt%	1.55	2,030	101
Evonik VESTAPE® PA12/CF	Nylon12	59 wt%	1.36	1,750	100
Evonik VESTAPE® PEEK/CF	PEEK	53 wt%	1.51	1,750	100
Maezio™ CF GP 1000T	PC	44 **vol%**	1.5	1,400	105
MaruHachi MCP1223	Nylon 6	52 **vol%**	–	1,900	120
Solvay APC-2 (AS4)	PEEK	66 wt%	–	2,070	138
Solvay APC-2 (IM7)	PEEK	68 wt%	–	2,900	172
Tenax®-E TPUD PEEK (HTS45)	PEEK	66 wt%	–	2,300	138
Toray Cetex® TC910	Nylon 6	60 wt%	1.45	1,900	100
Toray Cetex TC 1200 (AS4D)	PEEK	66 wt%	1.59	2,410	135
Toray Cetex TC 1200 (IM7)	PEEK	66 wt%	1.59	3,100	159
Toray Cetex® TC940	PET	60 wt%	1.53	1,272	94.2
Toray Cetex® TC1100 (AS4A)	PPS	66 wt%	1.59	2,020	134
Toray Cetex® TC1100 (IM7)	PPS	66 wt%	1.59	2,760	152

3.2.2 Woven fabric thermoplastic composite

Woven fabric is another common fiber architecture used in thermoplastic composites. The fibers in the woven fabric-reinforced thermoplastic composite are arranged by interlacing warp and weft fiber tows (including reinforcement fiber tows, commingled fiber yarns, powder impregnated fiber tows, as well as uni-tapes) in certain patterns. Warp is the longitudinal fiber tow that is held stationary in tension on a loom, and weft (also called fill) is the transverse fiber tow that is interlaced over and under the warp fiber tow. Woven fabric-reinforced thermoplastic composites have fibers in multi-axes and their mechanical properties are more balanced in their longitudinal and transverse directions compared to unidirectional fiber-reinforced thermoplastic composites. Different weave patterns, including plain weave, twill weave, and harness satin weave, are used in the woven fabric-reinforced thermoplastic composite. Those patterns can be differentiated by tracing one fiber tow and counting the number of the fiber tows in the other direction that are above and under it. Figure 3.5 shows the common weave patterns used in thermoplastic composites. One fiber tow is highlighted in a different color in each weave pattern for conveniently tracking the number of the fiber tows in the other direction that are above and under it. A side view of each weave pattern is also shown in the figure for better illustration of the weave pattern.

The fiber tows in the woven fabrics are interlocked mechanically. The extent to which the fiber tows are interlocked varies among plain weave, twill weave, and satin weave. The fabric shows higher stability but lower drapeability when there is more mechanical interlock. Higher stability benefits handling of the fabrics but lower

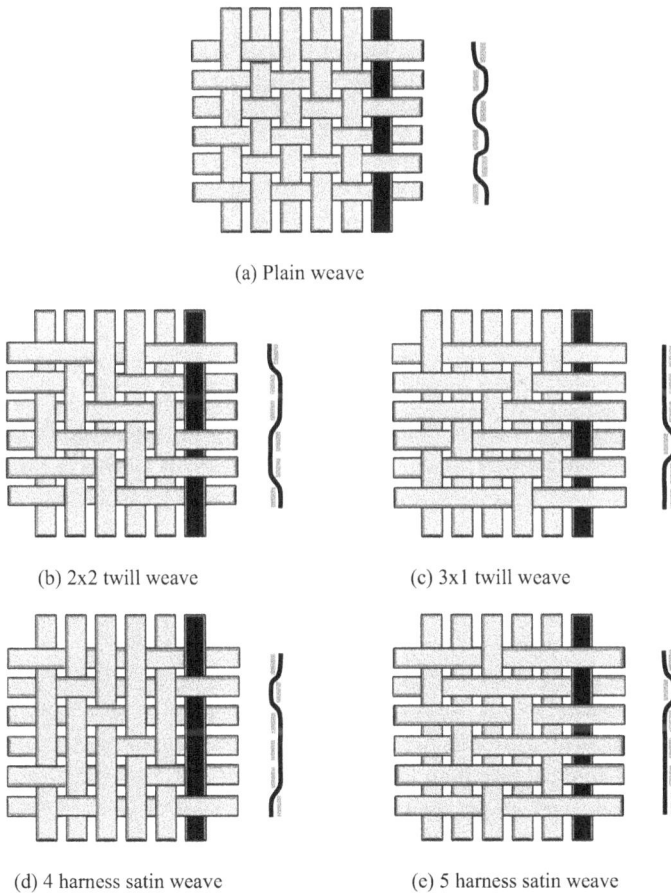

(a) Plain weave

(b) 2x2 twill weave

(c) 3x1 twill weave

(d) 4 harness satin weave

(e) 5 harness satin weave

Fig. 3.5: Weave patterns in woven fabric-reinforced thermoplastic composites: (a) plain weave; (b) 2 × 2 twill weave; (c) 3 × 1 twill weave; (d) 4 harness satin weave; and (e) 5 harness satin weave.

drapeability is not in favor of molding thermoplastic composites with complex geometries. In addition, the weave pattern of a fabric affects the mechanical properties of its thermoplastic composite. The following sections discuss each weave pattern and its effect on the drapeability and mechanical properties of the composite.

1. **Plain weave**. Plain weave fabrics are produced by passing each warp fiber tow alternately over and under each weft fiber tow (Fig. 3.5a). Thermoplastic composites reinforced with plain weave fabrics possess similar mechanical properties in its 0° and 90° direction if the same fiber type and tow size are used in both directions. The plain weave fabric is relatively stable because of the interlocking between the warp and weft fiber tows. The fiber tows in the plain weave fabrics have a high level of crimp, which results in lower efficiency in load bearing and lower mechanical properties compared to other weave patterns. In addition, the plain weave fabric has a low drapeability, and it is often challenging to produce components with complex geometries using this type of fabric.

2. **Twill weave**. Twill weave fabrics are produced by passing one weft fiber tow over and under two or more warp fiber tows and passing the next weft fiber tow in the same manner except that it is stepped one tow to the left or right of the preceding tow. The shift of one tow shows up as a "step" on the twill weave fabric. Typical twill weave patterns include twill 2 × 2 (Fig. 3.5b) and twill 3 × 1 (Fig. 3.5c). They have less crimp than the plain weave fabric. Therefore, the thermoplastic composite reinforced with twill weave fabrics is able to offer higher mechanical properties than the plain weave fabric-reinforced composite.

3. **Satin weave**. Satin weave is similar to twill weave but has fewer intersections between the warp and weft fiber tows. Harness number, the total number of fiber tows that one fiber tow (in the other direction) crosses over and passes under before the pattern repeats, is used to describe the satin weave pattern. Typical harness numbers are 4 (Fig. 3.5d), 5 (Fig. 3.5e), and 8. Satin weave fabrics have relatively low crimp and good drapeability because of fewer warp and weft intersections created in the pattern. However, the fabric is not symmetrical and has low stability.

Thermoplastic matrix can be introduced into woven fabrics before or after the weaving. Preforms such as commingled fiber tows and uni-tapes have the thermoplastic matrix integrated with the fabric before weaving. Those preforms have the thermoplastic matrix in a different form. The thermoplastic matrix in the commingled fiber tow is in a fiber form and the uni-tape has fibers impregnated with thermoplastic through a melt impregnation process. The commingled fiber tow has great flexibility and is best suited for weaving because both fiber and matrix are in fiber form. After weaving, the preform is then molded into woven fabric-reinforced thermoplastic composites. The thermoplastic matrix can also be introduced after weaving. Reinforcement fibers are woven into fabrics with a desired weave pattern first, and the thermoplastic matrix is introduced to the fabric through film stacking, powder impregnation, or solvent impregnation to form thermoplastic composite preforms for consequent molding processes.

3.2.3 Braided fabric thermoplastic composite

Braided fabric is another fiber arrangement used in thermoplastic composites. It is different from the woven fabric architecture even though fiber tows cross each other in both cases. In woven fabrics, the warp and weft fiber tows are interlaced perpendicularly and the interlacement angle is 90°. However, the interlacement angle in the braided fabric can range from 10° to 170°. Braided fabric can be biaxial and triaxial, based on the number of fiber tows/yarns used in braiding. Biaxial braids consist of two fiber tows/yarns while triaxial braids have three yarns, one of which is oriented in the longitudinal direction (0° braid angle). Braided fabric-reinforced thermoplastic composites generally possess advantages over woven fabric-reinforced thermoplastic composites in impact resistance, efficiency in bearing torsional load, and interlaminar strength.

The braided fiber thermoplastic composite is mainly made for tubing products. The fiber tows braided in a tube shape are able to deform radially. The flexibility in changing the diameter by changing the braid angle during processing facilitates the manufacture of complex composite tubular structures. However, change of the angle can complicate the design as the mechanical properties of the composite structure are affected by the braid angle. A common process for making the tubular structure is described in Section 4.4.10. The fiber tows used for braiding are typically commingled fiber tows (consisting of reinforcement fibers and thermoplastic fibers) although other preforms such as powder impregnated fiber tows (reinforcement fibers impregnated with thermoplastic powders) or uni-tapes can also be used for braiding.

3.2.4 Noncrimp fiber thermoplastic composite

NCF-reinforced thermoplastic composites gain their reputation for their noncrimp fibers, in contrast to the wavy fibers in woven or braided fabric thermoplastic composites. The NCF consists of a series of unidirectional fibers with different orientations that are stitched together using stitching yarns. Two main methods are used to introduce thermoplastics to the noncrimp fibers: (a) the thermoplastic in a fiber form is commingled with the reinforcement fiber and stitched into a NCF composite preform; (b) the thermoplastic in a film form is placed between reinforcement fiber layers and stitched into a NCF composite preform. The thermoplastic fiber or film is then melted and impregnated with the reinforcement fibers in subsequent molding processes. The stitching yarn is normally an organic fiber that has adequate strength but also flexibility that allows bending and shaping during stitching or molding. The yarn material includes polyethylene terephthalate (PET), nylon, and so on. The stitching yarns normally melt and stay inside the composite after molding.

The NCF thermoplastic composite does not have the waviness that is commonly seen in the woven or braided fabric thermoplastic composites. Therefore, the NCF composite has better efficiency in load-bearing. NCF also offers advantages such as flexibility in fiber orientation arrangement, customizability, ease of handling, and good drapeability. Figure 3.6 shows a NCF preform that consists of commingled glass fibers and polypropylene (PP) fibers (preform dimension: 152 mm × 152 mm). Both the glass fibers and PP fibers are aligned in the vertical direction. Glass fibers reflect more light and show more brightness in the preform. Stitching yarns and the backing material for the preform are also noticeable.

Fig. 3.6: A noncrimp fiber preform consisting of commingled glass fibers and polypropylene fibers.

3.2.5 Self-reinforced composite

Self-reinforced composites are a group of thermoplastic composites that have the same type of polymers as both their reinforcement and matrix. The self-reinforced composite is sometimes called self-reinforced polymer composite, one polymer composite, or single polymer composite. Since fibers made of thermosets do not have the adequate strength and modulus as a reinforcement, self-reinforced composites are exclusively used for thermoplastic composites.

The self-reinforced composite has the same work principle as the conventional composite, that is, the reinforcement and the matrix working together synergistically. The reinforcement possesses great mechanical properties and strengthens the matrix while the matrix transfers the load to the reinforcement and provides protection to the reinforcement. The only difference between the self-reinforced composite and conventional thermoplastic composite is that the reinforcement and the matrix are from the same thermoplastic family for the self-reinforced composite.

Several thermoplastics, including PE, PP, PET, and liquid crystal polymer, have been used in the self-reinforced composite. PE-based self-reinforced composites are the first developed self-reinforced composites. High molecular PE fibers were added into lower molecular PE matrix to form PE-based self-reinforced composites in the 1970s [1]. The fibers used in self-reinforced composites are produced such that their molecular chains are highly crystalline and aligned in the fiber direction (Fig. 2.54). As mentioned in Section 2.3.1.3, a thermoplastic polymer has much higher modulus and strength when its molecular chains are highly aligned and crystalline from processes such as drawing. The elastic modulus and tensile strength of the fiber can be more than one order of magnitude greater than those of the bulk thermoplastic that does not have any alignment of the molecular chain in spite of the presence of crystalline regions in the bulk thermoplastic. In addition, the highly crystalline thermoplastic fiber possesses a higher melting temperature than the bulk thermoplastic in the matrix. The difference in the melting temperature allows melting of the bulk thermoplastic only while maintaining the integrity of the fiber during manufacture of self-reinforced composite preforms as well as molding of the self-reinforced composite.

Table 3.3 lists the elastic modulus, tensile strength, and melting temperature of the thermoplastic matrix and the fibers that are used to produce self-reinforced composites. The elastic modulus and tensile strength of the self-reinforced composites are also listed. The mechanical properties of the fibers and the composite are measured in the fiber direction. Among those composites with different polymer systems, PP-based self-reinforced composites are the most commercialized because of ease of processing, low cost, and superior creep resistance of PP. Commercial PP-based self-reinforced composites include CURV®, PURE®, and Tegris®. Table 3.4 lists those self-reinforced composites and their properties. More detailed information about the self-reinforced composites can be obtained from reference [2].

The self-reinforced composite has been used in a variety of applications because of its lightweight and excellent toughness. Figure 3.7a shows a suitcase made of CURV® that takes advantage of the superior impact resistance and great wear resistance of the self-reinforced composite. Figure 3.7b shows a CURV® door panel for freezers used to store biological samples at an extremely low temperature for aerospace application, owing to the retained toughness of the PP-based self-reinforced composite when exposed to extremely low temperatures.

In addition to the PP- and PE-based self-reinforced composites, there are other types of self-reinforced composites, such as PET-based self-reinforced composite, PMMA-based self-reinforced composite, and LCP-based self-reinforced composite. The properties of the PET-based and LCP-based self-reinforced composites are listed in Tab. 3.3. Both the composites show great mechanical properties. The PMMA-based

Tab. 3.3: Properties of thermoplastic matrices, reinforcement fibers, and their self-reinforced composites (adapted from Reference [2]).

Constituent material properties				Composite material properties		
Constituent material	Elastic modulus (GPa)	Tensile strength (MPa)	Melting temperature (°C)	Self-reinforced composite	Elastic modulus (GPa)	Tensile strength (MPa)
PE matrix	0.6–0.9	20–45	105–138	PE-based composite	6.2–79	173–1,500
PE fiber	66–124	2,000–2,900	142–150			
PP matrix	1.2	27–30	150–166	PP-based composite	2.4–13	42–385
PP fiber	6.9–19	450–650	176			
PET matrix	2.7	12–55	212–250	PET-based composite	6.9–10	89–350
PET Fiber	11.4–14	370–1250	265			
LCP matrix	8–17.9	50–185	212–319	LCP-based composite	18.2–71	300–1,300
LCP fiber	52–103	1,100–3,374	276–330			

self-reinforced composite has relatively low modulus and strength due to limited properties of PMMA fibers. Aramid fiber-reinforced nylon matrix composites have also been considered a type of self-reinforced composite sometimes. It should be noted that the aramid fibers and the matrix nylon belong to the nylon family but they have different chemical formulas. The aramid fiber is an aromatic nylon while the matrix nylon is an aliphatic or semiaromatic nylon.

The self-reinforced composite offers several advantages over conventional thermoplastic composites. First, the self-reinforced composite has a very low density. Both the PE-based and PP-based self-reinforced composites have a density less than 1 g/cm³. Second, the self-reinforced composite offers superb impact resistance even at low temperatures. Third, because the self-reinforced composite is made of only one thermoplastic, there is no need to separate the fiber from the matrix during recycling when the composite reaches the end of life. In addition, great bonding between the reinforcement and the matrix can be achieved because they have the same formula and are compatible with each other. This is especially beneficial for achieving better fiber/matrix bonding for the thermoplastics that possess low surface energy, such as PP and PE.

The self-reinforced composite has several disadvantages though compared to the conventional thermoplastic composite. First, there are only a limited number of thermoplastic polymers that can be used to produce fibers with adequate strength for bearing load in the self-reinforced composite. Those polymers mainly include PE, PP, PET, and LCP that can form highly aligned and crystalline molecular chains. Second, because the reinforcement fibers in the self-reinforced composite are made of

Tab. 3.4: Densities and mechanical properties of several commercially available self-reinforced composites (reprinted from manufacturer's data).

PP-based self-reinforced composite	ρ (g/cm³)	E (GPa)	Tensile strength (MPa)	Flexural modulus (GPa)	Flexural strength (MPa)	Elongation at failure	Notched Izod impact strength (KJ/m)		Methods for producing preforms
							20 °C	−40 °C	
CURV®	0.92	5	180	3.6	107	17%	4.75	7.5	Hot compaction
PURE®	0.78	5.5	200	4.8	50	9%	–	–	Co-extrusion
Tegris®	0.78	5.5	200	5.5	–	6%	4.8	–	Film stacking

(a)   (b)

Fig. 3.7: Components made from PP-based self-reinforced composites: (a) suitcase (adapted from Reference [3]); (b) door panel (500 mm × 300 mm) for freezers used in aerospace application.

organic materials and their performance is highly affected at elevated temperatures, the performance of the self-reinforced composite at those temperatures is limited. Additionally, the difference in the melting temperatures between the fiber and the matrix is small, resulting in a narrow processing window. The narrow processing window poses a great challenge in producing the self-reinforced composite preform and molding the self-reinforced composite. The temperature needs to be high enough to melt the matrix and induce adequate flow of the molten matrix under pressure but low enough to prevent any molecular chains in the fiber from relaxation and maintain the mechanical properties of the fiber as much as possible. This has been considered as the key issue for processing of the self-reinforced composite.

Self-reinforced composite preforms can be produced via thermal and solution methods. The thermal method involves melting a portion of the reinforcement fiber in a single-component process or melting the matrix in a bicomponent process followed by consolidation under pressure. It is dominantly used for producing self-reinforced composites.

The melting of a portion of the reinforcement fiber in the single-component process is called hot compaction process. This process starts with one-component thermoplastic fibers that are produced through processes such as drawing. The fibers are melted partially on their surface. Normally only 10% of their volume is melted. The molten polymer is then forced to flow among the fibers under pressure. The fibers are bonded with the matrix and the self-reinforced composite preform is then produced. Because the matrix has the same chemical formula as the fiber, the bonding is superb at the fiber/matrix interface. Figure 3.8 shows the processing steps of the hot compaction method.

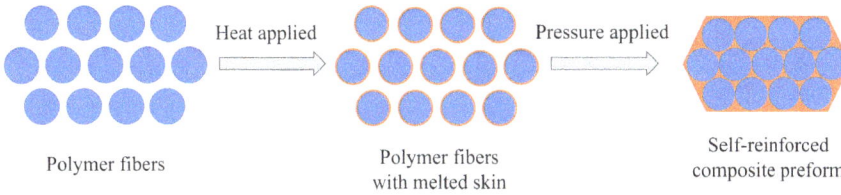

Fig. 3.8: A hot compaction process to produce self-reinforced composite preform by melting a small portion of the fibers followed by consolidation under pressure.

Other thermal methods for producing self-reinforced composite preforms include the bicomponent process. The thermoplastic matrix with a lower melting temperature than the fiber is introduced by different methods, such as co-extrusion, film stacking, commingling, or powder impregnation. The bicomponent material is heated above the melting temperature of the matrix but below the melting temperature of the fiber. Pressure is then applied to produce the self-reinforced composite preform. Figure 3.9(a–d) show the bicomponent methods used to introduce the thermoplastic matrix.

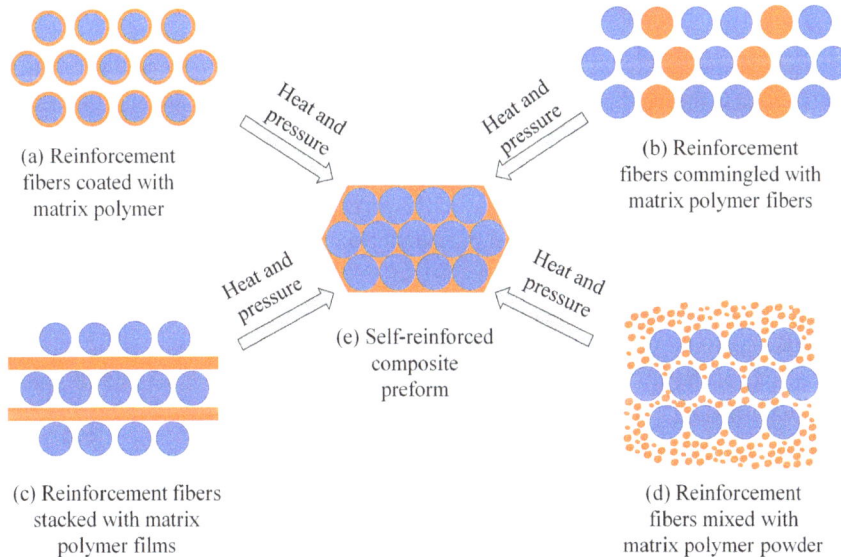

Fig. 3.9: Manufacture of self-reinforced composite preforms using bicomponent process: (a) co-extrusion method; (b) commingling method; (c) film stacking method; and (d) powder impregnation method; to produce (e) self-reinforced composite preform.

3.3 Discontinuous fiber-reinforced thermoplastic composite

A discontinuous fiber-reinforced thermoplastic composite is comprised of a thermoplastic matrix and discrete fibers. The discontinuous fiber thermoplastic composite does not have the same performance as the continuous fiber-reinforced thermoplastic composite because its fibers have much less aspect ratios than those in the latter. However, because of their low fiber length, the discontinuous fiber thermoplastic composites possess better processability.

Discontinuous fiber-reinforced thermoplastic composites include random fiber mat thermoplastic composite (or glass fiber mat thermoplastic composite (GMT)), SFT composite, and LFT composite. The following sections describe those discontinuous fiber-reinforced thermoplastic composites.

3.3.1 Random fiber mat thermoplastic composite

Random fiber mat thermoplastic composite is a composite comprised of a fiber mat pre-impregnated with a thermoplastic matrix. A typical random fiber mat thermoplastic composite is GMT composite. It was initially developed in the 1960s for the main purpose of replacing stamped sheet metals in automotive industries. The GMT composite offers a combination of good properties, lightweight, and good processability at a moderate cost. Those advantages have made GMT composites one of the prevalent materials for secondary structural components in the automotive industry.

"G" in GMT composites stands for glass fibers because of their use for the first random fiber mat thermoplastic composite material in the 1960s. Although glass fibers are still dominantly used in GMT composites, GMT composites reinforced by other fibers, such as carbon and cellulose fibers. are also available. The matrix in GMT composites can be any thermoplastic. However, PP is the initial thermoplastic used in the early GMT composites and has remained as the dominant matrix material in GMT composites, accounting for 95% of commercially available GMT products. Other thermoplastics such as PC, PE, and nylons have also been used in GMT composites.

Processing of the GMT composites involve combining fibers with a thermoplastic to produce a GMT preform first and molding the GMT preform under heat and pressure. The processing of GMT composites is described in Section 4.5.3.

The GMT technology has been advancing with more versatility in material combination, enhanced impact, and structural performance by hybridizing with continuous fiber thermoplastic composites. GMTex is one of the examples. Continuous fabric layers are added to the conventional GMT. The fabric layers are positioned at different locations through the thickness to provide versatile material options. Both the fabric layers and glass fiber mat layers undergo the similar process as shown in

Fig. 4.33 to form a hybrid composite preform. The composite preform is then heated in an oven, transferred into a mold, and molded into products. The production rate is high and cycle time can be less than one minute. Because of the integration of continuous fibers into the GMT, the impact resistance, modulus, and strength are significantly improved. The advanced GMTex has found its use mainly in automotive application. Typical components made of GMTex include front-end modules, back door inner panels, bumper brackets, and spare wheel well.

3.3.2 Long fiber thermoplastic composite

3.3.2.1 Introduction

LFT composites are a group of discontinuous fiber-reinforced thermoplastic composites with a fiber length more than critical fiber length. Their fiber length normally ranges from 1 mm to more than 10 mm. The LFT composite has a good combination of great mechanical properties and ease of processability. Because of its relatively high fiber length, the LFT composite has considerably increased modulus, strength, and impact properties than SFT composites. Meanwhile, it possesses much better processability than the continuous fiber thermoplastic composite. The combination of good modulus and strength, ease of processability, and relatively low cost of the LFT composite has prompted its increasing use not only in the automotive application, but also in other applications, such as electronics, military, and mass transit. Figure 3.10 shows the global market share of LFT composites across different industry sectors in 2016.

Common thermoplastics used in LFT composites include PP, nylon, and PPS. Among all the LFT composites, PP matrix LFT composites have more than 65% of the market share. Among all types of PP matrix LFT composites, long glass fiber-reinforced

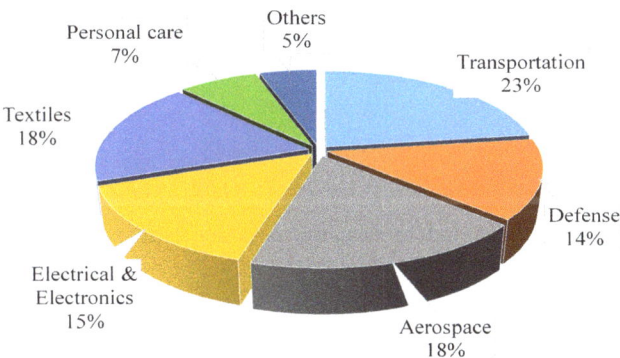

Fig. 3.10: The global market share of long fiber thermoplastic composite across different industry sectors reported in 2016 (adapted from Reference [4]).

PP composite dominates the market because of its extensive use in the automotive industry. It has been mainly used for semi-structural automotive components such as seat, bumper, underbody panels, door panels, dashboard panels, front end modules, roof modules, battery trays, and spare-wheel pans.

3.3.2.2 Critical fiber length

When a discontinuous fiber-reinforced thermoplastic composite is under loading, the load is transferred from the thermoplastic matrix to the fiber through the fiber/matrix interface. Depending on the length of the fiber, the peak stress in the fiber can be lower than or equal to its tensile strength. Critical fiber length is the minimum fiber length required for the maximum fiber stress to equal the tensile strength of the fiber at its center (mid-length). When the fiber length is lower than the critical fiber length, the center of the fiber has a stress less than its tensile strength. When the fiber length is no less than the critical fiber length, the center of the fiber has a stress equal to its tensile strength.

When a fiber in the fiber-reinforced thermoplastic composite is under loading, the stress condition of the fiber is assumed to be symmetrical with respect to its center, and the stress increases linearly from its ends. An element with a length of dx in the fiber showing all of applied stresses is sketched in Fig. 3.11a. The fiber has a length L and a diameter d. The element undergoes three stresses, σ, $\sigma + d\sigma_x$, and τ (Fig. 3.11b). σ is the normal stress applied on the element in the fiber direction, and τ is the shear stress acting on the interface between the fiber and the thermoplastic matrix. The geometry of the fiber can be considered to be symmetrical about its center as most fibers, such as the glass or carbon fiber, are cylindrical and have a consistent cross section along the length direction (x direction). At the center of the fiber, x is equal to 0, or, $x = 0$.

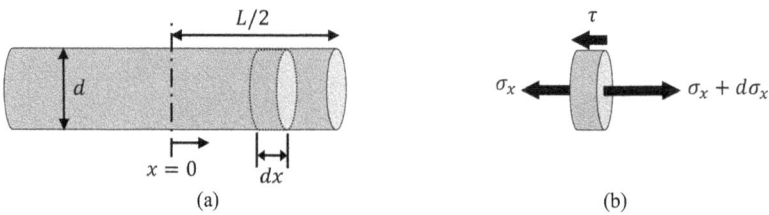

Fig. 3.11: (a) An element in a fiber in fiber-reinforced thermoplastic composite; (b) stresses, σ_x, $\sigma_x + d\sigma_x$, and τ, on the element. The fiber has length L and diameter d.

Equation (3.1) is obtained by balancing all of the forces applied on the element in the fiber length direction (x direction):

$$\sigma_x A + \tau A' = (\sigma_x + d\sigma_x)A \qquad (3.1)$$

where A is the cross-sectional area of the fiber, $\pi(d/2)^2$; A' is the interface area between the element and the thermoplastic matrix, $2\pi(d/2)dx$:

$$\sigma_x \times \pi(d/2)^2 + \tau \times 2\pi(d/2)dx = (\sigma_x + d\sigma_x) \times \pi(d/2)^2 \tag{3.2}$$

After simplifying,

$$\tau dx = \frac{d}{4}d\sigma_x \tag{3.3}$$

The terms in Eq. (3.3) are integrated for x from the center $(x = 0)$ to the end of the fiber $(x = L/2)$, and σ_x from 0 to the tensile strength of the fiber, σ_f. σ_f is the upper limit that the stress in the fiber σ_x can achieve:

$$\int_0^{L/2} \tau dx = \int_0^{\sigma_f} (d/4) d\sigma_x$$

Therefore,

$$\frac{L}{d} = \frac{\sigma_f}{2\tau} \tag{3.4}$$

The critical fiber aspect ratio, $(L/d)_c$, is defined when the stress applied on the interface, τ, reaches the maximum interfacial bonding strength between the fiber and the matrix, τ_y:

$$\left(\frac{L}{d}\right)_c = \frac{\sigma_f}{2\tau_y} \tag{3.5}$$

For a fiber with a fixed diameter, its critical fiber length, L_c, is defined by the following equation:

$$L_c = \frac{\sigma_f d}{2\tau_y} \tag{3.6}$$

where d is the fiber diameter, τ_y is the interfacial strength between the fiber and the thermoplastic matrix, and σ_f is the tensile strength of the fiber. For a fiber with a given diameter, its critical fiber length is dependent on the fiber tensile strength and the bonding strength between the fiber and the thermoplastic matrix. A higher tensile strength and a weaker bonding strength result in a higher critical fiber length.

Figure 3.12 shows different fiber lengths in the long fiber-reinforced thermoplastic composite that result in different stress scenarios. Figure 3.12a shows the stress condition when the fiber length is less than the critical fiber length (e.g., in SFT composites). The peak stress in the fiber, which occurs at the fiber center, is lower than its tensile strength. In this case, the fiber does not fracture due to the fact that the stress in the fiber is less than its tensile strength. The contact area between the fiber and the thermoplastic matrix is not large enough to transfer enough load from the matrix to the fiber for the fiber stress to reach its tensile strength. As a result, the fiber can be debonded from the thermoplastic matrix through pullout instead of fracture.

Figure 3.12b shows that the fiber length reaches the critical fiber length and the stress at the fiber center equals to its tensile strength. Figure 3.12c shows that the fiber length is more than the critical fiber length. The stress reaches the fiber tensile strength at the location $(L_c/2)$ from each end. There is a section of the fiber with a length of $(L - L_c)$ that has a stress equal to the fiber strength. Therefore, the fiber can potentially fracture within this section. This principle has been used in fiber fragmentation testing to determine the fiber critical length (see Section 6.9.2).

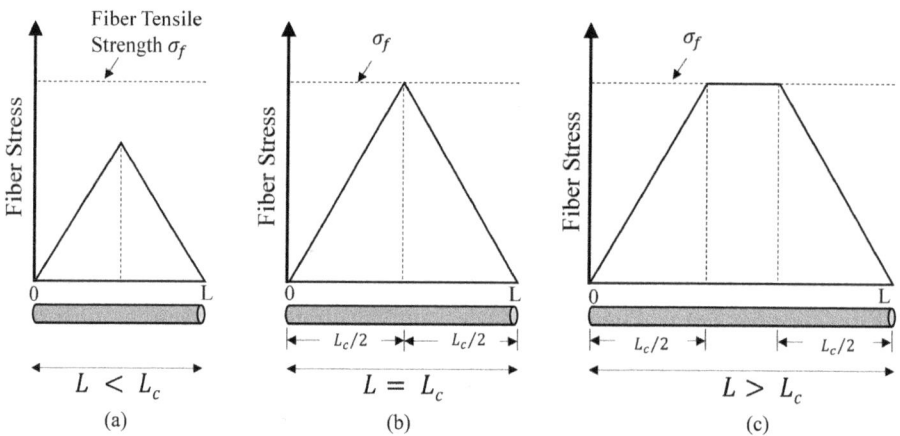

Fig. 3.12: The change of the fiber stress with fiber length, L, that is (a) less than, (b) equal to, and (c) more than the critical fiber length, L_c.

The interfacial bonding strength between the fiber and the thermoplastic matrix considerably affects the critical fiber length as indicated in Eq. (3.6). The critical fiber length increases with decreasing interfacial bonding strength. If the interfacial bonding strength, the diameter, and tensile strength of the fiber are known, the critical fiber length is determined. The critical fiber length can vary if the same fiber is added to different thermoplastic matrices due to the difference in the interfacial bonding strength between the fiber and different thermoplastics. Other conditions that can change the interfacial bonding strength and the critical fiber length include the surface treatment of fibers and the addition of coupling agents in the thermoplastic matrix, both of which can enhance the bonding at the fiber/matrix interface and, therefore, increase the interfacial bonding strength and decrease the critical fiber length.

When a fiber has the length greater than the critical fiber length, it is indicated that the load can be efficiently transferred from the matrix to the fiber and the maximum load-bearing capacity of the fiber can be reached. It is beneficial to have a lower critical fiber length. A lower critical fiber length indicates a larger probability for the fiber length staying above the critical fiber length in a composite after fiber attrition resulted from processing.

Example question 3.1 The glass fiber in a discontinuous glass fiber-reinforced PP composite has interfacial bonding strength of 12 MPa with the PP matrix. The glass fiber has an average diameter of 15 μm and tensile strength of 3,200 MPa. What is the critical fiber length of the glass fiber? If the glass fiber is treated with a different size, which improves its bonding strength with PP to 15 MPa, what is the critical fiber length?

Solution: Critical fiber length can be calculated using Eq. (3.6). The values of σ_f, d, and τ_y are already given. Therefore, the critical fiber length of the glass fiber L_c can be calculated.

Since the critical fiber length is equal to $\dfrac{\sigma_f d}{2\tau_y}$

$$L_c = \frac{3,200\ \text{MPa} \times 15\ \mu m}{2 \times 12 \text{MPa}} = 2\ \text{mm}$$

When the interfacial bonding strength is improved to 15 MPa, the critical fiber length of the glass fiber is calculated as follows:

$$L'_c = \frac{3,200\ \text{MPa} \times 15\ \mu m}{2 \times 15\ \text{MPa}} = 1.6\ \text{mm}$$

3.3.3 Short fiber thermoplastic composite

Thermoplastic composites with a fiber length less than the critical fiber length are called SFT composites. SFT composites normally have fiber length less than 1 mm. SFT composites were developed in the 1960s for the purpose of enhancing the performance of neat thermoplastics while still maintaining ease of processability. The SFT composite is still one of the commonly used materials for secondary structural components in applications such as automotive. Glass and carbon fibers are the common reinforcements added in the SFT composite. Other fibers such as basalt and cellulose fibers are also used in the SFT composite.

The addition of short fibers to thermoplastics improves the mechanical properties, such as elastic modulus, tensile strength, fatigue strength, and impact strength, of the thermoplastic. With increasing fiber content, the mechanical properties of the SFT composite increase. In addition, dimensional stability of the SFT composite is improved compared to neat thermoplastics. Its lower coefficient of thermal shrinkage resulting from the addition of short fibers can help mold products with less warpage.

Manufacture of SFT composite products involves multiple processing steps, namely, manufacture of SFT composite pellets and molding of the pellets. The processing details for producing SFT composite pellets are discussed in Section 4.5. Injection molding is a typical process used to manufacture SFT composite products. It can result in severe fiber length attrition because of the high shear stresses involved in the process. Fig. 3.13 shows the fiber length distribution in glass fiber-reinforced PET composites with different fiber contents (15 wt% and 30 wt%) [5]. Fiber length decreases with the increase of fiber percentage due to increased fiber-to-fiber interactions. The average fiber length in the molded composite decreases from an initial fiber length of 4.5 mm to 0.71 mm and 0.55 mm for composites with a fiber content of 15 and 30 wt%, respectively [5].

The fibers in SFT composites generally have a length less than the critical fiber length. Therefore, the maximum tensile stress that a fiber in the SFT composite undergoes is less than its tensile strength. Because of the short fiber length in the

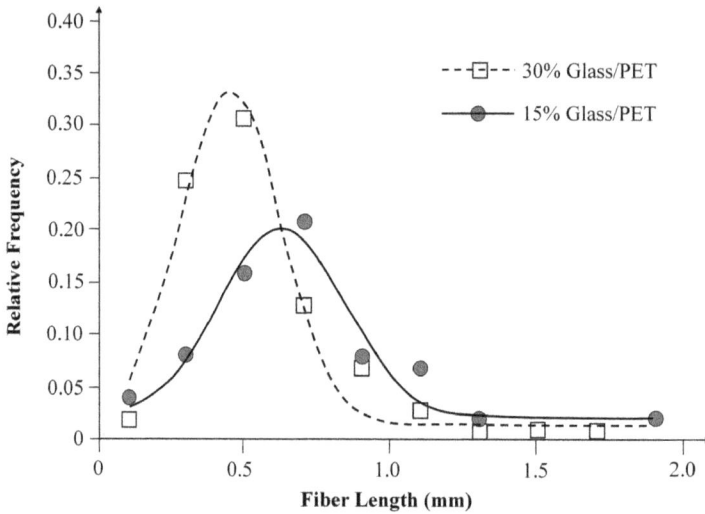

Fig. 3.13: The fiber length distribution of injection molded glass fiber PET composites with different fiber contents (reprinted from Reference [5] with permission).

SFT composite, the contact area or the interfacial area between the fiber and the matrix is not large enough to transfer adequate load for the fiber to reach its tensile strength. As a result, the fiber does not normally fracture but debonds with the matrix instead under loading. The phenomenon that the fiber debonds from the matrix is called fiber pullout. It is one of the main failure mechanisms in the SFT composite. Matrix fracture is another main failure mechanism in SFT composites. Both the mechanisms can be observed in the fracture surface of a fractured SFT composite as shown in Fig. 3.14. Fiber pullout is indicated by those exposed fibers and the holes resulted from the fiber pullout.

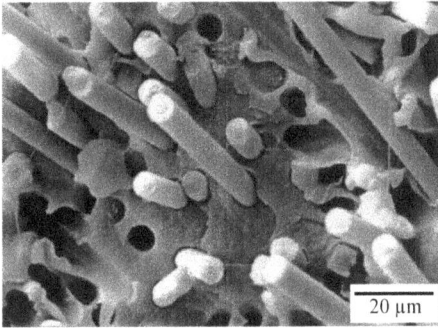

Fig. 3.14: The fracture surface of a short carbon fiber-reinforced polypropylene composite showing fiber pullout and matrix fracture, typical failure mechanisms in SFT composites (reprinted from Reference [6] with permission).

The SFT composite has also been widely used in additive manufacturing for its balanced combination of relatively good mechanical properties and great printability. Short fibers can be readily blended with a thermoplastic matrix to produce filaments for additive manufacturing. One example is the filament feedstock made of a short carbon fiber-reinforced nylon 6 composite (Fig. 5.2c) for the fusion deposition modeling process. The details are provided in Section 5.2.1.

3.4 Thermoplastic composite-based hybrid material

A hybrid material consists of different types of materials in an integrated manner. Each material in the hybrid material possesses different functions. Thermoplastic composites have been increasingly hybridized with other thermoplastic composites. For example, continuous fiber-reinforced thermoplastic composites are hybridized with discontinuous fiber-reinforced thermoplastic composites. Thermoplastic composites can also be hybridized with other materials that are not composites, such as foams and metals. The following sections describe several thermoplastic composite-based hybrid materials.

3.4.1 Overmolded thermoplastic composites

Continuous and discontinuous fiber-reinforced thermoplastic composites are often hybridized to gain the benefits from both thermoplastic composites. Continuous fiber thermoplastic composites offer great mechanical properties while their processability is relatively low, especially for the products with complex geometries (ribs, nonuniform part thickness, etc.). On the other hand, discontinuous fiber-reinforced thermoplastic composites have good processability but their mechanical properties are relatively low. Hybridizing these two types of thermoplastic composites provide combined benefits in both processability and mechanical properties.

Hybridization can be achieved through an overmolding process, a common process in hybridizing thermoplastic composites with other types of thermoplastic composites or even metals. The continuous fiber thermoplastic composite is placed into a mold (normally heated), on which a discontinuous fiber thermoplastic composite melt is then placed. With the aid of high pressure, forming, consolidation, and joining are done in one single step. The overmolding can take place in a setting of compression molding or injection molding, The details of the overmolding process are described in Section 4.7.

A variety of thermoplastic composites have been used in the overmolding process. Typical material combinations include organosheet and LFT (Fig. 4.11), unitape and LFT, and woven fabric composite and SFT. The same matrix material is commonly used to achieve good bonding between the composites. Figure 3.15 shows the microstructure of the cross-sectional view of an overmolded hybrid material made of continuous carbon fiber-reinforced PPS composites and a long glass fiber-reinforced PPS composite. The LFT composite is sandwiched between the continuous fiber composites.

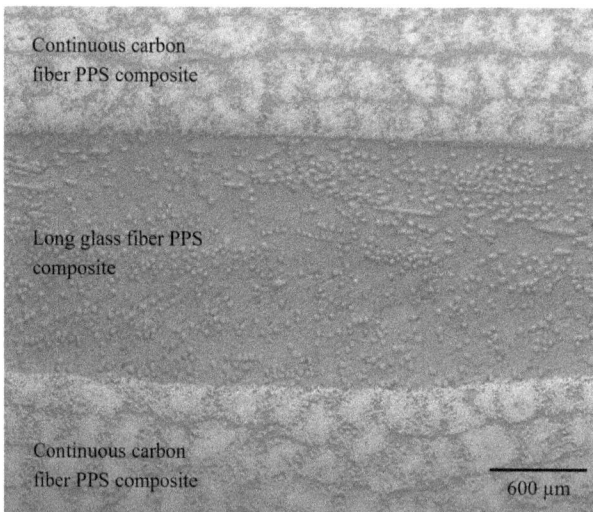

Fig. 3.15: A hybrid material consisting of continuous carbon fiber-reinforced PPS composite and long glass fiber-reinforced PPS composite.

Case study 3.1: Hybridizing thermoplastic composite with metal
Metallic materials can be hybridized with thermoplastic matrix composites although they are distinctly different materials. Mechanical locking using Z-pins can efficiently induce good bonding between the metal and the composite. Figure 3.16 shows a hybridized material consisting of an aluminum alloy and a long glass fiber PP matrix composite [7]. The aluminum with Z-pins is placed into a compression mold before a LFT extrudite is molded with the metal (Fig. 3.16a). The Z-pins facilitate great bonding at the metal/composite interface (Fig. 3.16b and c).

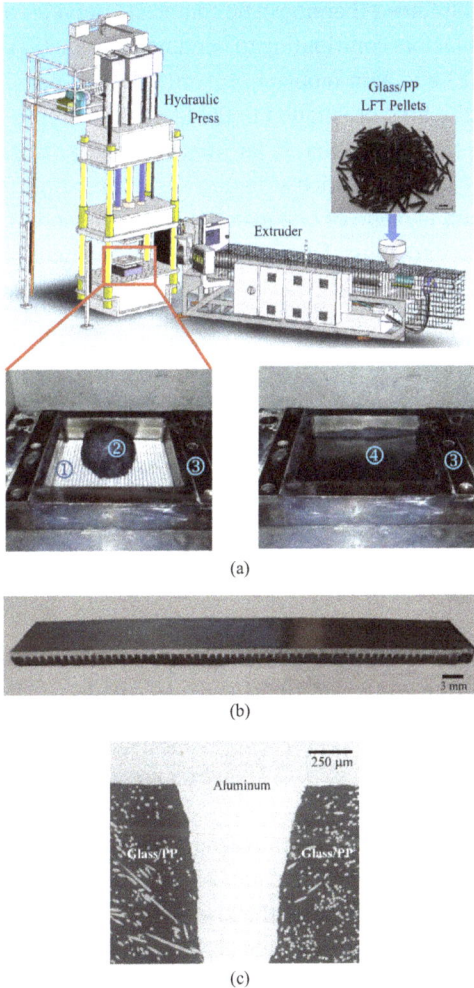

Fig. 3.16: (a) A compression molding process used to overmold LFT composite onto aluminum sheet with integrated Z-pins (①-aluminum sheet with Z-pins; ②-glass/PP melt; ③mold; and ④molded hybrid material plate); (b) overmolded LFT/metal hybrid material; (c) microstructure of the hybrid material at the interface (adapted from Reference [7]).

3.4.2 Thermoplastic composite sandwich material

A thermoplastic composite sandwich material is another hybrid material. It is generally comprised of two facesheets and one core, as shown in Fig. 3.17. The facesheet is made of a continuous fiber thermoplastic composite while the core material is made of low-density materials, such as foams, and makes up most of the volume of the sandwich structure. The combination of the facesheet and the core offers great bending stiffness and geometric stability.

The facesheet can be made of any of the continuous fiber-reinforced thermoplastic composites. The continuous fiber-reinforced thermoplastic composite provides a high elastic modulus (E), one of the major factors contributing to bending stiffness (EI), which is the product of elastic modulus (E) and area moment of inertia (I). It is noted that the elastic modulus is dependent on the fiber orientation in the facesheet composite. In addition, the facesheet, located on the surface that is far from the neutral axis, results in a large area moment of inertia (I). The sandwich structure has a similar principle as steel I-beams. However, the I-beam is made of one material. Its high bending stiffness is achieved through its high elastic modulus of steel and high area moment of inertia due to its flange that is far from the neutral axis.

Fig. 3.17: A thermoplastic composite sandwich comprised of two thermoplastic composite facesheets and one core.

The equivalent bending stiffness of the thermoplastic composite sandwich in the length direction can be expressed by the following equation:

$$(EI)_{eq} = E_{facesheet} \frac{bt^3}{12} \times 2 + E_{facesheet} bt \left(\frac{c+t}{2}\right)^2 \times 2 + E_{core} \frac{bc^3}{12}$$

where $(EI)_{eq}$ is the equivalent bending stiffness of the thermoplastic composite sandwich, $E_{facesheet}$ is the elastic modulus of the thermoplastic composite facesheet in the length direction, b is the width of the sandwich, c is the thickness of the core, t is the thickness of the facesheet, and E_{core} is the elastic modulus of the core.

The elastic modulus of the facesheet, $E_{facesheet}$, is considerably greater than the elastic modulus of the core, or, $E_{facesheet} \gg E_{core}$. Therefore, the contribution from the core to the sandwich stiffness, $E_{core}(bc^3/12)$, is much smaller and is negligible. If the facesheet has much smaller thickness than the core or the overall sandwich thickness ($t \ll c$), the term $E_{\text{facesheet}}\left(bt^3/6\right)$ can be neglected, and the equivalent bending stiffness equation can be simplified below:

$$(EI)_{\text{eq}} \approx E_{\text{facesheet}} \frac{btc^2}{2}$$

Figure 3.18a displays several core materials that can be used in the thermoplastic composite sandwich structure. Those sandwich cores include foam-filled honeycomb, PP honeycomb, syntactic foam, balsa, and foam core (from bottom to top in the figure). PP is a versatile material and has been used in different forms in thermoplastic composites, including bulk matrix material, fibers, and honeycomb core. The PP honeycomb core is a low-density material for sandwich structures because of the large percentage of air in the core and the low density of PP. Figure 3.18b shows a sandwich panel (150 mm × 150 mm) consisting of PP honeycomb core and continuous glass fiber-reinforced PP composite facesheets. The sandwich is lightweight and provides excellent sound damping, low moisture absorption, and good impact resistance. The bonding between the core and facesheet can be realized via fusion bonding because of the same type of thermoplastic (PP) in both core and facesheet.

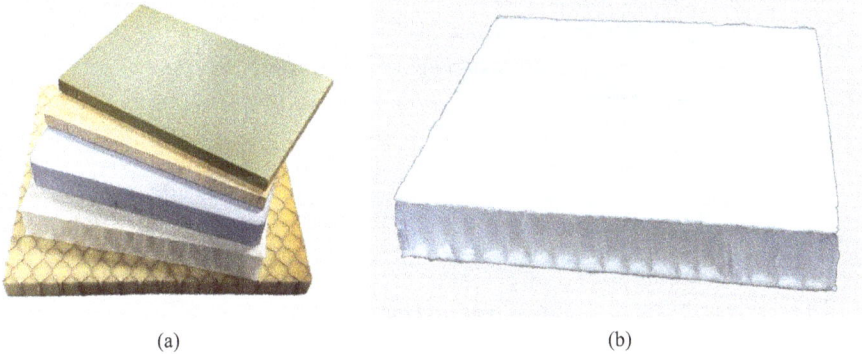

(a)   (b)

Fig. 3.18: (a) Core materials used in thermoplastic composite sandwich structure; (b) a thermoplastic composite sandwich structure made of a polypropylene honeycomb core and continuous glass fiber-reinforced polypropylene composite facesheets.

3.5 Multifunctional thermoplastic composite

It has become highly desirable to integrate multiple functions in one material, one component, or one device. A material that possesses multiple attributes and has more than one function simultaneously is called a multifunctional material. Multifunctional thermoplastic composite materials possess not only a structural function but

also at least one of the nonstructural functions, such as sensing, actuation, energy harvest and conversion, electromagnetic shielding, electrical conductivity, thermal conductivity, structural health monitoring, shape memory, or self-healing capability. Several multifunctional thermoplastic composite examples are listed below:
– micro-sized and/or nano-sized particles with tailored magnetic and dielectric properties are blended into thermoplastic composites to achieve electromagnetic shielding;
– fine bronze-woven mesh is integrated on the surface of thermoplastic composites used in aircraft structures to provide lightning strike protection;
– micro-sized and/or nano-sized pores are intentionally introduced to thermoplastic composites to render the composite structure lighter, more thermal insulating, and better vibration damping;
– glass fiber thermoplastic composites are added to the surface of carbon fiber-reinforced thermoplastic composites that make contact with aluminum alloys to prevent galvanic corrosion of the aluminum alloy;
– suspector materials are added to thermoplastic composites for fusion joining;
– materials such as carbon black and nanographene are added to thermoplastic composites to induce thermal and electrical conductivity;
– piezoelectric sensor-actuator arrays are integrated in thermoplastic composites for health monitoring/sensing function;
– shape memory thermoplastic polymers such as polystyrene, polyurethane, poly (ethylene oxide), acrylonitrile butadiene styrene, polytetrafluoroethylene, and polylactic acid are used in composites to induce shape memory function.

3.6 Summary

– The processability of thermoplastic composites decreases with increasing fiber length while mechanical properties such as modulus, strength, and impact resistance increase with fiber length.
– The fiber architecture in continuous fiber-reinforced thermoplastic composites and their preforms include unidirectional fiber, woven fabric, braided fiber, and NCF.
– Thermoplastic composites reinforced with unidirectional carbon fibers are one of the engineering materials with the highest specific strength and modulus in the fiber direction due to their excellent strength, modulus, and low density.
– Self-reinforced composites are a group of thermoplastic composites that have the same thermoplastic polymer as both their reinforcement and matrix. The self-reinforced composites mainly include polypropylene-based self-reinforced composite and polyethylene-based self-reinforced composite.

- Self-reinforced composite preforms are produced through a thermal method that involves melting a portion of the reinforcement fiber in a single-component process or melting the matrix in a bicomponent process followed by consolidation under pressure.
- Discontinuous fiber-reinforced thermoplastic composites can be classified into random fiber mat thermoplastic composites, LFT composites, and SFT composites.
- Fiber aspect ratio is defined as the ratio of fiber length to fiber diameter. It is one main parameter that determines the mechanical properties of discontinuous fiber-reinforced thermoplastic composites.
- Critical fiber length is the minimum fiber length required for the maximum fiber stress to equal the tensile strength of the fiber at its center (mid-length). When a fiber has the length greater than the critical fiber length, the load can be efficiently transferred from the matrix to the fiber, and the maximum load-bearing capacity of the fiber can be reached.
- Long fiber-reinforced thermoplastic composites possess a combination of great mechanical properties and good processability. It is increasingly used in applications that require a high production rate and great mechanical properties.
- Fiber length in short fiber-reinforced thermoplastic composites is less than critical fiber length. Short fiber-reinforced thermoplastic composites show typical failure modes such as fiber pullout and matrix fracture.
- Thermoplastic composites can be overmolded with other thermoplastic composites and metallic materials. The overmolded hybrid materials gain benefits from each material.
- Thermoplastic composite sandwich structures provide great bending stiffness and geometry stability. The large elastic modulus of the continuous fiber thermoplastic composite facesheet and its large distance from the neutral axis result in great bending stiffness (EI).

References

[1]   Capiati NJ, Porter RS. The concept of one polymer composites modelled with high density polyethylene. Journal of Materials Science. 1975;10(10):1671–7.
[2]   Alcock B, Peijs T. Technology and development of self-reinforced polymer composites. Polymer Composites–Polyolefin Fractionation–Polymeric Peptidomimetics–Collagens. 2011: 1–76.
[3]   https://www.luggagedirect.com.au/brands/samsonite-luggage/samsonite-lite-cube/samson ite-lite-cube-hardside-medium-76-cm-spinner-suitcase-graphite-58624.html. Accessed in Feb 2021.
[4]   https://www.marketresearchfuture.com/reports/long-fiber-thermoplastics-market-4889. Accessed in April 2021.

[5]   Pegoretti A, Penati A. Recycled poly (ethylene terephthalate) and its short glass fibres composites: Effects of hygrothermal aging on the thermo-mechanical behaviour. Polymer. 2004;45(23):7995–8004.

[6]   Fu S-Y, Lauke B, Mäder E, Yue C-Y, Hu X. Tensile properties of short-glass-fiber-and short-carbon-fiber-reinforced polypropylene composites. Composites Part A: Applied Science and Manufacturing. 2000;31(10):1117–25.

[7]   Yeole P, Ning H, Hassen AA. Development and characterization of a polypropylene matrix composite and aluminum hybrid material. Journal of Thermoplastic Composite Materials. 2021;34(3):364–81.

Chapter 4
Processing of thermoplastic composites

4.1 Introduction

Processing of thermoplastic composites refers to combining thermoplastic matrix and reinforcement to produce thermoplastic composite preforms and molding the preform into composite products. Elevated temperature and high pressure are normally involved in the processing of thermoplastic composites. In addition, the characteristics of their constituents, including the viscosity of the matrix, the volume percentage of the reinforcement, the form of the reinforcement, among others, considerably affect processing of the thermoplastic composite.

Unlike processing of thermoset composites that starts with monomers in a liquid state, processing of thermoplastic composites has to cope with thermoplastics that are already polymerized generally. The polymerized thermoplastic is solid at room temperature, and processing of the thermoplastic composite can be achieved after the thermoplastic matrix is heated to a certain temperature. Semicrystalline thermoplastics are heated above their melting temperature, and amorphous thermoplastics are heated to a temperature significantly higher than their glass transition temperature. However, even when heated to the elevated temperature, the thermoplastic has a high viscosity, which causes the difficulty in combining with fibers. It is worth pointing out a special case that processing of thermoplastic composites occurs at a very low viscosity because monomers of their thermoplastic matrix are used. The processing is discussed in Section 4.4.12. Other exceptions include the solution impregnation process in which the thermoplastic is dissolved and fibers are impregnated with the low viscosity solution (see Section 4.3.5).

Since thermoplastics have much higher viscosities, it is difficult to achieve adequate fiber impregnation and consolidation in one processing step. Instead, two steps, namely, pre-impregnation step and consolidation step, are used to produce thermoplastic composite products. The pre-impregnation step includes combining the thermoplastic matrix and the reinforcement to form a pre-impregnated material, or a preform, through various techniques. In the consolidation step, the preform is heated to its processing temperature, and pressure is applied to consolidate the preforms into a composite product.

4.2 Effect of viscosity on processing of thermoplastic composites

One of the main material variables that affect the processing of the thermoplastic composite is the viscosity of the thermoplastic matrix. The viscosity of a liquid polymer, such as thermoset resin or thermoplastic melt, is defined as the resistance of

https://doi.org/10.1515/9781501519055-004

the resin or the melt to flow. Viscosity is one major variable that determines the processability of composites. A lower viscosity of the matrix indicates better processability. For example, thermoset resins used in processing thermoset composites are liquid and have a low viscosity at room temperature, because the resin is in the monomer stage. Due to their low viscosity, it is relatively easy to process thermoset matrix composites. The low viscosity enables wet out and impregnation of fibers readily. Polymerization of the monomers occurs during the consolidation of the thermoset composite. Therefore, the impregnation and consolidation of the thermoset composite can be completed via one single processing step due to the low viscosity of thermosets. One exception is thermoset composite prepreg, an intermediate material that, has fibers pre-impregnated with partially cured (B-stage) thermoset resin and is subsequently consolidated via molding processes, such as autoclave molding.

When a thermoplastic is heated to its processing temperature, it still has a significantly higher viscosity than the thermoset. Complex viscosity, the resistance to flow as a function of dynamic shear (angular frequency ω), of the thermoplastic melt can range from 100 Pa·s to more than 10,000 Pa·s, when the angular frequency is less than 10^3 rad/s (Fig. 4.1). For example, nylon 66 melt has a viscosity of 200 Pa·s at an angular frequency less than 10^3 rad/s. Other thermoplastics, such as ABS and PEI, have even higher viscosities. The viscosity of ABS at 270 °C can reach 10,000 Pa·s at low angular frequencies. In contrast, a typical thermoset resin, for example, epoxy, has a viscosity of about 4 Pa s, which is 1/40 that of the nylon 66 melt, when the angular frequency is between 1 and 1,000 rad/s. Other thermoset resins such as vinyl ester and polyester have a similar or even lower viscosity, compared to epoxy.

The complex viscosities of some thermoplastics and thermoplastic composites as a function of angular frequency at elevated temperatures are compared in Fig. 4.1. All the composites in the figure are reinforced by discontinuous fibers. Different colors are used to indicate different types of thermoplastic polymer systems. The viscosity curve of an epoxy at room temperature is also included to demonstrate the enormous difference in the complex viscosity between epoxy and thermoplastic or discontinuous fiber thermoplastic composite melts. It is noted that both the axes are in a log scale.

Figure 4.1 shows the drastic complex viscosity decrease of thermoplastic melts, such as nylon 66 at 285 °C, with increasing angular frequency at high angular frequencies. For example, when the angular frequency increases from 10^3 to 10^5 rad/s (injection molding range) for nylon 66 at 285 °C, its complex viscosity decreases by one magnitude from 10^2 to 10^1 Pa s. The complex viscosity–angular frequency relationship indicates decrease of apparent viscosity with shear rate, according to the Cox–Merz rule and the shear-thinning nature of the thermoplastic melt. Shear thinning is a characteristic phenomenon of some non-Newtonian fluids whose viscosity decreases with increasing shear rate. Thermoplastic melts or discontinuous fiber thermoplastic composite melts are usually non-Newtonian fluids that show a shear-thinning behavior. Their viscosity decreases with shear rate. When the shear rate increases, polymer chains are more aligned in the melt, and their viscosity decreases accordingly. The decreased viscosity of the melts indicates a greater ability to flow under an applied pressure. This is more apparent in some processes that involve high shear rates, for example, injection molding.

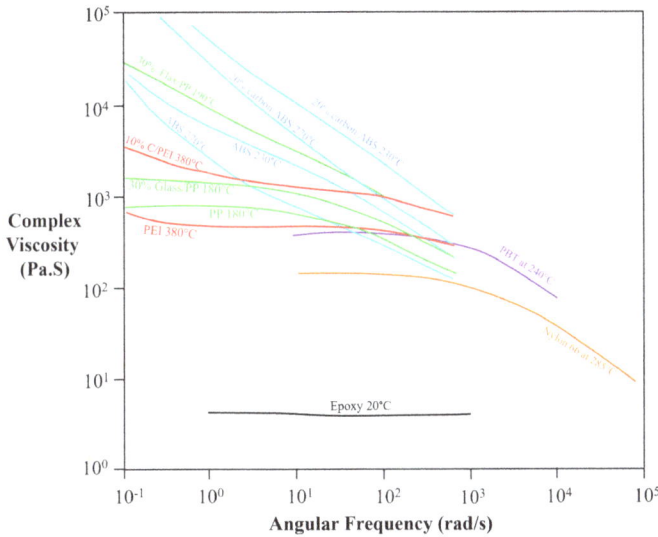

Fig. 4.1: Complex viscosity – angular frequency curves of various thermoplastic and discontinuous fiber-reinforced thermoplastic composite melts (adapted from References [1–5]).

The viscosity of thermoplastic melts or discontinuous fiber thermoplastic composite melts depends on their temperature. The viscosity decreases with increasing temperature, to a certain extent. When temperature is increased from 230 to 270 °C, the viscosities of both neat ABS and 20 wt% carbon fiber-reinforced ABS composite decrease, as shown in Fig. 4.1. The dependence of the melt viscosity on temperature can be described by mathematical models such as the Arrhenius model and the William-Landel-Ferry model. Lowering viscosity by increasing processing temperature can be beneficial to induce improved flow; however, a higher temperature can cause thermal degradation of the thermoplastic.

The viscosity of thermoplastics increases when fibers or other fillers are added. High fiber content results in a higher viscosity of the melt and considerably affects the flow and filling during processing. Figure 4.1 shows the viscosity difference between neat thermoplastics and discontinuous fiber-reinforced thermoplastic composites. For example, 20% carbon fiber added to ABS results in a significantly higher viscosity than neat ABS; the viscosity of 10 wt% carbon fiber-reinforced PEI composite increases by almost one magnitude compared to neat PEI at the same temperature (Fig. 4.1). However, the difference in viscosity reduces when the angular frequency or shear rate increases, which benefits processing of the discontinuous fiber-reinforced thermoplastic composite in high shear-rate processes, such as injection molding and extrusion.

Processing of continuous fiber-reinforced thermoplastic composite involves pre-impregnation and consolidation. Preforms are firstly produced through different preforming techniques to introduce thermoplastics to fibers. The composite preform is then consolidated into final products under heat and pressure through various molding methods. One main goal from processing of thermoplastic composites is to realize fiber impregnation, or coating fibers with the thermoplastic matrix. It can occur at different stages for different preforms. For example, fiber impregnation is achieved at the pre-impregnation stage for melt impregnation, while it is realized only at the consolidation stage for commingled preforms. The fiber impregnation takes place at both the pre-impregnation and consolidation stages for the thermoplastic composite produced from powder impregnation and film stacking preforms. These different preforming techniques are described in Section 4.3.

Fiber impregnation is a critical step for processing polymer matrix composites, including thermoset and thermoplastic composites. The impregnation of fibers relies on the flow of a liquid polymer, either liquid thermosets or thermoplastic melts, during processing of the composites. The flow follows Darcy's law:

$$Q = - \frac{KA*\nabla P}{\mu} \tag{4.1}$$

where Q is the polymer flow rate (unit: m^3/s), K is the permeability of the fibers (unit: m^2), A is the cross-sectional area of flow, ∇P is the pressure gradient (pressure difference over a certain length; unit: Pa/m), and μ is the viscosity (unit: Pa·S).

Darcy's law defines that the flow rate of a liquid polymer inside fabrics is proportional to the permeability of the fabric, the cross-sectional area of flow, and the pressure gradient. The permeability is defined as the ability of the fabric to allow the polymer to pass through. It is dependent on the available space in the fabric, the orientation of the fibers, and so on. Higher permeability results in higher flow rate of the liquid polymer, and, thus, faster fiber impregnation.

The viscosity (μ) of the liquid polymer adversely affects the flow rate and fiber impregnation. As thermoplastic melts have significantly higher viscosities than thermosets (Fig. 4.1), the impregnation of the thermoplastic melt onto fibers is much more difficult. Increasing processing temperature can reduce the viscosity of the thermoplastic melt, which can induce a higher flow rate and faster fiber impregnation. However, the thermoplastic can suffer from thermal degradation at a high processing temperature. Besides processing temperature, pressure gradient, ∇P, is another practical variable that can be controlled to induce a higher flow rate for the thermoplastic melt.

During processing of thermoplastic composites, a high processing pressure can be applied to induce a higher ∇P and greater flow rate. This approach has been adopted in processing thermoset composite by using high pressure to reduce the processing time and increase the production rate. For example, high-pressure resin transfer molding (HP RTM) can significantly reduce cycle times compared to a regular RTM process.

There are several pressures involved in the flow of the thermoplastic melt and impregnation of the fibers, during processing. These pressures include capillary pressure, gravitational pressure, and mechanical pressure:

$$P = P_c + P_g + P_m \tag{4.2}$$

where P is the total pressure, P_c is the capillary pressure, P_g is gravitational pressure, and P_m is the mechanical pressure or applied external pressure (including vacuum pressure). The gravitational pressure of the polymer melt in a capillary tube is normally much less than the mechanical pressure and capillary pressure. Therefore, it can be negligible.

Capillary effect plays a critical role during processing of thermoset and thermoplastic composites at the microscopic level. Capillary effect or capillary mechanism is a phenomenon that commonly occurs at the interface of a liquid contacting a solid surface in narrow spaces. The combination of surface tension and adhesive forces between the liquid and the solid wall propels the liquid to move and fill the space. The microscale narrow spaces among fiber bundles form capillary tubes, and polymer liquid flows under capillary pressure in the tube. The capillary pressure, P_c, is defined by the Young–Laplace equation below:

$$P_c = \frac{2\gamma\cos\theta}{r} \tag{4.3}$$

where γ is the liquid surface tension, θ is the contact angle between liquid and solid, and r is the capillary tube radius. For fiber bundles consisting of microscale fibers, the size of the capillary tube can be in a small scale, which generate a high capillary pressure, P_c, that can favor the impregnation by increasing the overall pressure, P.

In spite of a high capillary pressure generated in a small capillary tube, the flow distance over time is limited in the small capillary tube. The Lucas–Washburn equation describes the flow distance square over time that is directly proportional to the capillary tube size and inversely proportional to the viscosity of the polymer liquid, as shown in Eq. (4.4). It is indicated that the small size of the space in the fiber bundle results in a low travel distance square over time. The equation also shows that a polymer liquid with a higher viscosity, for example, thermoplastic melt, will need a longer time to fill the space between the fibers or impregnate the fibers under the capillary pressure only:

$$\frac{L^2}{t} = \frac{\gamma r \cos\theta}{2\mu} \tag{4.4}$$

where L is the travel distance by the thermoplastic melt, t is the travel time, and μ is the viscosity of the liquid.

As aforementioned, the impregnation of thermoplastic melts onto fibers generally involves externally applied pressure, or mechanical pressure, in addition to the capillary pressure. The effect of externally applied pressure on the flow of the thermoplastic melt within the fiber bundles needs to be considered. Modification of the Lucas–Washburn equation can be made to consider the effect of the mechanical pressure. Based on Eqs. (4.2) and (4.3),

$$P = \frac{2\gamma\cos\theta}{r} + P_m \tag{4.5}$$

Poiseuille's law defines the flow of a liquid media in a capillary tube with radius, r. It is expressed in the following equation:

$$Q' = \frac{\pi r^4 P}{8\mu(L(t))} \tag{4.6}$$

where Q' is the flow rate of the thermoplastic melt and $L(t)$ is the travel distance of the liquid inside the capillary tube, and it is a function of time. By combining Eqs. (4.5) and (4.6),

$$Q' = \frac{\pi r^4}{8\mu(L(t))} \left(\frac{2\gamma\cos\theta}{r} + P_m \right) \tag{4.7}$$

The flow rate is proportional to the mechanical pressure, P_m. For the thermoplastic melt that normally has a very high viscosity, a large amount of P_m is required to achieve a reasonable flow rate. Therefore, processing of thermoplastics generally involves use of mechanical pressure, which is applied through different ways, such as hydraulic press, vacuum pump, autoclave, and so on.

For a capillary tube that is cylindrical, the flow rate in the cylinder can be expressed as

$$Q' = \pi r^2 \frac{d(L(t))}{dt}$$

Therefore,

$$\frac{d(L(t))}{dt} = \frac{r^2}{8\mu(L(t))} \left(\frac{2\gamma\cos\theta}{r} + P_m \right)$$

After rearranging,

$$\int L(t)d(L(t)) = \int \frac{r}{4\mu} \left(\gamma\cos\theta + \frac{rP_m}{2} \right) dt$$

After solving the differential equation by assuming that γ, θ, and P_m are not a function of time:

$$\frac{1}{2}L^2(t) = \frac{r}{4\mu}\left(\gamma\cos\theta + \frac{rP_m}{2}\right)t + C$$

There is an initial condition for the flow of the polymer liquid in the capillary tube. When $t = 0$, $L(0) = 0$. After the boundary condition is applied to the equation, it is found that $C = 0$.

Therefore,

$$L^2 = \frac{r\gamma\cos\theta}{2\mu}t + \frac{r^2 P_m}{4\mu}t \tag{4.8}$$

or,

$$\frac{L^2}{t} = \frac{r\gamma\cos\theta}{2\mu} + \frac{r^2 P_m}{4\mu} \tag{4.9}$$

Equation (4.8) defines the travel distance of a polymer liquid, including thermoplastic melts, inside the fiber bundles in relation to space size, surface tension of the liquid, contact angle between the liquid and the fiber surface, liquid viscosity, externally applied pressure, and P_m. The viscosity of the thermoplastic melt has an adverse effect on its travel distance. When the viscosity of the thermoplastic melt increases, its travel distance shortens in a specific time. It is also seen that the externally applied pressure can enhance the flow in the fiber bundle. For thermoplastic melts that have a large viscosity, a large mechanical pressure, P_m, helps the melt travel a longer distance in a shorter time.

Equations (4.8) and (4.9) can also be applied to the flow in other capillary tubes with the aid of externally applied pressure. When there is no externally applied pressure, or $P_m = 0$, Eq. (4.9) becomes $\frac{L^2}{t} = \frac{r\gamma\cos\theta}{2\mu}$, which is the Lucas–Washburn equation described in Eq. (4.4).

Various preforming and molding methods have been developed to produce thermoplastic composite preforms and manufacture high-quality thermoplastic composite products, respectively. The following sections discuss those methods.

4.3 Manufacture of continuous fiber thermoplastic composite preform

Several pre-impregnation or preforming processes have been developed to produce thermoplastic composite preforms, an intermediate composite material for consequent molding of final thermoplastic composite products. Pre-impregnation is also the main processing step that controls the fiber content and determines the mechanical properties of final composite products. It is indispensable to go through the intermediate pre-impregnation stage for most thermoplastic composites in order to achieve sufficient impregnation and facilitate adequate consolidation in the final composite product. The main pre-impregnation processes are melt impregnation, powder impregnation, commingling, film stacking, and solution impregnation.

Figure 4.2 shows the schematics of different pre-impregnation techniques. The melt impregnation process results in a preform that is completely impregnated, as the thermoplastic covers the fibers completely. The powder impregnation process and the film stacking process produce preforms that are semi-impregnated. The preform from the commingling process does not have any fibers impregnated with thermoplastics, and thermoplastic fibers are only melted and impregnated with reinforcement fibers during molding. The following sections detail each pre-impregnation technique.

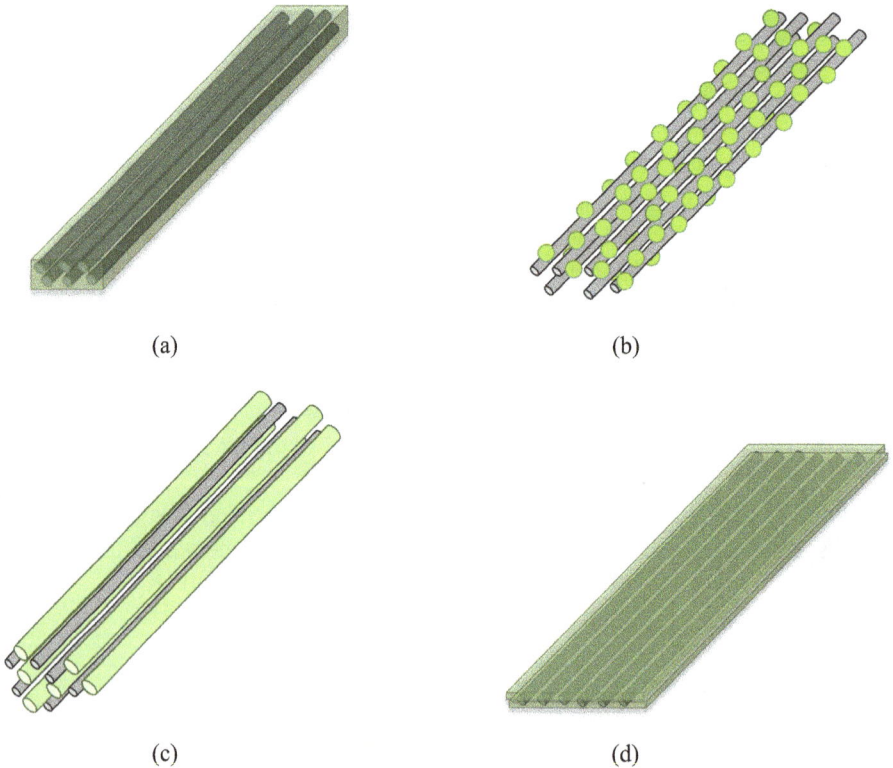

(a)   (b)

(c)   (d)

Fig. 4.2: Different techniques used for producing thermoplastic composite preforms: (a) melt impregnation; (b) powder impregnation; (c) commingling; and (d) film stacking.

4.3.1 Melt impregnation process

Melt impregnation process, or hot melt impregnation process, is one of the most commonly used pre-impregnation processes to produce thermoplastic composite preforms. It is a process in which reinforcement fiber tows pass through a thermoplastic melt in a die and gets impregnated with the thermoplastic. The preform produced through the melt impregnation process generally has complete fiber impregnation.

Figure 4.3 shows a typical schematic of the melt impregnation process for producing uni-tapes, one of the most commonly used preforms. Neat thermoplastic pellets are firstly fed into an extruder that melts the thermoplastic and supplies the molten thermoplastic to an impregnation die. Continuous fiber tows from fiber creels are pulled through a guide plate and a preheating station and finally into the impregnation die filled with the thermoplastic melt. The purpose of preheating the fiber tow is to increase the temperature of the filaments in the tow to prevent any sudden freezing of the molten thermoplastic when the filaments make contact with the molten thermoplastic. Fiber impregnation takes place when the fibers meet the molten thermoplastic in the impregnation die.

Impregnation pins are normally used and arranged in a certain way to assist with the impregnation of the fibers. The fiber tows are wet out through both macro impregnation (flow between fiber bundles) and micro impregnation (flow between individual filaments). It is desirable to spread the fiber tows onto the impregnation pins. Travel distance of the thermoplastic melt is significantly shortened when the fiber tow is spread, and better impregnation can be expected (Eq. (4.9)). The melt impregnation process produces a fully impregnated thermoplastic composite preform, as each individual filament is coated with thermoplastic.

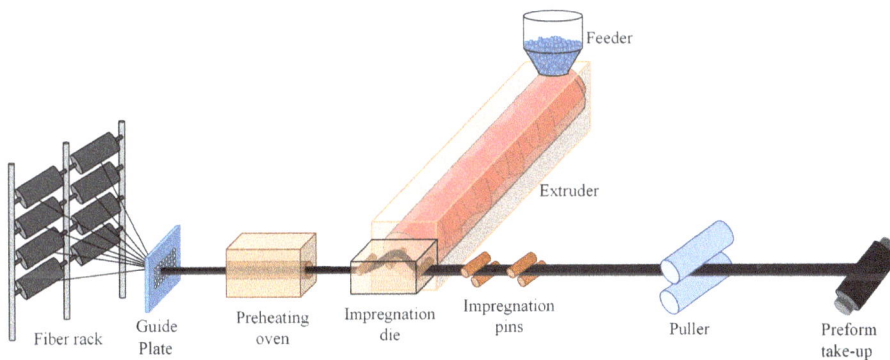

Fig. 4.3: Melt impregnation process for manufacturing unidirectional fiber-reinforced thermoplastic composite preforms.

A tension force is applied throughout the process to ensure that the fibers are pulled through the die continuously and helps reduce the slackness/waviness and crossover of the fibers. Both the slackness/waviness and crossover can lead to a preform with reduced mechanical properties. Under the tension load, the fibers are maintained straight, and the fiber tows also spread on the impregnation pins. The straightness of the fibers can maximize their load bearing capacity. The proper amount of tension load applied also helps spread the fiber tows open on the impregnation pins and improve the impregnation of the fibers with the molten thermoplastic. The melt-impregnated preform normally has good impregnation and minimal defects (Fig. 4.4b), which makes it feasible to manufacture high-quality thermoplastic composite products via the final molding process. The melt impregnation process is also used to produce continuous fiber thermoplastic composite filaments for additive manufacturing.

After the fiber tows are impregnated with the molten thermoplastic, the cross section of the impregnated fibers is controlled to a desired shape, dependent on the ensuing application. Typical cross sections are thin tape shapes or rod shapes. For example, the tape is mainly used for processing continuous fiber-reinforced thermoplastic composites, while the rod shape is used in additive manufacturing (Fig. 5.3) as well as producing long fiber-reinforced thermoplastic composite pellets (see Section 4.5.2).

The fiber weight fraction of the melt-impregnated preform can be controlled by several processing parameters, including the feed rate of the thermoplastic melt, the number of fiber tows, pulling speed, the amount of tension force used, aperture diameter, and so on. The fiber weight fraction of the melt-impregnated composite preform or any other preforms can be computed by a method that measures the mass of a segment of dry fiber with a certain length (m_f) and the mass of a segment of a pre-impregnated composite preform with the same length (m_c). The fiber weight fraction (W_f) of the preform can be calculated in accordance with the following equation:

$$W_f = \frac{m_f}{m_c}$$

The uni-tape is a versatile preform material for various molding processes. It can be used in several molding or forming processes, including compression molding, autoclave, filament winding, automated tape placement, pultrusion, and so on. Figure 4.4a shows a uni-tape roll (with a width of 300 mm) consisting of carbon fibers and a nylon 6 matrix. The microstructure of the cross section of a carbon fiber uni-tape in Fig. 4.4b shows complete fiber impregnation and minimal voids.

4.3.2 Powder impregnation

Powder impregnation process involves deposition of thermoplastic powders onto fibers, for the purpose of producing thermoplastic composite preforms for subsequent molding processes. It is a process that produces semi-impregnated preforms, because individual filament is not completely coated with thermoplastic. The powder

(a)

100 μm

(b)

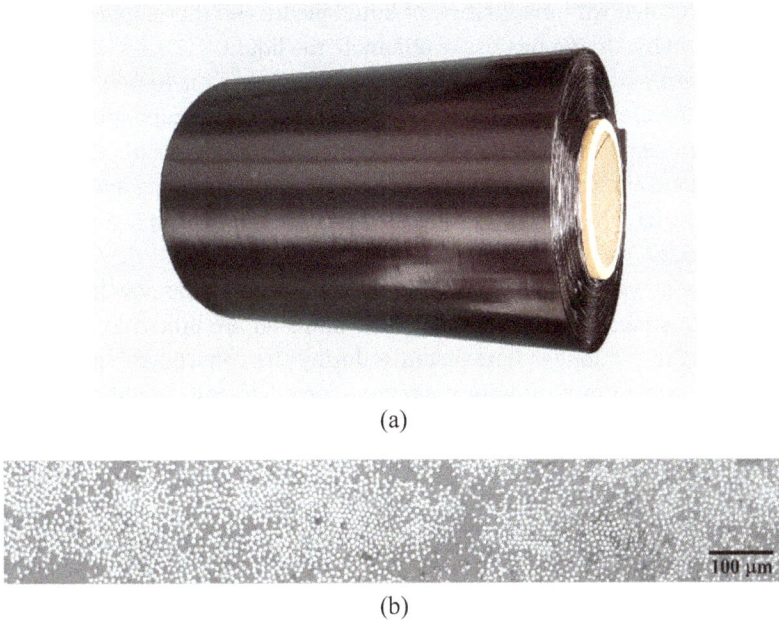

Fig. 4.4: (a) A uni-tape roll comprised of carbon fibers and nylon 6 matrix and (b) the microstructure of a uni-tape cross section showing complete fiber impregnation and minimal voids.

impregnation process is mainly applicable for thermoplastics that are difficult to be processed through other methods.

Different powder impregnation approaches, including dry impregnation and wet impregnation, have been developed. Dry impregnation takes place in a chamber in which thermoplastic powders are loosely packed or fluidized through flow of air or nitrogen. The powders are electrically charged, and deposition of the powders on the fibers is achieved through electrostatic forces. When the fibers pass through the chamber, normally, after being spread, the thermoplastic powders deposit onto the fibers and impregnation is achieved. In this approach, the friction between the fibers and impregnation pins is relatively high, and fiber breakage can occur. Separation of powders from the fibers after the impregnation can be another issue.

The wet impregnation approach uses slurry, that is, thermoplastic powders are suspended into a liquid media, normally water. The fibers are guided into the slurry and pins are used to spread and impregnate the fibers inside the slurry. The friction between the fibers and the impregnation pins is significantly reduced because the slurry acts as a lubricant. Figure 4.5 shows a schematic of the wet powder impregnation process. However, the wet impregnation approach has several drawbacks. The powder concentration in the slurry can decrease because of the consumption of the thermoplastic powder, which can result in an inconsistent fiber volume fraction. Constant measurement of the powder concentration needs to be carried out. In addition, the fibers coated with the mixture of liquid media and thermoplastic powders have to go through a drying process to eliminate the liquid.

After the powder impregnation step, thermoplastic powders are loosely adhered to the fibers, after the electrostatic charge is removed (in the dry impregnation approach) or the liquid media is dried (in the wet impregnation approach). The adhesion between the fibers and the powders can be strengthened by heating and melting the thermoplastic powder. The powder-impregnated preform is then supplied for molding composite products.

The powder size plays an important role in the uniformity of the powder on the fibers. Powders with similar diameters as the fiber diameter are preferred. A larger powder size may require a longer flow distance during the consequent consolidation step. It is also critical to uniformly impregnate powders onto the fibers to reduce voids and fiber-rich areas in the final thermoplastic composite product.

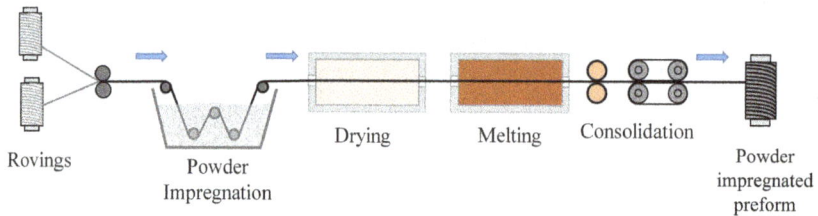

Fig. 4.5: Powder impregnation process used to produce thermoplastic composite preforms (adapted from Reference [6]).

Figure 4.6a shows a roll of a woven carbon fabric PPS preform (TowFlex®, Hexcel) that is approximately one meter wide. It consists of carbon fibers impregnated with

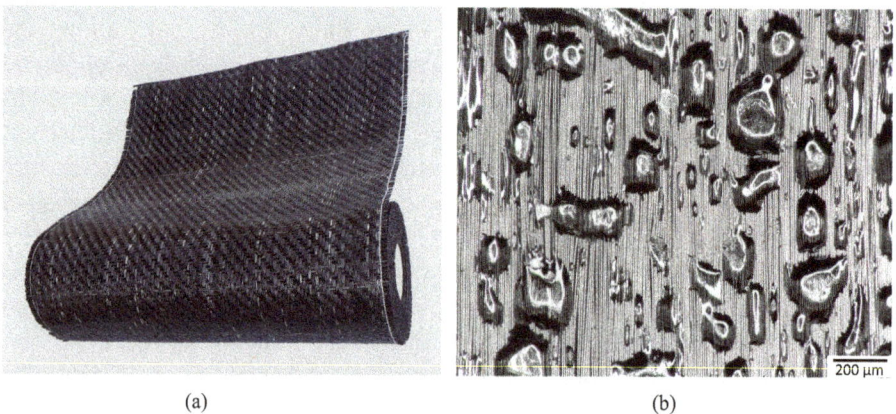

(a)   (b)

Fig. 4.6: (a) A roll of powder-impregnated preform consisting of woven carbon fabrics and PPS and (b) one tow of the preform showing PPS (from molten PPS powders) partially impregnated on carbon fibers.

PPS powders that are melted onto the fibers. The micrograph of one tow of the preform (Fig. 4.6b) shows the PPS (from molten PPS powders) on carbon fibers (in a vertical orientation). It can be seen that the powder-impregnated preform has the fibers partially impregnated.

4.3.3 Commingling

Commingling is one of the techniques developed in textile engineering in which different types of continuous fibers are mixed to form commingled-filament yarns. This technique is adopted by thermoplastic composite industries to mix reinforcement fibers and thermoplastic fibers and produce commingled yarns, another thermoplastic composite preform. The commingled filaments can be subsequently stitched, woven, or braided into preforms and processed into thermoplastic composites products through processes such as pultrusion and compression molding. Because both the reinforcement and the thermoplastic polymer are in a fiber form, the commingled yarns are flexible and suited for stitching, weaving, or braiding. The commingled preform also has a great drapeability because the flexibility of both reinforcement and thermoplastic polymer fibers.

A common commingling process involves guiding thermoplastic fiber yarns, such as PP fibers, nylon fibers, PPS fibers, and reinforcement fiber yarns, typically, glass fiber or carbon fibers through a sound box, before high pressure air is blown from a nozzle onto the fibers and forces the fibers to blend in the sound box. The commingled fibers are then collected by a spool. Figure 4.7a shows a spool of commingled carbon fibers and nylon fibers and Fig. 4.7b shows a close view of the carbon fibers (black color) and nylon fibers (white color).

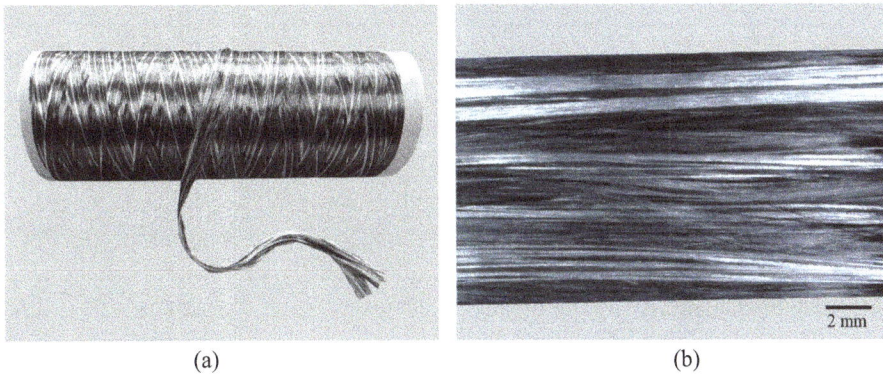

(a)   (b)

Fig. 4.7: (a) A spool of commingled fiber preform consisting of carbon fibers and nylon fibers; (b) a closer view of the commingled fibers showing carbon fibers (in black color) and nylon fibers (in white color).

The commingling technique is continuously advanced. An online commingling process has been developed for more efficient processing. The online commingling process is schematically shown in Fig. 4.8. Glass filaments and thermoplastic filaments are produced and commingled while passing a sizing applicator. This process is able to commingle the fiber yarns before they have pronounced fiber integrity, which usually happens after sizing is applied. It also skips the steps in traditional commingled processes, namely, winding of the polymer and glass fiber yarns, packing and transferring the yarns, and unwinding the fiber yarns for commingling.

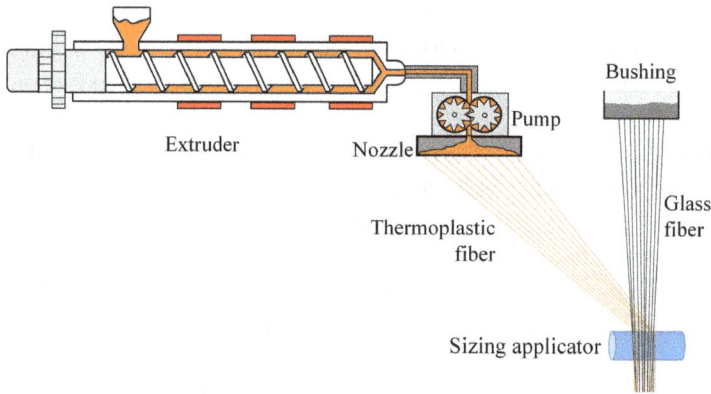

Fig. 4.8: A schematic of online commingling process for producing glass fiber thermoplastic composite preforms (adapted from Reference [7]).

The blending quality of the commingled fibers is determined by the uniformity of distribution of the reinforcement and thermoplastic fibers. It varies with processing parameters, including air pressure, fiber travel speed, nozzle design, fiber tow size, fiber flexibility, sizing applied on the reinforcement fiber, and so on. Two extreme cases of commingling are shown in Fig. 4.9. Figure 4.9a illustrates perfectly well commingled thermoplastic polymer fibers (in green color) and reinforcement fibers (in gray color), and Fig. 4.9b shows side-by-side commingled fibers that are not well commingled. The actual blending in any commingled fibers falls between these two extreme cases. The blending quality can affect the following consolidation process and the quality of final thermoplastic composite products. During the molding process, the commingled fiber preform is heated above the melting temperature of the thermoplastic fiber. The thermoplastic melt flows under capillary pressure and externally applied mechanical pressure and impregnates the reinforcement fiber. It is desirable that the melt has a shorter flow distance (Eq. (4.9)), so that the consolidation process happens at a faster rate with lesser possibility of developing defects such as dry fibers and voids. The blending quality can be evaluated by using a combination of microscopy and image analysis. Segments of commingled fibers are firstly prepared and mounted. The fiber distribution should not be disturbed during sample preparation. The cross sections of the fibers are then polished, and micrographs are finally taken for image analysis and evaluation of the blending quality [8, 9].

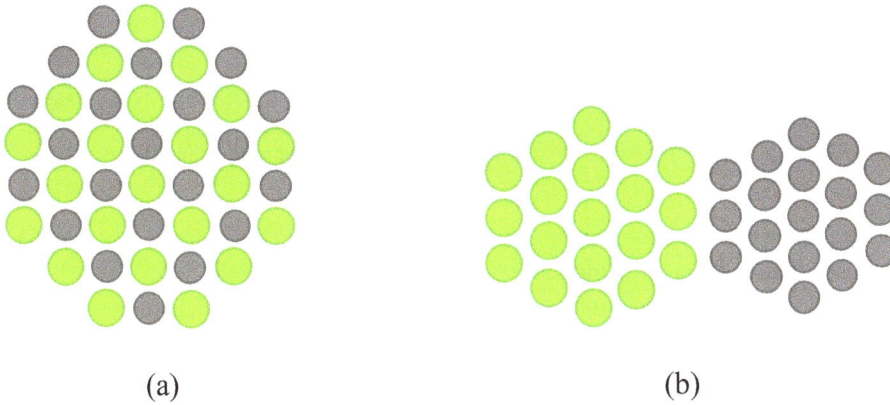

(a)   (b)

Fig. 4.9: Two extreme cases of commingled thermoplastic polymer fibers (in green color) and reinforcement fibers (in gray color): (a) perfectly commingled fibers and (b) side-by-side commingled fibers.

4.3.4 Film stacking

Film stacking is another pre-impregnation method used to produce thermoplastic composite preforms. The pre-impregnated preforms produced by this method are shown in Fig. 4.2d. The film stacking method is capable of producing unidirectional fiber preforms, fiber mat preforms, fabric preforms, laminate preforms, and so on. Commercially available preform products include Cetex® (Toray Advanced Composites and formerly TenCate Advanced Composites) and Tepex® (Bond-Laminates, Germany).

Figure 4.10 schematically shows the film stacking process used to produce preforms. Fibers in different forms (such as roving fiber, mat, or fabric) and thermoplastic films continuously go through a double belt press in which the films are melted and impregnated with the fibers. Pressure is applied in the double belt press to enhance flow of the thermoplastic melt and attain fiber wet out and impregnation to a certain extent.

The film stacking impregnation process has been used to produce organo sheets or organosheets, a thermoplastic composite preform comprising continuous fabrics pre-impregnated with a thermoplastic matrix. The organosheet offers an excellent combination of cost-effectiveness and superior material properties. In addition, the preform allows for customized design for the layup meeting required specifications. The organosheet can be made of different combinations of fibers and thermoplastics.

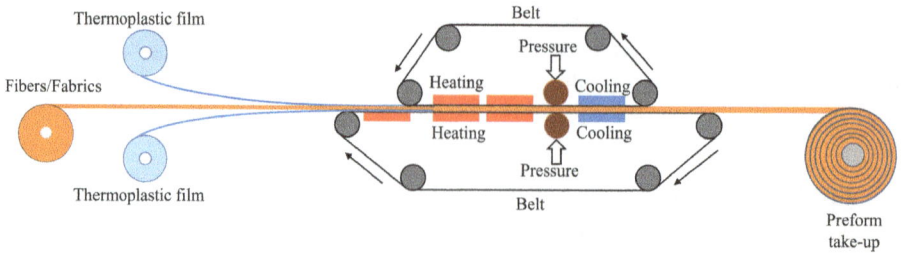

Fig. 4.10: The film stacking process for producing thermoplastic composite preforms.

However, the most common fibers used are glass fibers and carbon fibers. Thermoplastics used in the organosheet include PP, nylons, PC, PPS, and PEEK.

Production of organosheets involves introducing thermoplastic films onto fabrics and pre-impregnating the fabric with the thermoplastic. Thermoplastics can also be introduced through other forms, such as melts (in melt impregnation process) or powders (in powder impregnation process). The fabric and the thermoplastic film go through a heating process with pressure applied. The thermoplastic is melted and forced to penetrate the fabric and impregnate fibers under the applied pressure. The pre-impregnated fabric is then cooled and organosheet is formed. The process is continuous and it is, therefore, an efficient processing method. Preforms can also be produced with a desired number of organosheets stacked and semiconsolidated with desired layup sequence for specific applications. However, overall thickness of the organosheet should not pose any issue in its heating for consequent molding processes, such as thermostamping process (see Section 4.4.4), a typical process for molding the organosheet.

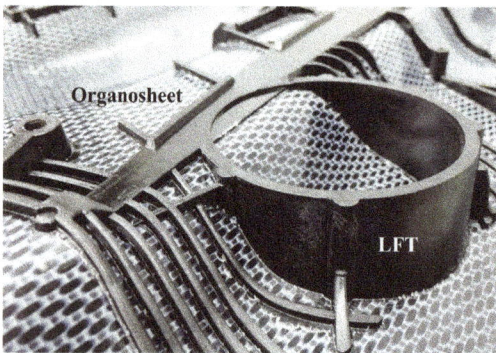

Fig. 4.11: Organosheet overmolded with long fiber thermoplastic (LFT) composite (reprinted from Reference [10] with permission).

Forming and consolidating the organosheet into a final product are normally done through a set of matched molds. The organosheet is firstly heated through infrared heating to its processing temperature. The material is then transferred to the mold, formed, and consolidated after the top and bottom molds are closed. The organo-sheet can be overmolded with other materials. The nature of the organosheet does not accommodate thickness variation and rib formation. Introducing other materi-als, such as GMT, LFT, SFT, or neat thermoplastic allows formation of features, such as ribs and standoffs, as shown in Figs. 4.11 and 4.45.

4.3.5 Solution impregnation

Solvent-based impregnation method, or solution impregnation method, uses a solu-tion consisting of a thermoplastic and its solvent to impregnate fibers. The thermo-plastic is firstly dissolved in the solvent and the fibers are impregnated with a low viscosity solution. The solvent is removed after completion of the impregnation. Sol-vent-based impregnation is normally suited for amorphous thermoplastics that are more susceptible to dissolution by solvents, due to their noncrystalline structure and less dense polymer chain arrangement, compared to semicrystalline thermo-plastics. For the thermoplastic with polar groups, polar solvents are normally used for efficiently dissolving the thermoplastic.

Normally, the thermoplastic used in solvent impregnation is of powder form for easier dissolution in its solvent. It should be pointed out that the wet powder im-pregnation method (see Section 4.3.2) also uses thermoplastic powders and a liquid medium; however, the powder is not dissolved but only physically suspended in the liquid medium. The thermoplastic powders are dissolved in the solvent and form a solution with a low viscosity. The low viscosity significantly promotes wet out and impregnation of fibers. The preform is then dried to remove the solvent for consequent molding processes. Any residual solvent in the preform can affect the performance of the composite, although molding at a higher temperature and long time is able to remove the residual solvent to a certain extent. In addition, the com-posite produced by the solvent-based impregnation method is less chemical-resistant because the thermoplastic is susceptible to solvent attack.

4.4 Consolidation of continuous fiber thermoplastic composite preform

Consolidation of the continuous fiber thermoplastic composite preform is the second processing step in which the thermoplastic composite preform is shaped or molded with the help of heat and pressure to achieve consolidation. Various methods have been developed for consolidating the preforms. Those methods include pultrusion, double belt molding, continuous compression molding, thermostamping, roll form-ing, filament winding, tape placement, compression molding, autoclave molding, diaphragm forming, and bladder molding. Some of the methods are originated from processing thermoset composites and have been adopted for consolidating thermoplastic composite preforms.

The abovementioned methods can produce continuous fiber thermoplastic composite parts with different geometric complexity. Certain consolidation methods can produce quite complex geometries, while some other methods are only capable of coping with simple geometries. The cost of producing composite parts is inversely proportional to the geometric complexity. Pultrusion and double belt pressing are continuous processes and are the most cost-effective methods; however, both the methods can only manufacture thermoplastic composite parts with simple geometries/profiles. On the other hand, diaphragm forming and bladder molding processes can produce complex thermoplastic composite parts; however, these processes are associated with high processing costs. Figure 4.12 illustrates the processing cost in relation to the part complexity for the methods used in consolidating continuous fiber-reinforced thermoplastic composite preforms.

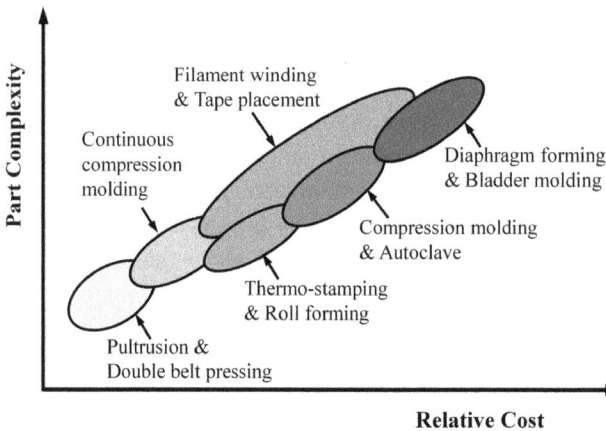

Fig. 4.12: Processing cost in relation to part complexity for various methods that are used to consolidate continuous fiber-reinforced thermoplastic composite preforms (adapted from Reference [11]).

4.4.1 Pultrusion

Pultrusion is known to be a continuous process that produces structures with a constant cross section. It is capable of efficiently producing profiles with simple geometries. In the pultrusion process different types of preforms are used, including commingled fibers and uni-tapes. The die used in the thermoplastic composite pultrusion process has a heating zone and a cooling zone. The heating zone ensures the heating and melting of the thermoplastic matrix, while the cooling zone is adequately long to dissipate the heat from the composite and allow its exit out of the die with good dimensional stability. The die remains closed throughout the pultrusion process. Figure 4.13 shows a schematic of the pultrusion process for producing thermoplastic composite profiles.

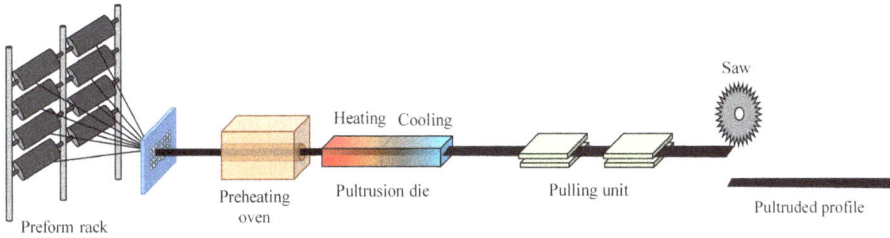

Fig. 4.13: A schematic of pultrusion process for producing thermoplastic composite profiles.

4.4.2 Double belt pressing

Double belt pressing is a method that applies pressure and heat to impregnate or consolidate thermoplastic composites through a series of belts. This process is versatile and has been adopted for producing thermoplastic composite preforms in which thermoplastics are introduced to fibers through different methods, including melt impregnation, powder impregnation, and film stacking. The heat and pressure applied through the belts force the molten thermoplastic onto the fibers and achieve fiber wet out and impregnation. It is also a continuous process that is used to consolidate thermoplastic composite preforms into simple geometries. Figure 4.14 shows a schematic of the double belt pressing method that consolidates several layers of preforms into thermoplastic composite products.

In addition to producing continuous fiber thermoplastic composite preforms and continuous fiber thermoplastic composite products, the double belt pressing method can also be used for consolidating discontinuous fiber thermoplastic composite mat preforms. The preforms are unwound from bobbins and fed into the double belt press. The preforms are heated to their processing temperature and pressure is applied to achieve consolidation.

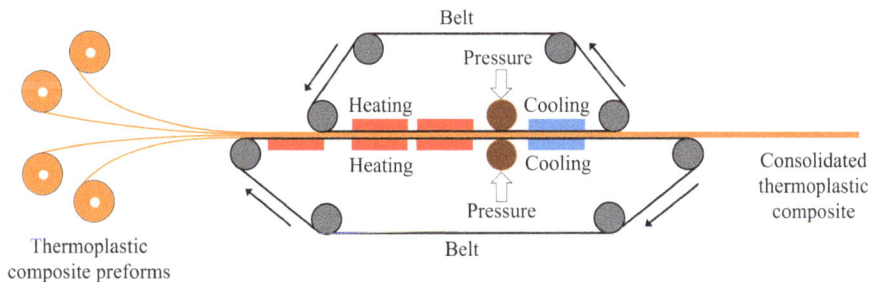

Fig. 4.14: Double belt pressing method for consolidating thermoplastic composite preforms.

4.4.3 Continuous compression molding

Continuous compression molding is a process combining a continuous feeding process that is commonly seen in the pultrusion or double belt process with an intermittent process (compression molding) to consolidate and shape continuous fiber thermoplastic composite preforms into relatively simple profiles. It has been used to manufacture stringers with a variety of profiles for aircraft applications, such as C-channels, H-beams, U-sections, L- and T-stringers, and hollow trapezoidal/hat stringers.

Figure 4.15a illustrates the continuous compression molding process. Several layers of preforms are fed into a preheating and pre-shaping station that applies heat and pressure to shape the preforms. The pre-consolidated preforms are pulled into a compression molding station, where a set of compression molds close and consolidate the preforms into parts with desired profiles. Figure 4.15b shows the front view of the cross section change of the composite from flat preforms to a U-shaped profile after molding. A cooling section in the mold (in blue color) ensures that the consolidated composite part cools and retains its shape after the mold opens. The molded profile is then pulled out of the compression mold and cut to a desired length.

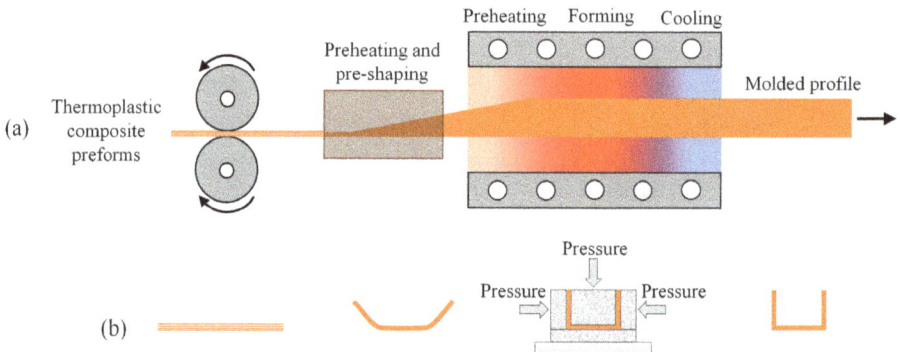

Fig. 4.15: (a) A schematic of continuous compression molding process (adapted from References [12, 13]) and (b) a front view of the cross section change of the composite from preforms in a sheet form to a U-shaped profile after molding.

4.4.4 Thermostamping

Thermostamping is defined as a group of compression molding-like processes that do not require an extensive flow of the thermoplastic composite material and take place at a high cycle rate. Heating of thermoplastic composite preforms is required before the molding process. The heated preform is then transferred between a set of matched molds for forming and consolidation. The forming and consolidation pressure may not be as high as that used in compression molding. This process is similar to the metal stamping process that shapes products at a high production rate. The process is, sometimes, referred to as thermoforming, stamp forming, hybrid molding, hot-drape forming, or stamping.

Figure 4.16 shows a schematic of the thermostamping process. Preforms are firstly clamped and heated in an infrared or convention oven to a temperature that creates a tacky thermoplastic matrix. The preforms are then moved to a forming and consolidation station and molded between the matched molds. Consolidation of the preforms is realized after the mold is closed and pressure is applied. As the mold temperature is much lower than the preform temperature, the consolidated composite can be sufficiently cooled to maintain its geometric stability before demolding. The consolidated composite is finally demolded for finishing operations such as trimming, machining, and so on. For parts that require a specific layup sequence and thickness at certain areas, preforms can be cut into a desired size and orientation and spot-welded together prior to heating and molding. The thermostamping process can achieve a relatively high production rate (cycle time in minutes). Good consolidation can also be realized. It is necessary to use preforms that already have good fiber impregnation, for example, uni-tapes, since there is limited flow and fiber impregnation during the thermostamping process. Scraps may be generated from trimming of the edges used to clamp and hold during heating.

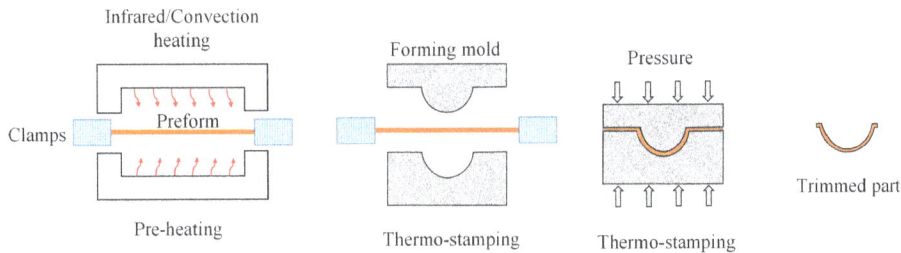

Fig. 4.16: Thermostamping process for forming and consolidating thermoplastic composite preforms (adapted from Reference [14]).

4.4.5 Roll forming

Roll forming is a cost-effective process used to convert flat thermoplastic composite preforms into a profile by passing the preform through a series of matched rolls. The process was followed originally for forming sheet metals and later adopted for thermoplastic composite preforms because of the characteristic of pliability when heated to the processing temperature, similar to heated sheet metals. The roll forming process is a semicontinuous forming process that is used to produce profiles with different shapes, such as C-shaped, T-shaped, and hat-shaped profiles.

Figure 4.17 shows the roll forming process. Thermoplastic composite preforms, including woven fabric and unidirectional fiber-reinforced thermoplastic composite preforms, are heated to the processing temperature of the preforms, before being fed between a series of matched rolls for consolidation and shaping. The composite is consolidated and gradually deformed when passing through the rolls, until formation of the composite profile with a desired cross section is achieved.

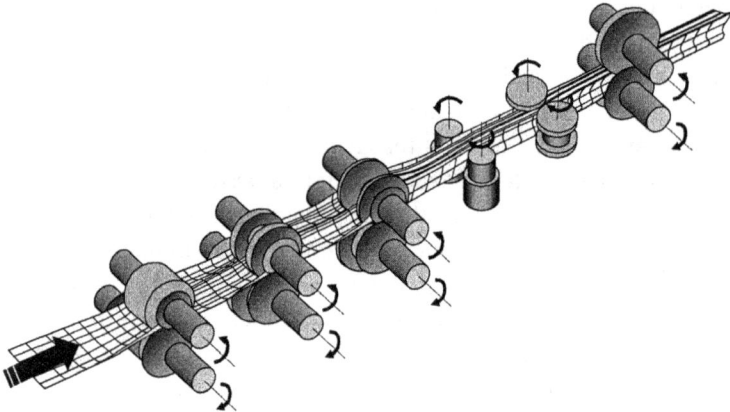

Fig. 4.17: Roll forming process for producing profiles from thermoplastic composite preforms (reprinted from Reference [15] with permission).

4.4.6 Autoclave molding

Autoclave molding is a process initially developed for manufacture of high-quality thermoset composite structures for aerospace application. The process is known for its capability of producing continuous fiber composite structures with minimal voids, high fiber contents, and highly tailorable fiber orientation. The urge for producing low-void and high-quality thermoplastic composite structures has driven the adoption of the autoclave molding process by thermoplastic composite industries. Although the void content cannot be reduced to the same level as seen in thermoset composites (<1%) mainly because of the high viscosity of the thermoplastic, the process has enabled the production of high-performance thermoplastic composites with a void content below 2%.

Figure 4.18 shows a schematic of the autoclave molding process. Preforms with a desired layup sequence are firstly placed in a single-sided mold. A vacuum bag is used to seal the preform and a vacuum port is installed to remove the air inside the vacuum bag through a vacuum pump. The preforms and the mold are then placed into an autoclave and heated to above the melting temperature of the thermoplastic matrix. A high positive pressure is applied, in the meantime, to consolidate the composite.

Autoclave molding is a common process for producing material testing samples, for example, unidirectional fiber-reinforced thermoplastic composite samples. Unidirectional fiber-reinforced thermoplastic composite preforms are stacked onto a flat mold and the molding steps previously mentioned are used to produce a consolidated plate from which testing samples are prepared. Samples are commonly cut in two directions, 0° and 90°, and tested to obtain mechanical properties in both directions.

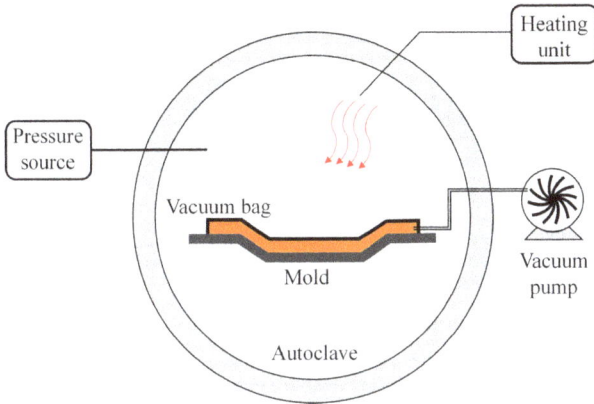

Fig. 4.18: Schematic of the autoclave molding process for manufacturing high-quality thermoplastic composite structures.

4.4.7 Compression molding

Compression molding is another processing method that consolidates thermoplastic composites preforms using a set of matched molds. The compression molding process can be used to mold both continuous fiber and discontinuous fiber-reinforced thermoplastic composites. This section only focuses on the compression molding process for the continuous fiber-reinforced thermoplastic composite. All types of preforms, including commingled fiber preform, powder-impregnated fiber preform, and melt-impregnated tape preform, can be molded using this process.

Figure 4.19 shows a schematic of the compression molding process. A stack of continuous fiber thermoplastic composite preforms is placed in the cavity of the bottom mold before the top mold is lowered down and closed. Shear edge, a telescopic feature along the circumference of the mold cavity, allows air to escape from the cavity but seals the material inside the mold cavity. The preform is then heated above the melting temperature of the thermoplastic matrix. A high pressure is applied to force the thermoplastic melt to flow and fill any gaps or cavities in the composite to achieve consolidation. The composite is then adequately cooled before it is demolded. A set of ejector pins is normally used for demolding (only one ejector pin is shown in the figure).

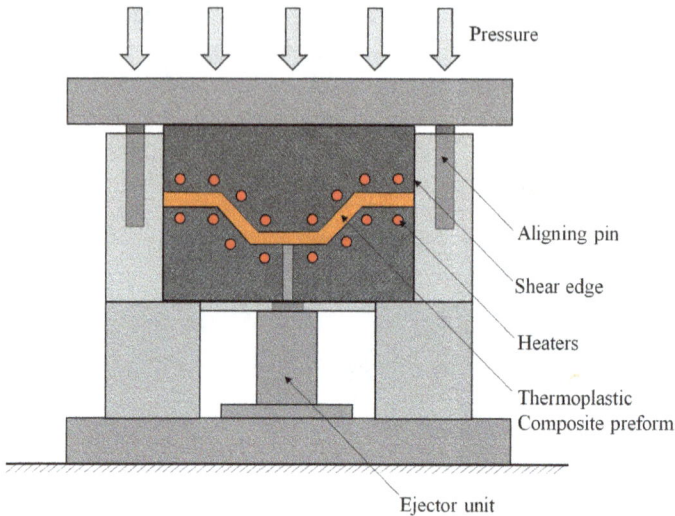

Fig. 4.19: Schematic of the compression molding process for consolidating continuous fiber-reinforced thermoplastic composite preforms.

A large amount of flow of thermoplastic melt can occur during compression molding. The flow significantly benefits the consolidation of the composite, but can cause issues such as fiber misalignment and flash (extra material outside the part due to excessive flow into the shear edge). The flow may shift fibers out of position, causing fiber misalignment. Fiber misalignment can happen to thermoplastic composite preforms with any fiber architecture. Fiber misalignment in woven fabric-reinforced thermoplastic composite preforms is generally less than in other types of preforms, because the fibers are interlocked and have a low freedom to move out of position when the flow of the molten thermoplastic occurs, whereas the thermoplastic composite with unidirectional fiber architecture can have the most fiber misalignment.

Compression molding can be used to produce unidirectional fiber-reinforced thermoplastic composites from different types of preforms, such as uni-tapes or stitched, commingled, and unidirectional fiber fabrics (Fig. 3.6). Unidirectional fiber thermoplastic composite is normally produced for the purpose of material testing and generating basic material property data. During compression molding of the unidirectional fiber thermoplastic composite, the fibers tend to shift with the flow of the thermoplastic. The fiber movement can cause fiber misalignment, often known as fiber wash. When 0°-degree samples with misaligned fibers are tested, inaccurate property data (lower strength and modulus) can result from misaligned fibers. Care has to be taken to use optimized pressure and temperature in order to minimize any excessive fiber movement. The following method describes a setup that can further reduce the fiber misalignment.

Figure 4.20 shows a setup for compression molding a unidirectional fiber-reinforced thermoplastic composite with minimal fiber wash. Unidirectional fiber composite preforms are prepared such that they have the same width as the mold cavity width, while its length is greater than the mold in the fiber direction. When the mold is heated above the melting temperature of the matrix during processing, the matrix of the preform stack inside the mold melts, but the overhanging sections are maintained below the melting temperature. The two mold walls perpendicular to the fiber direction are removed, so that the flow in the fiber direction has less resistance. On the contrary, the flow in the transverse direction has more resistance because of the two mold walls. This setup minimizes the fiber wash by promoting melt flow in the fiber direction and hindering the flow in the width direction. In addition, the overhanging sections normally do not reach the melting temperature and act as anchoring points at both ends to keep the fibers straight during processing and minimize fiber misalignment. A tension force can also be applied to stretch the fibers during molding, to further minimize fiber misalignment. Steel caul plates are normally used to sandwich the preforms to achieve even heating and less fiber wash. High temperature polymer films (for example, polyimide film) can be inserted next to the preform to help molding and provide good surface finish.

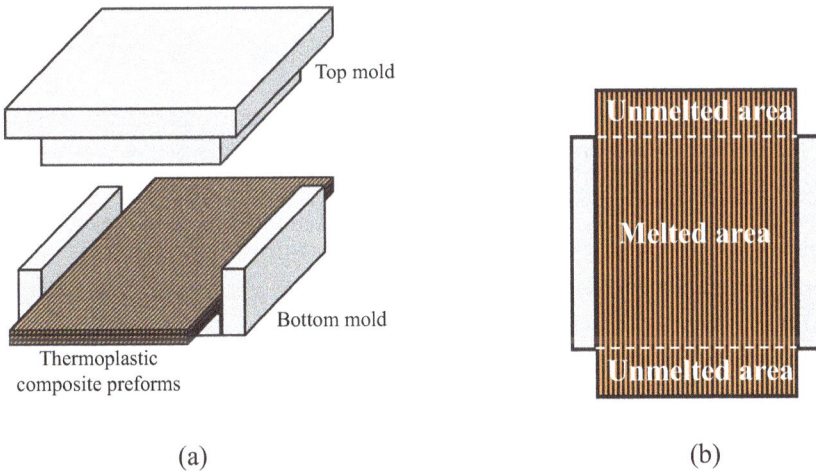

(a)   (b)

Fig. 4.20: (a) An isometric view of a setup for compression molding unidirectional fiber thermoplastic composite preforms with minimized fiber misalignment and (b) a top view of the bottom mold and preform assembly, showing overhanging of the preform stack.

4.4.8 Filament winding

Filament winding is a highly automated processing method for manufacturing tu-bular composite products with either open or closed ends. The filament winding process is used to produce thin-walled thermoplastic composite products, such as pipes, pressure vessels, tanks, and so on. A compressed gas tank made of thermo-plastic composites is shown in Fig. 1.3b. Thermoplastic composites are increas-ingly used for the filament-wound pipes or tanks because of their relatively low material cost, great mechanical properties, and high production rate.

Figure 4.21 schematically shows the filament winding process for manufactur-ing thermoplastic composite structures. A thermoplastic composite preform, typi-cally uni-tapes, is heated in a pre-heating station and wound on a mandrel. Heating is required to raise the preform temperature at the consolidation point. Consolida-tion is achieved by applying pressure through a consolidation roller.

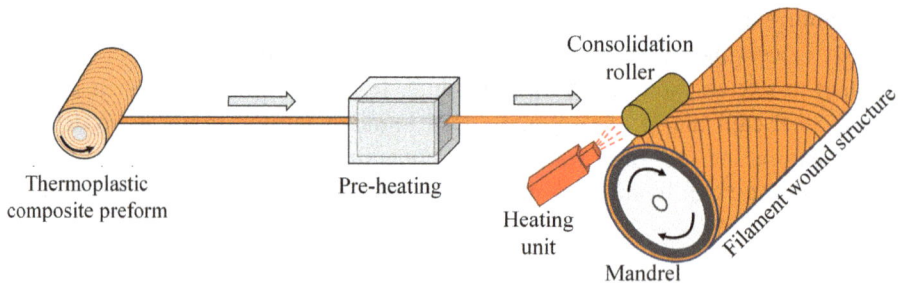

Fig. 4.21: Filament winding process for producing thin-walled thermoplastic composite structures.

The filament winding process is similar to the automated fiber placement process, because both processes use similar thermoplastic composite preforms, typically uni-tapes, and both heat and pressure are applied for in situ consolidation of a layer to the previous layer. However, the preform used in the filament winding process is constantly under tension during the winding process, while the preform in the auto-mated fiber placement process can be under no stress. The preforms are not cut dur-ing the filament winding process until the winding process is completed. However, the preform can be cut during the automated fiber placement process.

Thin-walled structures such as pipes and tanks have a diameter-to-wall-thickness ratio of equal to or more than 20 and experience a plane stress condition. For a thin-walled pipe or vessel with a thickness, t, as shown in Fig. 4.22, an axial stress, σ_{axial}, and a hoop stress, σ_{hoop}, are resulted when an internal gas or fluid pressure, P, is ap-plied. The stress in the radial direction (or thickness direction) is considered to be zero, as the thin-walled structure is under a plane stress condition. The axial stress can be calculated in relation to the pressure and the dimensions of the structure (radius r and thickness t) as follows:

$$\sigma_{axial}(2\pi r)t = P(\pi r^2)$$

Therefore,

$$\sigma_{axial} = \frac{Pr}{2t} \tag{4.10}$$

The hoop stress resulted from the pressure can also be expressed according to the pressure and the dimensions of the thin-walled structure:

$$2\sigma_{hoop}Lt = PL(2r)$$

$$\sigma_{hoop} = \frac{Pr}{t} \tag{4.11}$$

From Eqs. (4.10) and (4.11), the stress in the hoop direction is twice that in the axial direction. Therefore,

$$\sigma_{hoop} = 2\sigma_{axial}$$

The winding angle in a filament-wound composite pipe/vessel is defined as the angle between the center line and the tape/fiber-laying angle. It determines the load-bearing capability in its axial and hoop directions. Figure 4.22a shows a thin-walled thermoplastic composite structure with the winding angle, α. Considering a unit body from the structure with a right angle and a unit length on the bottom side, as shown in Fig. 4.22b, the lengths of the other two sides are $(\tan \alpha)$ and $(1/\cos \alpha)$, respectively.

Figure 4.22b shows the unit body that the load, T, is applied to each fiber. Therefore, the force in the axial direction is $T(\cos \alpha)$, and the force in the hoop direction is $T(\sin \alpha)$. If the number of fibers in the unit body is n, the total force in the axial direction is $nT(\cos \alpha)$, and the force in the hoop direction is $nT(\sin \alpha)$.

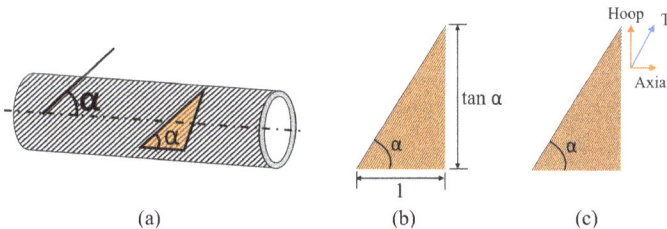

(a) (b) (c)

Fig. 4.22: (a) A thin-walled thermoplastic composite structure with a unit body with a unit length on the bottom side; (b) the unit body with side lengths specified; and (c) the unit body with a tension force, T, applied in the fiber direction.

In the hoop direction, the force balance can be expressed in the following equation:

$$\sigma_{hoop}(1)t = nT(\sin \alpha)$$

In the axial direction, the force balance can be described below:

$$\sigma_{axial}(t)(\tan \alpha) = nT(\cos \alpha)$$

After the stresses are replaced with the terms from Eqs. (4.10) and (4.11), the force balance equations in the hoop and axial directions can be rewritten in Eqs. (4.12) and (4.13), respectively.

$$\frac{Pr}{t}(1)t = nT(\sin \alpha) \tag{4.12}$$

$$\frac{Pr}{2t}(t)(\tan \alpha) = nT(\cos \alpha) \tag{4.13}$$

After combining Eqs. (4.12) and (4.13),

$$(\tan \alpha)^2 = 2$$

Therefore,

$$\alpha = 54.7°$$

The angle 54.7° is known as magic winding angle or magic angle. It is commonly used for manufacturing filament-wound structures. When the filament-wound structure with the magic winding angle is pressurized, the combination of its hoop stress and axial stress results in a stress acting along the fiber direction, resulting in the highest efficiency in its load- bearing capacity.

4.4.9 Automated tape placement

Automated tape placement (ATP) is also called automated tape laying or automated tow placement. It is a process that can produce thermoplastic composite products with customized layup sequences and thicknesses at local areas. Uni-tapes are typically used for in the process. The uni-tape generally is placed onto the previous layer by a tape placement unit. Consolidation between the layers is achieved through heating the local area and applying pressure at the nip point through a compaction roller. Various heating methods, including flame, ultrasonic, infrared, induction, and laser heating, have been used to provide the heat to melt the thermoplastic matrix around the nip point. Among those different heating methods, laser heating is one of the most effective methods and provides high energy density, rapid response time, uniform heating, and excellent efficiency with minimal heat loss.

A schematic of the ATP process is shown in Fig. 4.23. Each layer of the uni-tape can be placed in a well-controlled angle and that angle can be varied from layer to layer, based on the design requirement. After each layer is placed, the uni-tape is cut and the next layer is placed. This process also allows different thickness at desired areas, for example, areas with high stresses, according to product design. The laying angle and path of each uni-tape are controlled through software, which makes the tape placement a highly automated process. The process is sometimes considered an additive manufacturing processes as it has the similar layer-by-layer construction principle (see Chapter 5).

The bonding strength between the layers is determined by the tape-laying speed, tool temperature, heating temperature, consolidation pressure, and so on. Additional consolidation operations, such as ultrasonic vibration, can be conducted to improve the fracture toughness of the ATP thermoplastic composite. Post-processing on the ATP parts can also be done to improve the bonding strength and other mechanical properties. For example, autoclave molding is typically used for reducing the voids and improving interlaminar shear strength in the ATP composite.

ATP is able to produce highly customized products at a relatively high production rate with good repeatability. Other advantages of the ATP process include efficient material utilization and labor savings. Due to these benefits, ATP has been used in various applications, including aerospace, oil and gas exploration, and military. Common thermoplastic composites used in the ATP process are carbon/PEEK, glass/PEEK, carbon/PPS, and so on.

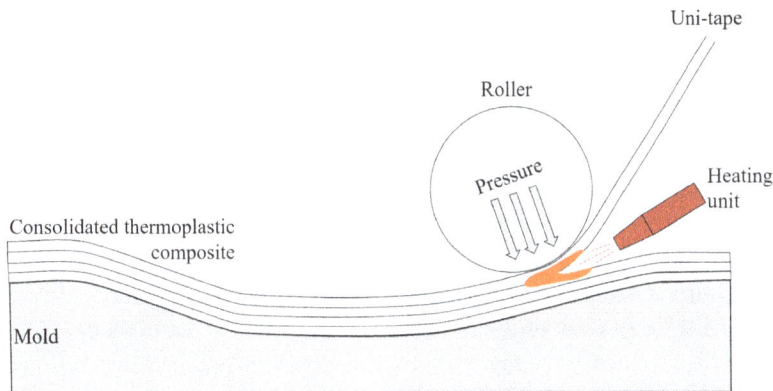

Fig. 4.23: A schematic of the automated tape placement (ATP) process (adapted from Reference [16]).

4.4.10 Bladder molding

Bladder molding is another process that can be used to produce hollow thermoplastic composite structures. Figure 4.24 shows the bladder molding process. Braided
tubular preforms made of pre-impregnated fabrics such as commingled yarn-based
fabrics or powder- impregnated fabrics are placed inside a mold. An expandable
bladder, normally made of high temperature silicone rubber, is inserted inside the
preform. When compressed air is pumped to inflate the bladder, the pressurized
bladder pushes the preform against the mold. The stretched bladder can induce a
tensile force to the fibers, which maintains the straightness of the fibers and minimizes fiber misalignment. Heat is applied to melt the thermoplastic matrix and consolidation of the preform is achieved under the pressure. The consolidated composite
is then cooled and demolded.

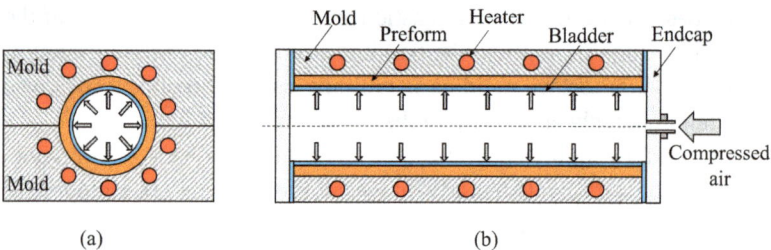

Fig. 4.24: (a) The front view and (b) side view of the bladder molding process for consolidating
tubular thermoplastic composite preforms.

4.4.11 Diaphragm forming

Diaphragm forming is a low production rate process that is capable of producing
high-quality thermoplastic composite products. The process uses diaphragms made
of stretchable materials such as silicone rubber to apply pressure onto thermoplastic composite preforms. When one diaphragm is used, the process is called single
diaphragm forming process. When two diaphragms are used, the process is called
double diaphragm forming process.

Figure 4.25 shows a schematic of the single diaphragm forming process. Vacuum
pressure is applied through a set of vacuum ports connected to vacuum pumps. The
vacuum can effectively remove the air entrapped in the preform and between preform
layers. Positive pressure, which can be much higher than the vacuum pressure, is applied to consolidate the preform. Both air and fluid can be used as the media to apply
the positive pressure. Figure 4.47 shows a schematic of a double diaphragm forming
process that uses a fluid to apply the positive pressure.

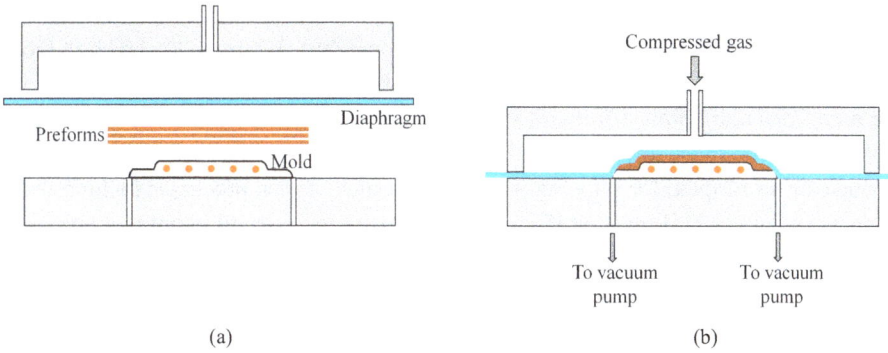

Fig. 4.25: Schematic of the single diaphragm forming process: (a) before consolidation and (b) after consolidation (reprinted from Reference [17] with permission).

4.4.12 In situ polymerization process

Thermoplastics have high viscosities even when they are melted because of their polymerized structure. Their viscosities are significantly higher than the viscosity of thermoset resins. For instance, the viscosity of some thermoplastic melts, such as ABS, can reach more than 1,000 Pa·s at low shear rates, as compared to the viscosity of approximately 4 Pa·s (Fig. 4.1) for epoxy resins. The high viscosity of thermoplastic melts results in difficulty in their flow and impregnation of fibers. According to Darcy's law (Eq. (4.1)), the viscosity of a liquid, such as molten thermoplastics, is a parameter that adversely affects the flow of the liquid in a permeable media, such as fiber tows or fabrics. A higher viscosity results in a lower flow rate which considerably hinders impregnation of fibers.

In situ polymerization process is a process that can address the high viscosity issue by using monomers as the starting material for processing thermoplastic composites. It avoids the use of thermoplastics that are already polymerized, which have a high viscosity even when melted. The process involves melting the monomer and mixing the molten monomer with an initiator, in the presence of a catalyst, to form a solution that is infused into fiber tows or fabrics. The solution has a very low viscosity (Fig. 4.26) even compared to thermoset resins. The viscosity of the melted monomer can be one magnitude lower than that of thermoset resins and several magnitudes lower than that of thermoplastic melts, which tremendously facilitates the flow and fiber impregnation. Polymerization of the thermoplastic monomer takes place after the solution is infused into the fabric under controlled temperature and moisture conditions. This chemical reaction can be analogized to individual snakes (repeated molecular units or monomers) connecting together (being polymerized) into snake chains (polymer with the structure shown in Fig. 1.1b).

One main advantage of this process is the low viscosity of the molten thermoplastic monomer, which induces easy infusion into the fiber tow or fabric and, therefore, great fiber impregnation. The viscosity can also be controlled through varying temperature and concentrations of the initiator and catalyst. Figure 4.26 shows the low viscosity at the beginning of polymerization of nylon 6 from ε-caprolactam at 150 °C for a caprolactam/catalyst/initiator ratio of 100/4/4 [18]. Due to the low viscosity of the melted monomer, any liquid molding process that is used in infusing thermoset resins can be adopted for infusing the monomer melt. Those processes include resin transfer molding (RTM) process (Fig. 4.27), pultrusion process, filament winding, and so on. In addition, the in situ polymerization can occur at a very high rate and complete within a few minutes.

A common thermoplastic used in molding thermoplastic composites via the in situ polymerization approach is nylon 6. Its monomer, ε-caprolactam is polymerized into nylon 6 through a ring-opening polymerization process (Fig. 2.17). Figure 4.28 shows a woven glass fabric-reinforced nylon 6 matrix organosheet that is made through the in situ polymerization process. Other thermoplastics that can be polymerized in situ include PMMA (polymerization of methylmetacrylate monomer with the help of peroxide initiator), nylon 12 (ring opening of ω-laurolactam with the help of activator and initiator), PET (ring-opening metathesis polymerization of macrocyclic oligomers), PBT (polymerization of a macrocyclic oligomer mixture from depolymerization of linear PBT), thermoplastic polyurethane (di-isocyanates reacting with diol), and PC (polymerization of macrocyclic bisphenol-A with the help of initiator). An example is Elium® (ARKEMA company), a commercially available product

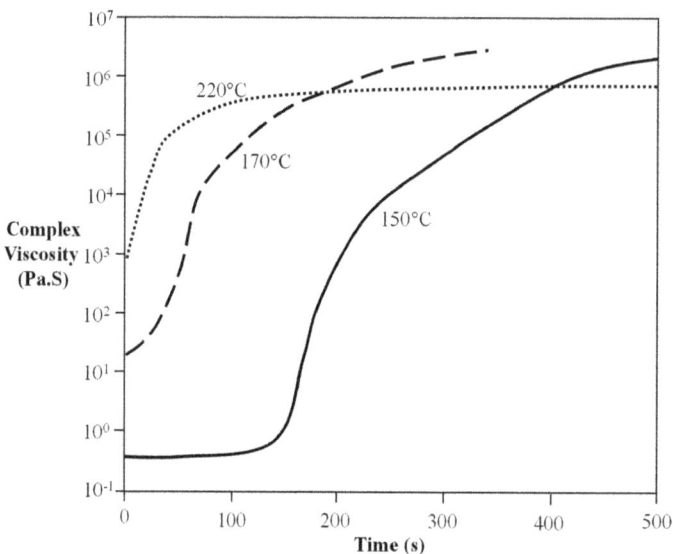

Fig. 4.26: Complex viscosity as a function of time at different temperatures for in situ polymerization of nylon 6 from ε-caprolactam (adapted from Reference [18]).

made of methyl methacrylate (MMA) monomers that can be polymerized with the aid of peroxide initiators through free radical polymerization [19].

The in situ polymerization process, however, has its limitations in processing thermoplastic composites. It is only suited for some thermoplastic polymers that can be polymerized from monomers without any by-products detrimental to composite properties. It also has a strict temperature range in order for the polymerization to initiate and complete. Initiator and/or catalyst are often required. In addition, the in situ polymerization process may be highly susceptible to moisture and oxygen exposure. Both moisture and oxygen can hinder and even stop the polymerization. In order to minimize the exposure to moisture and oxygen, the monomer, initiator, and catalyst are often dried to eliminate the moisture absorbed before melting and, furthermore, inert gas such as nitrogen is used to maintain an inert and protective atmosphere throughout the melting, mixing, and infusion steps.

Figure 4.27 shows an RTM process that infuses a mixture of melted monomer, initiator, and catalyst into a stack of fabrics inside a mold. Vacuum is commonly used to assist the infusion. The chemical composition, for example, the ratio among the monomer, the initiator, and the catalyst, and the processing conditions, such as mold temperature, are well controlled, such that in situ polymerization starts right after the infusion is finished. When the polymerization completes, an in situ polymerized thermoplastic composite is produced. Figure 4.28 shows an organosheet consisting of woven glass fabric and in situ polymerized nylon 6 matrix. The low viscosity also allows fiber impregnation at a high rate, making it possible to manufacture large scale components such as wind turbine blades. The first thermoplastic composite wind turbine blade is made of an in situ polymerized PBT composite [20].

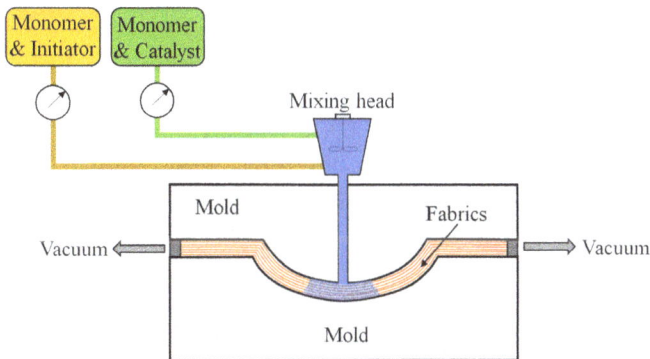

Fig. 4.27: In situ polymerization process for infusing a low viscosity mixture of monomer, initiator, and catalyst into fabrics and producing thermoplastic composites.

Fig. 4.28: An organosheet (15 × 10 mm) consisting of woven glass fabric and in situ polymerized nylon 6 matrix.

4.5 Manufacture of discontinuous fiber thermoplastic composite preform

Discontinuous fiber-reinforced thermoplastic composites include short fiber, long fiber, and glass mat-reinforced thermoplastic composites. The preform for the discontinuous fiber- reinforced thermoplastic composite is an intermediate material consisting of fibers pre-impregnated with a thermoplastic and ready for molding (or forming in the extrusion process). Pellets are the preform material for molding (or forming) short and long fiber-reinforced thermoplastic composite products. The pellet consists of discontinuous fibers pre-impregnated with a thermoplastic matrix. Figure 4.29 schematically shows short fiber thermoplastic (SFT) and long fiber thermoplastic (LFT) pellets. The SFT pellet is normally several millimeters long, and the fibers in the pellet generally are less than 1 mm long (Fig. 4.29a). On the other hand, the LFT pellet is comprised of thermoplastic and unidirectional fibers that are aligned in the pellet length direction. The pellet length can be up to 25 mm. The fibers in the LFT pellet can be fully impregnated (Fig. 4.29b) or wire coated (partially impregnated) by the thermoplastic matrix (Fig. 4.29c). The preform for molding GMT composites is a fiber mat pre-impregnated with a thermoplastic matrix. The processing of the GMT is described in Section 4.5.3.

It is considerably easier to mold the discontinuous fiber thermoplastic composite, because of the discontinuity of the fiber and its low fiber aspect ratio compared to continuous fiber-reinforced thermoplastic composites. A number of methods, including injection molding, compression molding, extrusion, and so on, are used to mold (or forming) the discontinuous fiber-reinforced thermoplastic composite. Table 4.1 summarizes the preforms for discontinuous fiber thermoplastic composites and their common molding (or forming) methods.

Tab. 4.1: Preforms of discontinuous fiber thermoplastic composites and their molding (or forming) processes.

Discontinuous fiber thermoplastic composite	Preform	Starting fiber length	Common molding or forming processes
Short fiber thermoplastic (SFT) composite	Pellet	Normally less than 1 mm	– Injection molding – Extrusion – Overmolding
Long fiber thermoplastic (LFT) composite	Pellet	Normally between 6 and 25 mm	– Compression molding – Injection molding – Injection-compression molding – Extrusion – Overmolding
Glass mat thermoplastic (GMT) composite	Sheet	Normally between 10 and 50 mm	– Compression molding – Thermostamping – Overmolding

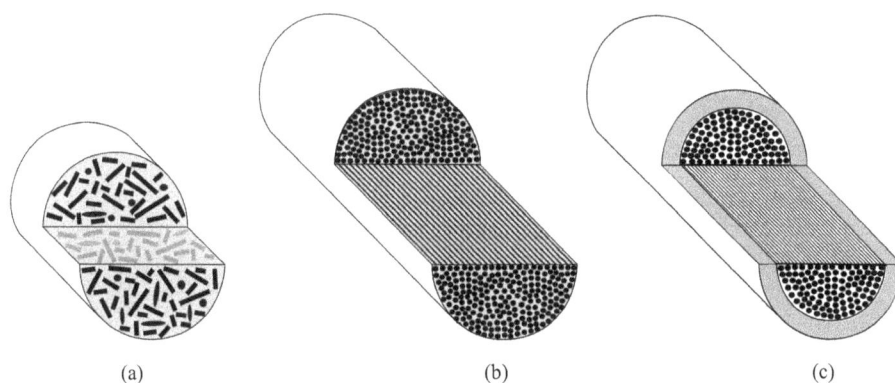

(a) (b) (c)

Fig. 4.29: (a) Short fiber thermoplastic composite pellet; (b) long fiber thermoplastic composite pellet (complete fiber impregnation); and (c) wire-coated long fiber thermoplastic composite pellet (partial fiber impregnation).

4.5.1 Short fiber thermoplastic composite pellets

SFT composite pellets are the preform for molding SFT products. The SFT pellet is commonly produced through an extrusion process that compounds its constituent materials, such as fibers, thermoplastic matrix, and additives (fiber retardants, UV additives, etc.) with desired ratios.

Figure 4.30a shows the extrusion process for producing SFT composite pellets through a twin screw extruder. Neat thermoplastic pellets are fed into the barrel of the extruder through a feeder. The temperature of the barrel is well controlled, such that the thermoplastic is heated above its melting temperature but below its degradation temperature. Additives can be pre-mixed with the neat thermoplastic pellets but can also be fed through a port, in case there is separation or segregation between the additive and the pellets during pre-mixing. Fiber roving can be introduced into the extruder through another port. The high shear stress created between the twin screws as well as between the screw and the barrel wall breaks the fibers into small lengths, typically less than 1 mm. After the fiber is sufficiently mixed and compounded with the molten thermoplastic, the SFT composite is extruded into a filament through a rounded aperture on an extrusion die. When the extruded filament is adequately cooled, a chopper or pelletizer cuts it into pellets with a length of several millimeters. Figure 4.30b shows the SFT pellets consisting of short carbon fibers and

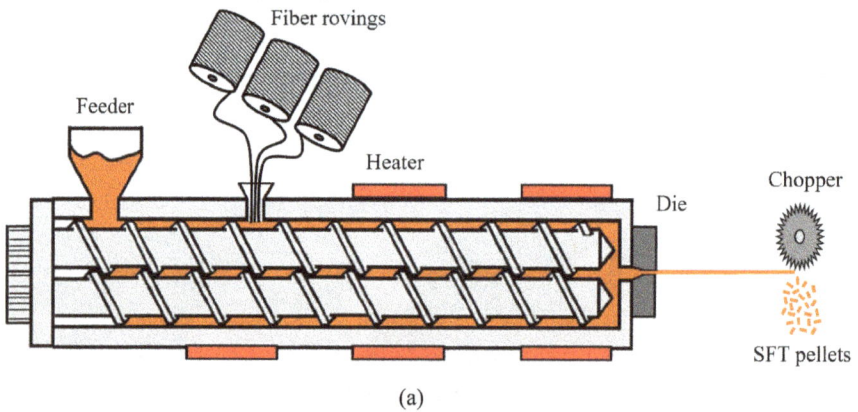

(a)

(b)

Fig. 4.30: Manufacture of SFT composite pellets via twin screw extrusion: (a) a schematic of the extrusion process and (b) pellets produced from the twin screw extrusion process consisting of short carbon fibers and polyetherimide matrix.

polyetherimide produced through the twin screw extrusion process. The pellets have a length of approximately 2 mm.

Twin screw extruders are dominantly used in producing SFT pellets; however, single screw extruders can also be used for producing SFT pellets. When the single screw extruder is used, a high backpressure is normally set to ensure adequate mixing and dispersion of the fibers in the thermoplastic matrix.

4.5.2 Long fiber thermoplastic composite pellets

Long fiber-reinforced thermoplastic composite pellets, abbreviated as LFT composite pellets or LFT pellets, are the preform used to mold LFT composite products. The LFT pellet can be categorized into fully impregnated pellets and partially impregnated pellets (Fig. 4.29). The fully impregnated pellet is normally produced through a melt impregnation process and is the most commonly used (Fig. 4.31). Fiber roving is firstly introduced into an impregnation die after being preheated. Molten thermoplastic is supplied to the impregnation die through an extruder. The wet out and impregnation occur when the fibers make contact with the molten thermoplastic in the die. Impregnation pins improve the fiber wet out and impregnation by spreading the fibers. Finally, the impregnated fibers are pulled through a cylindrical aperture and shaped into a rod before being cooled and cut into pellets by a chopper or pelletizer as shown in Fig. 4.31. Typical lengths for LFT pellets are 6 mm (¼″), 12 mm (½″), and 25 mm (1″). It is to be noted that the fiber length in LFT products molded from these pellets is shorter than the original fiber length due to fiber attrition during molding. Both Figs. 4.32a,b show 12-mm-long pellets consisting of long glass fibers and polypropylene. The LFT pellets in Fig. 4.32a have a natural color, while the pellets in Fig. 4.32b have a black color due to added carbon black as UV additives (see Section 2.3.2).

In the melt impregnation process, the temperatures of the extruder and the impregnation die have to be properly controlled to ensure that the thermoplastic is heated above its melting temperature but below its degradation temperature. It is desirable for the impregnation to take place at a temperature that is approximately 40 °C above the melting temperature of the thermoplastic. At that temperature, the thermoplastic has a low viscosity that can help fiber impregnation tremendously, and the thermal degradation of the thermoplastic can be minimized in the meantime. The melting temperature and thermal degradation temperature of the thermoplastic can be obtained by using differential scanning calorimetry and thermogravimetric analysis, respectively. Those thermal analysis methods are described in Chapter 6.

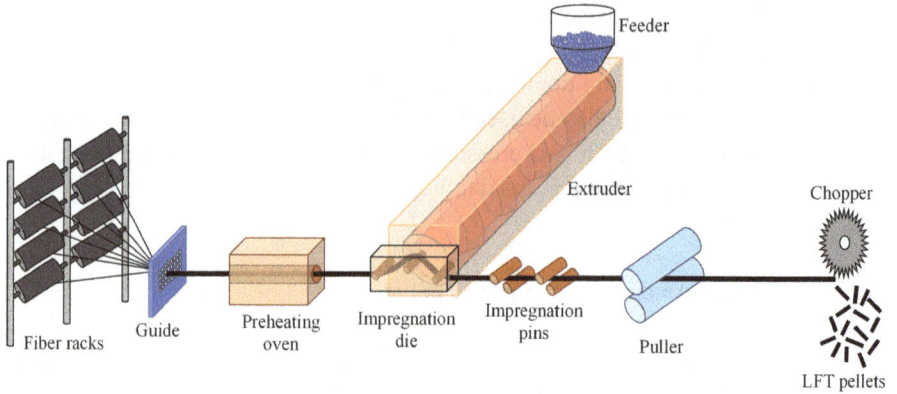

Fig. 4.31: Melt impregnation process for producing long fiber-reinforced thermoplastic composite pellets.

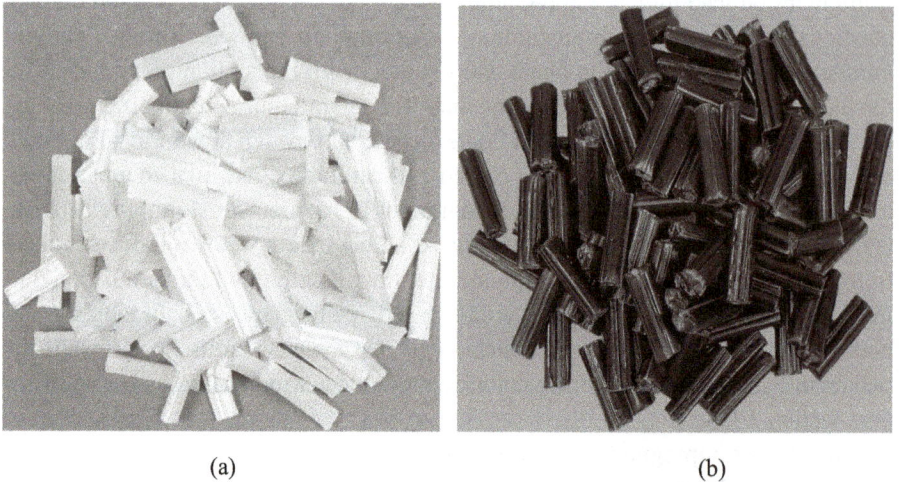

(a)   (b)

Fig. 4.32: Long glass fiber-reinforced polypropylene pellets with (a) natural color and (b) carbon black added. The pellet length in both images is about 12 mm.

The long fiber-reinforced thermoplastic pellets are subsequently used in molding processes such as compression molding or injection molding for producing LFT composite products. The molding processes are described in Section 4.6.

4.5.3 Glass fiber mat thermoplastic composite

Glass fiber mat thermoplastic composite, or GMT composite, consists of discontinuous fibers, mainly discontinuous glass fibers, in a mat form and a thermoplastic matrix.

The fiber mat is pre-impregnated with the thermoplastic matrix to form a preform that is a partially consolidated and semifinished composite for consequent molding processes. The fibers normally have a length ranging from 10 to 50 mm and are randomly oriented. In addition, continuous fibers such as unidirectional fiber can be integrated in the GMT composite.

Semifinished GMT preforms are firstly produced by partially impregnating the fiber mat with thermoplastics. The thermoplastic introduced to the fiber mat can be in different forms, such as powder, fiber, and melt. For example, chopped reinforcement fibers are mixed with thermoplastic powders in an aqueous solution to form slurry. The slurry is pumped onto a moving screen equipped with vacuum that is used to remove most water from it. The mat is then dried before being partially consolidated under heat. The thermoplastic can also be in a fiber form. A wet lay process or air lay process is used to lay reinforcement fibers and thermoplastic fibers into a mat (Fig. 8.12b). Furthermore, thermoplastics can be melted and extruded onto the fiber mat. Figure 4.33 shows the extrusion approach. A thermoplastic melt is supplied through an extruder onto the fiber mat. Multiple fiber mats can be used in the process, although only one mat layer is shown in the figure. The molten thermoplastic is normally introduced between the mat layers when multiple mat layers are used. The fiber mat and the molten thermoplastic then pass through a double belt press, and partial consolidation is achieved via heating and pressure applied by the double belt press. The preform is then cooled under pressure in the double belt press before being cut into a desired size. GMT preform suppliers include AZDEL Inc. and Quadrant Plastic Composites AG.

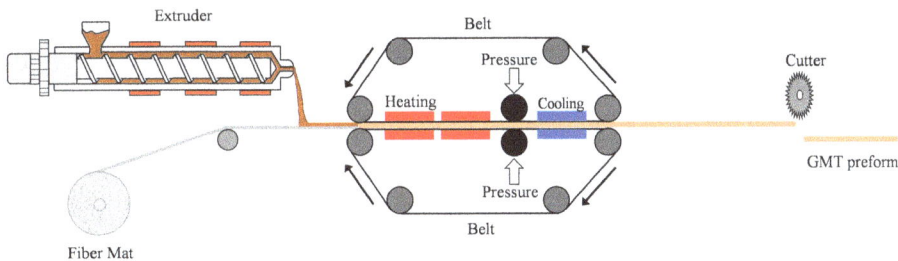

Fig. 4.33: A typical process for producing GMT composite preforms by partially impregnating fiber mat with molten thermoplastic.

The GMT preform is normally compression molded or thermostamped into final products through three processing steps: heating, transferring, and consolidation. The preform is firstly heated in an oven to a certain temperature for molding (for example, above the melting temperature of the thermoplastic matrix, if the thermoplastic is semicrystalline) before being transferred to a set of matched molds for compression molding or thermostamping. In compression molding, the preform experiences an extensive flow and fills the cavity when the top and bottom mold close. Minimal flash is resulted from precisely weighed preform before heating. The extensive flow resulting from high molding pressure enables filling of cavity with complex features such as standoffs, ribs, and so on. When thermostamped, however, the GMT is molded in its original sheet form, and the GMT sheet is preheated and placed to cover the mold cavity before the mold closes and applies pressure to consolidate the GMT composite. There is less flow and a constant thickness through the final product is resulted.

4.6 Molding of discontinuous fiber thermoplastic composite preform

Different processes are used to mold the discontinuous fiber thermoplastic composite preform, including pellets and random mats. These processes are described in detail below.

4.6.1 Injection molding

Injection molding process is one of the most common processes for molding discontinuous fiber-reinforced thermoplastic composites, typically SFT composites. The process is shown in the schematic in Fig. 4.34. SFT composite pellets are fed into a single screw extruder that melts the thermoplastic matrix. The thermoplastic composite melt is then transferred to the front of the extruder barrel by a rotating screw. The melt accumulates at the front of the extruder barrel and pushes back the reciprocating screw in the extruder. When the composite melt reaches a set length, the gate opens and the composite is injected into a mold cavity with a packing pressure up

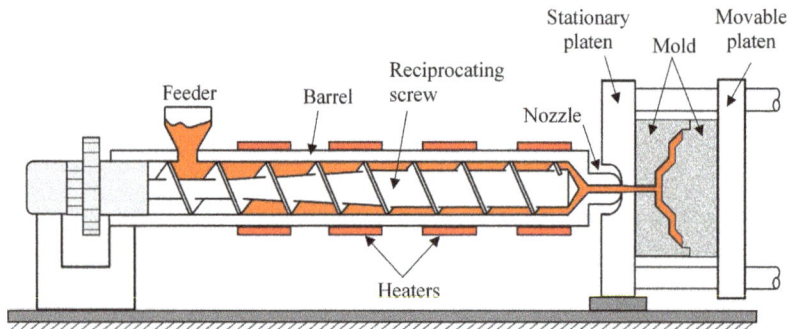

Fig. 4.34: Schematic of an injection molding unit for molding short or long fiber-reinforced thermoplastic composites.

to several dozens of megapascals (several thousand pound-force per square inch). The gate, sprue, and runner of the injection molding unit have to be designed properly to ensure filling of the mold cavity and minimal weld lines. After the mold is filled, the composite is then cooled in the mold under pressure and then demolded. Normally, a set of ejecting pins (not shown in the figure) is used to push out molded components during demolding.

Injection molding is a common process that has been extensively used to manufacture components at a high production rate for various applications, including automotive. The injection molded component can be produced with complex geometries. Figure 4.35 shows an automotive component (about 40 cm × 40 cm) that is injection molded from long carbon fiber-reinforced nylon composite. Due to the large shear stress involved in the process and fiber-fiber interaction, the length of the fibers in the molded component can be significantly degraded. For example, when LFT pellets (starting fiber length more than 6 mm) are used in the injection molding process, the attrition can be so severe that the average fiber length can be less than 1 mm in the molded component. The fiber degradation creates fiber fragments, which can be observed in injection molded components. Figure 4.36 shows the fiber fragments in an injection molded long carbon fiber PEEK composite for automotive application molded from 6-mm-long LFT pellets.

The significant flow involved in the injection molding process forces the fibers to align in the flow direction and induce preferable fiber orientation in that direction. Figure 4.36 shows the preferred fiber orientation in the flow direction. Mechanical properties, such as tensile strength and tensile modulus, are higher in the flow direction.

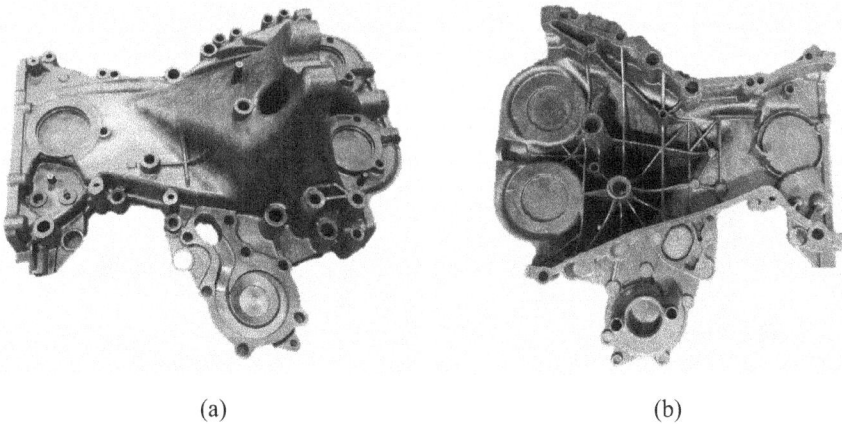

(a)   (b)

Fig. 4.35: (a) The front side and (b) the back side of an injection molded discontinuous carbon fiber-reinforced nylon composite component for automotive application.

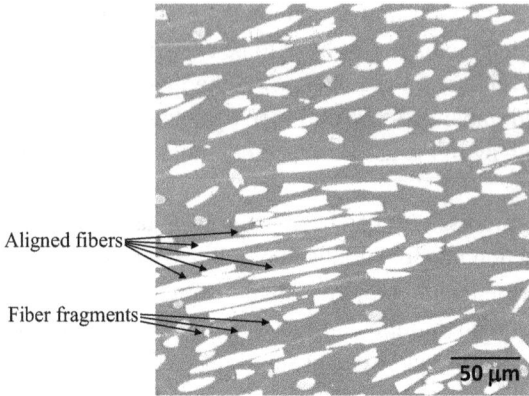

Fig. 4.36: The fibers in an injection molded carbon fiber-reinforced PEEK composite. Note the presence of aligned fibers and a large number of carbon fiber fragments.

4.6.2 Compression molding

Compression molding is another commonly used process for molding discontinuous fiber-reinforced thermoplastic composites, especially LFT composites. Figure 4.37 shows a schematic of the compression molding process. This process mainly comprises two processing steps, extrusion process and compression molding process. Firstly, pellets are fed into a low shear extruder, often known as a plasticator. The thermoplastic matrix in the composite pellets is melted in the plasticator and the fibers are dispersed in the thermoplastic by the rotation motion of the screw. The melt is then transferred to the front of the extruder before being extruded out with a specified length as an extrudite or charge for the compression molding process. Secondly, the charge is then placed into a set of matched molds, namely top and bottom molds. The top and bottom molds close and force the charge to flow and fill the mold cavity. The fibers can develop a preferred orientation due to the flow of the composite melt. The preferred orientation results in higher properties in that flow direction than the direction perpendicular to the flow direction. The difference in strength between these two directions can be more than 60%. As both extrusion and compression molding processes are involved, the compression molding process is also called extrusion–compression molding process.

The advantages of the compression molding process include retained fiber length and high production rate. The average fiber length can reach more than 9 mm when 25 mm long pellets are used in the compression molding process [21]. A greater fiber length leads to enhanced mechanical properties. The high production rate of the compression molding process is also favored in various applications, especially in the automotive industry. Figure 4.38 shows a long glass fiber-reinforced polypropylene composite door panel (approximately one meter wide) compression molded for mass transit application [21]. The fiber length distribution of the compression molded composite door panel is presented in Fig. 6.2.

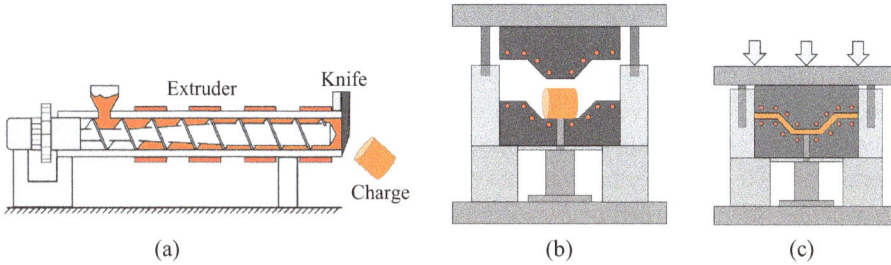

Fig. 4.37: Extrusion–compression molding of discontinuous fiber-reinforced thermoplastic composites: (a) extrusion process producing a charge; (b) the charge being placed into mold cavity; and (c) molding of the composite.

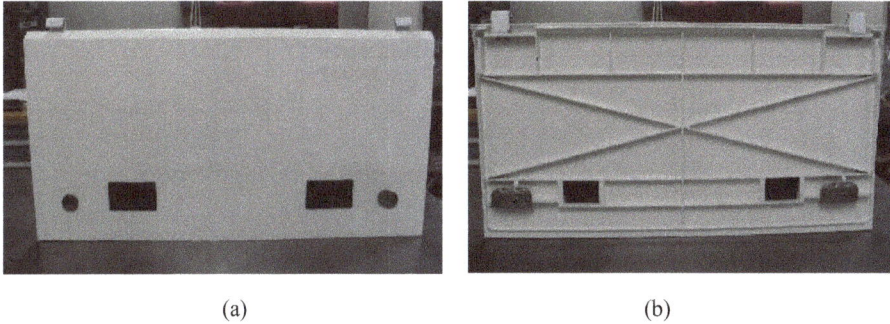

Fig. 4.38: (a) The front side and (b) the back side of a long glass fiber-reinforced polypropylene composite door panel (approximately one meter wide) produced via compression molding process after being painted and mounted with hardware (adapted from Reference [21]).

4.6.3 Injection-compression molding

Injection-compression molding process is a molding process that combines the injection molding and compression molding processes. Figure 4.39 shows a schematic of the injection-compression molding process. A thermoplastic composite melt is injected into a partially open mold followed by a compression molding process in which the mold closes and consolidates the composite. The injection pressure in this process is less than half of the pressure experienced in a typical injection molding process, because there is no need to totally fill the mold. The lower injection pressure results in a lower shear stress in the composite melt, which provides better retention of fiber length and improvement of mechanical properties as well as reduction in residual stresses and warpage, compared to the ones produced through the injection molding process.

The injection-compression molding process has reduced cycle time by running two processing steps in parallel, namely, closing of the mold in the compression molding step and rotation of the screw inside the extruder for preparing the thermoplastic melt for the next cycle. In addition, the pressure applied from the compression molding step is more evenly distributed and, therefore, the resulted shrinkage is more uniform compared to conventionally injection molded parts. The thermoplastic composite melt is directly injected into the mold, which reduces the production time by avoiding charge transferring from extruder to mold as seen in compression molding.

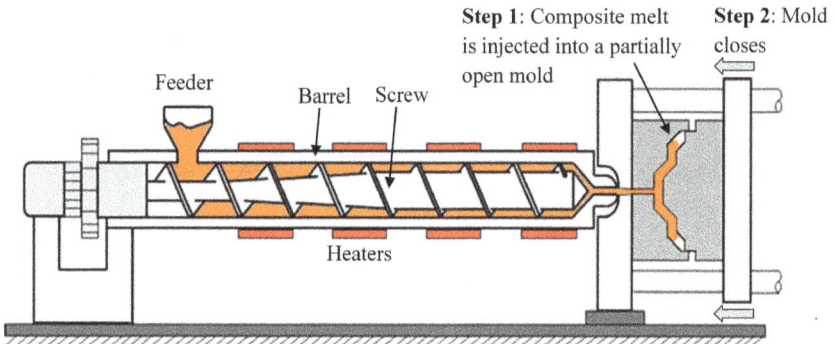

Fig. 4.39: Schematic of injection-compression molding process for molding discontinuous fiber-reinforced thermoplastic composites.

4.6.4 Direct in-line compounding

LFT composite components can be manufactured using direct in-line compounding, a process that combines the compounding process and a molding process (injection molding or compression molding). It is also referred to as direct LFT or D-LFT process. Its main purpose is to save cost by avoiding certain steps such as packing, handling, and shipping the long fiber-reinforced composite pellets. It is estimated that there is up to 30% raw material cost saving in D-LFT, compared to conventional processing of long fiber-reinforced composites.

The direct in-line compounding process (Fig. 4.40) involves three main steps, namely, compounding, mixing, and molding. A thermoplastic is firstly melted and compounded with additives (if required) in an extruder. The molten thermoplastic is then directed into to a twin screw extruder in which fiber roving is introduced for mixing. The high shear stress in the twin screw extruder cuts the continuous fiber roving into long fibers. It also induces impregnation of the fibers with the thermoplastic melt as well as dispersion of fibers in the melt. The composite melt is then molded into parts through compression molding or injection molding. Figure 4.40 shows the direct in-line compounding process with a compression molding unit. The extrudite produced from an extruder is cut into a desired length and transferred by a heated conveyor. An automated handling unit is used to place the extrudite into the mold cavity for compression molding. By directly converting raw materials (neat thermoplastics, additives, and fibers) into final products, the direct in-line compounding process eliminates additional heating as well as packing, handling, and shipping of pellets as required in conventional LFT molding processes. The material properties of the LFT composite produced from this process are comparable to those from the conventional compression molding process or injection molding process.

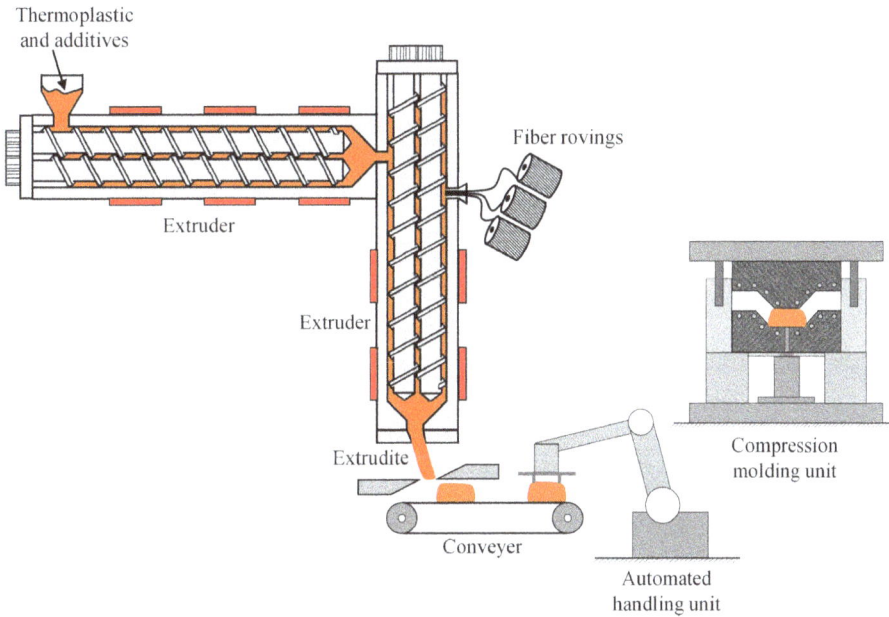

Fig. 4.40: Direct in-line compounding process that combines compounding, mixing, and molding of long fiber thermoplastic composites.

4.6.5 Extrusion

Extrusion process is a process in which a material is forced through a die to form a continuous profile with a fixed cross section. SFT composites have been traditionally used in the extrusion process because of their ease of processability. In addition, LFT composites are being used increasingly in the extrusion process for their better mechanical properties.

Figure 4.41 shows a schematic of the extrusion process. Thermoplastic composite pellets are fed into an extruder in which the composite is heated to its processing temperature and pushed through a die with a desired cross section. The composite material undergoes compressive stresses and shear stresses. Tension stresses may be involved when pulling force is applied for assisting the extrusion. The extruded profile is finally cooled to maintain its geometry and cut into a desired length. Cellulose fiber-reinforced composites with PP or PE matrices are one of the common thermoplastic composites used in the extrusion process. The composite material is extruded into a variety of geometries, including beams (Fig. 1.3d), rods, tubes, bars, sheets, and so on, for typical applications such as construction.

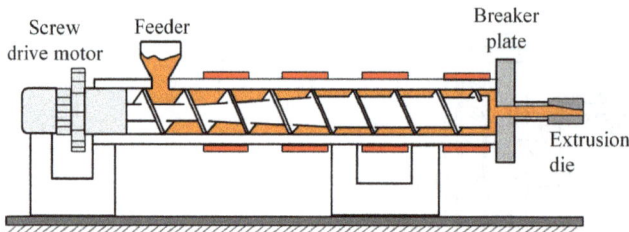

Fig. 4.41: Extrusion process for producing discontinuous fiber thermoplastic composite profiles.

4.7 Overmolding process

The abovementioned processes, such as compression molding and injection molding processes, are normally used for molding one single thermoplastic composite material, either continuous or discontinuous fiber-reinforced thermoplastic composite. However, the need for structurally efficient and cost-effective thermoplastic composite products has driven the advancements in processing technology and the development of hybrid molding processes. The hybrid molding process, often known as overmolding process, is a process that realizes the production of components consisting of two or more different materials.

Overmolding process is a typical hybrid molding process that combines a discontinuous fiber-reinforced thermoplastic composite (or neat thermoplastics) with other thermoplastic composites or even other materials, such as metallic materials, in a single molding process, for example, compression molding or injection molding process. Molding of the discontinuous fiber thermoplastic composite involves flow of the composite melt and filling of the mold cavity. The flow process has essentially enabled overmolding with other materials, such as continuous fiber thermoplastic composite and metal, provided that the material is secured inside the mold cavity during flow of the discontinuous fiber thermoplastic composite. Different combinations of materials can be molded together to produce multifunctional and cost-effective hybrid materials at a high production rate.

Common thermoplastic composites materials used in overmolding are discontinuous fiber-reinforced thermoplastic composites and continuous fiber-reinforced thermoplastic composites. Overmolding these different types of thermoplastic composites normally involves heating of continuous fiber-reinforced thermoplastic composite preforms, transferring the preform into a mold, injection molding or compression molding a discontinuous fiber-reinforced thermoplastic composite onto the preform. The overmolding process provides an excellent combination of design flexibility from discontinuous fiber-reinforced thermoplastic composites and improved strength and modulus from continuous fiber-reinforced thermoplastic composites. It is critical to achieve adequate bonding between these two different thermoplastic composite materials. Poor interfacial bonding often induces premature failure of the overmolded composite. Therefore, it is preferred to have the same thermoplastic matrix in both composites, because it is often easier to achieve better bonding between the same thermoplastic. In addition, it is indispensable for both the material surfaces to be in a melted status to achieve strong bonding. It is occasional that the composites have different matrices. When the matrix is different, it is normally required that the matrix in the continuous fiber thermoplastic composite has a lower melting temperature than that of the polymer in the discontinuous fiber composite or the neat thermoplastic.

Both injection molding and compression molding are the common processes used in the overmolding process. Below are the detailed descriptions of the overmolding process.

1. **Injection overmolding process**. Injection molding of discontinuous fiber-reinforced thermoplastic composites (or neat thermoplastics) is one of the most common molding processes. It is capable of producing secondary structural components with complex geometry designs, in a short cycle time. However, the molded component has limited strength and modulus because of its short fiber length. Adding continuous fiber- reinforced thermoplastic composites to the component can achieve improved strength and modulus with minimal sacrifice on cycle time and tooling cost. Figure 4.42 shows the steps in an injection overmolding process: (a) heating of a continuous fiber thermoplastic composite preform; (b) transferring of the heated preform to a mold; (c) thermostamping of the preform into shape; (d) injection molding of discontinuous fiber thermoplastic composite or neat thermoplastic onto the continuous fiber composite; and (e) demolding of the overmolded hybrid material.

   The injection overmolding process is widely used for producing components with complex features such as standoffs and ribs. Figure 4.43a shows an airbag module that is overmolded by injection molding a discontinuous fiber-reinforced nylon composite with a continuous fiber-reinforced nylon composite.

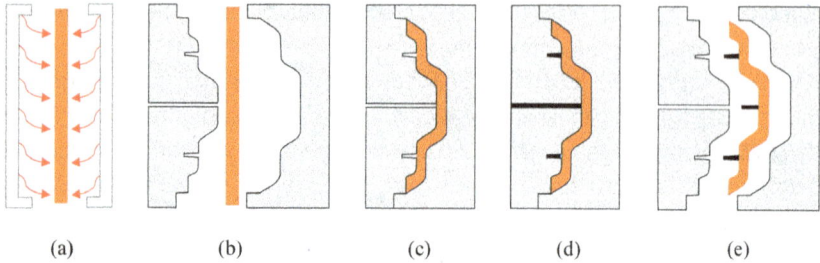

(a)   (b)   (c)   (d)   (e)

Fig. 4.42: Steps involved in an injection overmolding process: (a) heating of continuous fiber thermoplastic composite preform; (b) transferring; (c) thermostamping; (d) overmolding; and (e) demolding.

Thermoplastic composites can also be injection overmolded with metallic materials. Figure 4.43b shows an electric part that consists of brass inserts injection overmolded with a glass fiber PPS composite.

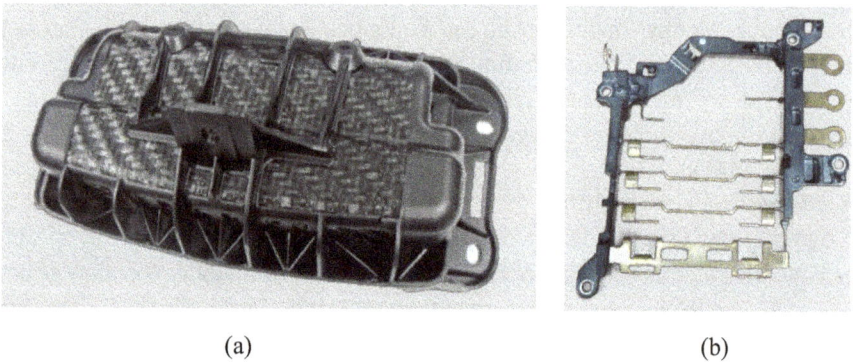

(a)   (b)

Fig. 4.43: (a) An airbag module consisting of woven glass fabric-reinforced nylon 6 composite and short glass fiber-reinforced nylon composite produced by injection overmolding process (adapted from Reference [22]); (b) An overmolded electric part consisting of discontinuous glass fiber PPS composite and metal inserts.

2. **Compression overmolding process.** LFT composites are gaining more and more attention in various applications because of their ease of manufacturing and great specific properties, along with other advantages. Compression molding process is commonly used to mold the LFT composite to minimize the fiber attrition and achieve enhanced strength and modulus. Compression overmolding process is able to integrate other materials, such as continuous fiber thermoplastic composites and metallic materials, with LFT composites.

In the compression overmolding process, a continuous fiber thermoplastic composite or a metallic material is firstly secured inside a set of compression molds that is normally heated. An LFT charge is then placed on the continuous fiber thermoplastic composite or the metal. Closing of the molds forces the LFT charge to flow, enabling not only filling of the mold cavity but also bonding with the continuous fiber thermoplastic composite or the metal. The bonding between the LFT and the continuous fiber thermoplastic composite is fusion bonding, while the bonding between the LFT and the metal is normally achieved through mechanical interlocking. Figure 4.44 shows a compression overmolded differential used for distributing torque to drive wheels in trucks. The hybrid differential consists of a steel skeleton overmolded with a long carbon fiber-reinforced PPS composite. The hybrid design of the differential allows a 40% weight saving compared to its cast iron counterpart [23].

Fig. 4.44: A compression overmolded differential consisting of a steel skeleton and a long carbon fiber-reinforced PPS composite (adapted from Reference [23]).

Case study 4.1: Overmolding of thermoplastic composite with multiple materials

Development of innovative and highly efficient thermoplastic composite structural components considerably benefits from advancement of materials technology, polymer chemistry, processing technique, and so on. Those components can, therefore, not only meet stringent structural requirements but also provide other advantages such as weight saving and use of recycled materials, among others. Integration of multiple materials in one component is one of the approaches that can achieve those targets while still maintaining low weight and high production rate. The following examples describe a thermoplastic composite component for automotive application that uses different combinations of materials to meet structural requirement and achieve weight saving.

Figure 4.45 shows a prototype of an automotive seat back that has integrated three different types of thermoplastic composites, that is, woven fabric thermoplastic composite, uni-tape, and discontinuous fiber-reinforced thermoplastic composite. The woven fabric composite is the main structure of the seat back. Uni-tapes with high strength and modulus are added strategically at high stress areas, such as the corners of the seat back (Fig. 4.45). A discontinuous fiber thermoplastic composite is injection molded to form ribs that provide extra rigidity to the seat back. In addition, the overmolding process also allows integration of metal inserts in the seat back for assembly purpose.

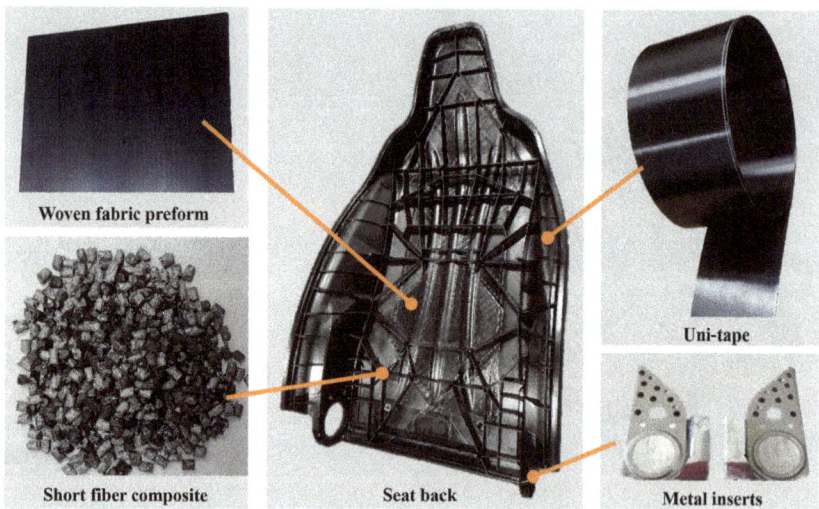

Fig. 4.45: A seat back consisting of different thermoplastic composite materials, including woven fabric, unidirectional fiber, and discontinuous fiber-reinforced thermoplastic composites (adapted from Reference [24, 25]).

Figure 4.46 shows another prototype of an automotive seat back with the same geometry design but with a different combination of materials: recycled carbon fiber thermoplastic composite, uni-tape, and neat thermoplastic. The recycled carbon fiber is in a nonwoven mat form and impregnated with nylon matrix through an in situ polymerization process. The recycled carbon fiber-reinforced nylon composite is the main structure of the seat back. Uni-tapes are added to high stress areas. A neat thermoplastic is molded as the ribbing structure that provides extra rigidity to the seat back. Metal inserts for mounting the seat back structure can also be overmolded with the composites. The hybrid thermoplastic composite seat back reduces weight by 40% over conventional metal construction [25].

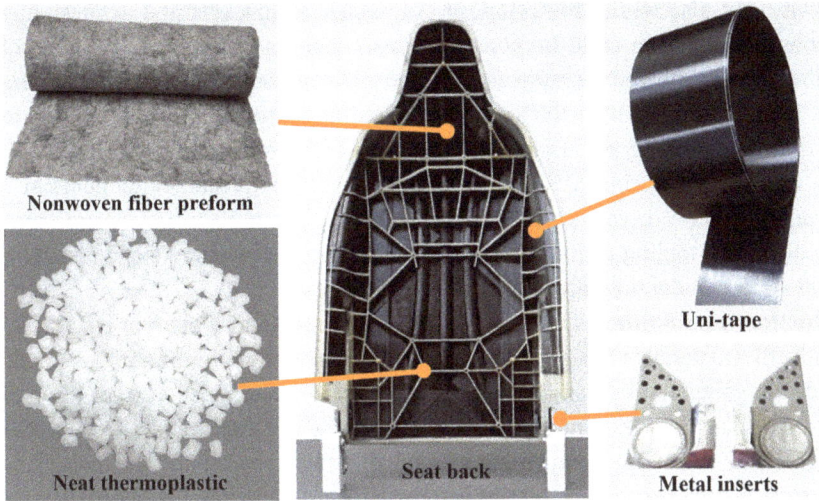

Fig. 4.46: (a) A seat back structure consisting of recycled carbon fiber mat composite, uni-tape, and neat thermoplastic (adapted from References [24, 25]).

Both the seat back prototypes have the same geometry design; however, different material combinations are used to achieve the structural requirement as well as weight saving, compared to conventional metal counterparts. The example demonstrates the excellent material design flexibility of thermoplastic composites and their great processability.

4.8 Tooling technology

A mold is an object with a cavity that is used to manufacture products by filling the cavity with pliable materials, such as thermoplastic composite melts. It is also known as a tool. The process of designing and engineering the molds for manufacturing products is called tooling. It is one of the essential elements in manufacture of thermoplastic composite components, as it determines the part geometry, dimension accuracy, surface appearance, material flow, fiber orientation, and so on. Common tooling materials include tool steels and aluminum alloys. Tool steel is commonly used for mass production for its high strength, excellent wear resistance, great thermal resistance, and fatigue resistance. Aluminum alloys are often used in prototyping tools due to its ease of machining and relatively good strength.

The mold in processing thermoplastic composites is normally subject to aggressive conditions, such as high molding pressure and a large number of thermal cycles with drastic temperature changes. High molding pressure is commonly required to force high viscosity thermoplastic composite melt to flow not only inside the mold to fill the mold cavity, but also within the composite to remove air pockets and achieve adequate consolidation. The mold for some processes such as compression molding of continuous fiber-reinforced thermoplastic composites experiences drastic temperature changes from heating to above the melting temperature of the thermoplastic matrix to cooling down below its recrystallization temperature before demolding.

The need for producing high-quality thermoplastic composite components at a high production rate drives advancements in tooling technology. Below are some cases of advanced tooling technology developed for molding well-consolidated composites at a high production rate.

1. **Induction heating tool**. Induction heating is a process in which an electrically conductive material is heated via electromagnetic induction. Alternating magnetic fields are generated by passing a high-frequency alternating AC current through an electromagnet. When an electrically conductive material is placed inside the magnetic field, an electric current is generated in the material, and the material is therefore heated because of its electrical resistance. This method is known to heat targeted areas in an electrically conductive material rapidly. It has been used in precisely heating a compression molding or injection molding mold in the area in contact with the thermoplastic composite that needs to be heated. RocTool technology is an example of integrating induction heating in molds. It is capable of heating mold surfaces to more than 400 °C within seconds. The high temperature achieved at the thermoplastic composite surface helps minimize surface defects, which is critically important for applications, such as automotive, that require excellent surface finish on exterior components. Class A finish is achievable using this tooling technology. In addition, cooling lines embedded in the tool can realize fast cooling and, eventually, high production rate. Due to the high heating rate, the cycle time can reach down to 2 min.

2. **Diaphragm assisted tooling** is another tooling technique used to reduce cycle time and increase production rate. Hydropressure is applied through a fluid to two flexible bladders or diaphragms between which a mold and composite preforms are placed, as shown in Fig. 4.47. Breathing cloth (not shown in the figure) is normally inserted between the bladder and the preforms to provide channels for removing air in the composite preform and ensures good vacuum throughout the preform. The pressure pushes the composite preform onto the mold to shape the part. The heated fluid conducts heat to the preform and raises its temperature to its processing temperature. Rapid heating and cooling are achieved by replacing fluids with different temperatures. The mold used in the process has much less volume and rigidity than that used in autoclave and other molding processes. The heating fluid has a high heat capacity that enables efficient heat conduction to the preform and the reduced mold volume facilitates fast heating. The heating rate is typically 8–15 °C/min. The single-sided tooling cost is also low. Diaphragm-assisted tooling technology eliminates the need for heating and cooling

lines inside the mold, lowers tooling cost, shortens cycle time, and reduces energy consumption.

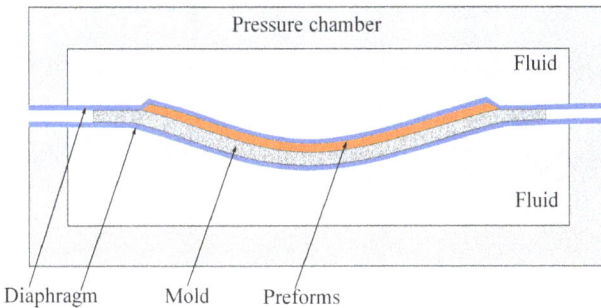

Fig. 4.47: Schematic of diaphragm-assisted tooling for molding thermoplastic composites.

3. **3D printed tooling**. One of the main hurdles in composite product development is the long lead time in prototype tooling as well as high prototype tooling cost. Both the long lead time and high cost stem from the time-consuming machining of metal molds. 3D printing of prototyping tool offers a solution to the issue by eliminating the machining process and reducing the tooling cost. 3D printed tooling also allows fast iterations of any modifications, such as corrections and improvements on the prototyping tool. With the availability of various materials for 3D printing, including fiber-reinforced thermoplastic composites with a high temperature capability, prototyping tools can be printed for design concept verification, geometry optimization, processing optimization, and so on.

The printed mold using thermoplastic composites is mainly used for molding thermoset composite parts because of low pressure and relatively low temperature involved. A variety of thermoplastic composites are available for 3D printing tooling for molding thermoset composite parts. When there is low molding pressure involved and no need for elevated temperature in the processes, such as in vacuum-assisted resin transfer molding (VARTM), it is not necessary to use high performance thermoplastic for the tooling. For example, a mold made of 20 wt% carbon fiber-reinforced ABS composite has been printed for molding thermoset composite for Shelby Cobra hood (Fig. 4.48). Large scale molds have also been printed using thermoplastic composites for producing thermoset composite structures such as windmill blades. When an elevated temperature is involved in processing, for instance, in autoclave molding, composites with high-end engineering thermoplastic matrix such as PPS and PSU can be selected as the printing material. These thermoplastics possess good strength and thermal stability. Carbon fibers are normally added as reinforcement to improve the strength and modulus and enhance the thermal conductivity of the mold. Furthermore, the addition of carbon fibers can dramatically enhance the heat deflection temperature of the mold (see Section 6.6.2).

Fig. 4.48: 3D printed thermoplastic composite mold for prototyping thermoset composite Shelby Cobra hood (reprinted from Reference [26] with permission).

4.9 Joining of thermoplastic composites

Joining of composites with other composites or other materials, such as metals, is commonly required during assembly. Traditional methods for joining fully cured thermoset composites are limited to mechanical fastening and adhesive bonding. On the other hand, the matrix in the thermoplastic composite can be reheated and reshaped, providing more joining options for the thermoplastic composite. Those joining options are fusion-based and therefore known as fusion joining, fusion bonding, or fusion welding.

Fusion bonding of thermoplastics or thermoplastic composites requires the polymer chains in both workpieces to diffuse, penetrate, and entangle at the interface. The bonding surfaces of workpieces are heated to near or above the melting temperature of their thermoplastic matrix, through different heating methods. At the same time, pressure is applied to create an intimate contact at the bonding surface to assist the diffusion, penetration, and entanglement of the polymer chains as well as migration of fibers in the thermoplastic composite. A combination of the heat and the pressure results in interchanging of materials, including the thermoplastic matrix and the fiber at the interface. Adequate cooling is required in all the fusion bonding methods. The interface is normally cooled down below the glass transition temperature of the thermoplastic matrix under pressure to achieve good bonding between the workpieces. Figure 4.49 schematically shows the polymer chains and the fibers penetrating the interface during joining of two workpieces made of short fiber-reinforced thermoplastic composites. The fibers at the bonding line function as z-pins and improve the bonding at the interface.

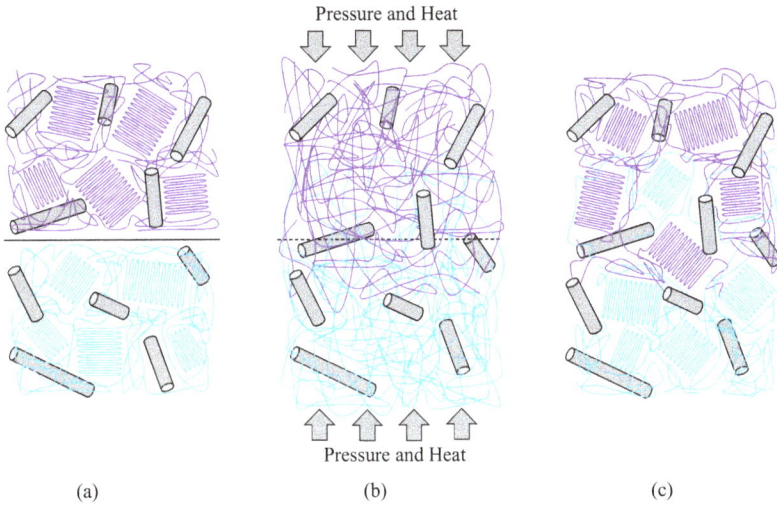

Pressure and Heat

Pressure and Heat

(a)   (b)   (c)

Fig. 4.49: A schematic showing polymer chains and fibers penetrating the interface during joining of two short fiber-reinforced thermoplastic composites; (a) two short fiber-reinforced thermoplastic composites to be joined; (b) heat and pressure applied to induce fusion bonding; and (c) fusion bonded thermoplastic composite.

Heat is generated from different energy sources for joining thermoplastic composites. Based on the heat source from which the thermoplastic composite is heated for fusion bonding, these are resistance welding, induction welding, ultrasonic welding, laser welding, friction welding, infrared welding, and microwave welding. Figure 4.50 shows these fusion bonding methods and conventional bonding methods, that is, adhesive bonding and mechanical fastening. Table 4.2 summarizes the working principles for the fusion bonding methods as well as their pros and cons.

4.9.1 Resistance welding

Resistance welding is a joining process that uses heat generated by resistive materials or resistors to join two thermoplastic composite workpieces. It is also known as implant resistance welding, resistive implant welding, electrical-resistance fusion, or electro-fusion. Heat is generated when electricity is passed through the resistor. The amount of heat can be well controlled, and the heated area is controlled by placing the resistor at desired locations. This allows melting of the thermoplastic matrix in contact with the resistor and joining of the composites under an applied pressure. The joining process can be completed at a high rate, normally within minutes, because of effective heating using electricity. However, the resistor will remain in the joined composites. Figure 4.51 shows the setup for resistance welding.

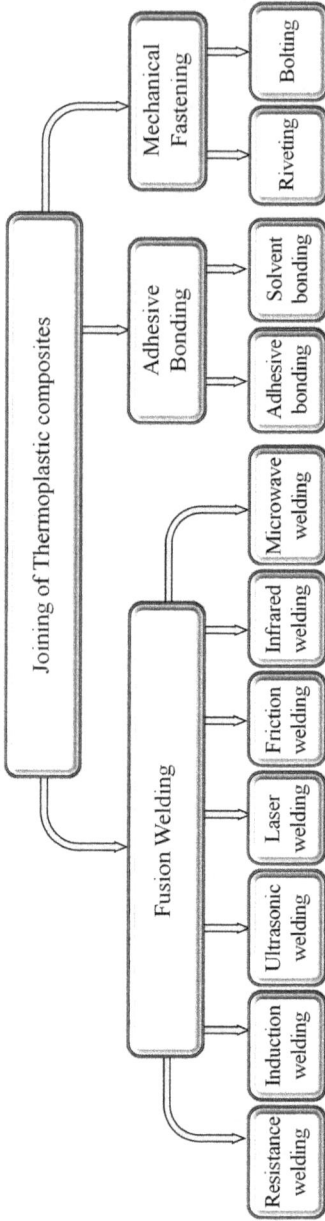

Fig. 4.50: Different methods for joining thermoplastic composites.

Tab. 4.2: Summary of fusion bonding methods for joining thermoplastic composites.

Joining method	Working principle	Advantages	Disadvantages
Resistance welding	Electric current passing resistive element and generating heat	– Good control on heating site – Time independent of welding length – No part thickness limit	– Resistive element such as carbon fiber or metal mesh is required – Resistive element stays in the weld
Induction Welding	Heat generated by Eddy current in conductive element caused by varying EM field	– Fast heating – High efficiency – No contact required	– Limited part thickness – Difficult to focus heat at weld line – Susceptor required for nonconductive composites – Susceptor stays in the weld
Ultrasonic welding	Heat generated at the interface by high-frequency vibration	– High welding speed – Good for spot welding – Minimal heat-affected zone – No additional material required	– Difficult to produce large joints – Limited to lap joints
Laser welding	The surface of one workpiece is heated by laser radiation	– Very high weld speed – Accurate control – No contact required – Minimal heat-affected zone	– Special weld design required – Material dependent – One workpiece is required to be laser transparent or semitransparent
Friction welding	Heat generated at the interface by friction	– Able to join different materials – No additional material required	– Limited to simple geometries – Low welding speed – Nonuniform weld line
Microwave welding	Heat generated by microwave at the interface	– Efficient heating – No contact required	– Material dependent – Susceptor required – Still under development
Conduction welding	Conduction heat used to heat the workpieces	– Joining of complex geometries – No additional material required	– Large heat-affected zone – Not efficient for materials with low thermal conductivity

Heat generated from the resistor follows the Joule's Law as stated in Eq. (4.14). It is proportional to the square of current, the resistance, and the time of current flow:

$$H = I^2 Rt \qquad (4.14)$$

where H is the heat generated, I is the current flowing through the resistor, R is the resistance, and t is the time of current flow.

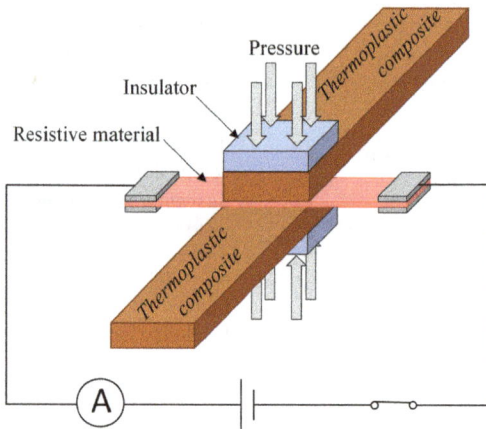

Fig. 4.51: A typical setup for resistance joining of thermoplastic composites.

The resistor is made of electrically conductive materials such as carbon fibers or metal (typically stainless steel) wires. Both unidirectional carbon fibers and carbon fabrics can be used as the resistor. The resistor remains in the composites, and that makes it advantageous for using carbon fibers as the resistor for carbon fiber-reinforced thermoplastic composites. Normally carbon fiber preforms with the same matrix as the composites to be joined are used. The matrix at both ends of the carbon fiber preform resistor is removed by flame or chemicals for connecting it to electricity supply.

Stainless steel meshes are another heating element used in the resistance welding. The metal mesh can be coated with the matrix material for better contact with composite workpieces. It provides a wider processing window because it is less sensitive to the welding parameters. However, the stainless steel mesh stays in the composite and can induce stress concentration at the bonding location. In addition, weight penalty because of the high density of the stainless steel and susceptibility to acidic corrosion are other disadvantages.

4.9.2 Induction welding

Induction welding is a noncontact welding method that is often used in joining thermoplastic composites. It is also referred to as electromagnetic welding. The Faraday effect is the principle for induction heating. An alternating electromagnetic field is created to induce eddy current in electrically conductive materials, such as carbon fibers that are placed at the surfaces to be joined. The eddy current results in heating of the conductive material due to resistance heating (see Section 4.9.1). When the heat is conducted to the thermoplastic matrix and melts the thermoplastic, pressure is applied to ensure a good contact between the workpieces and induce diffusion, penetration, and entanglement of molecular chains. The fusion bonding is realized when the composites are cooled under pressure. For a long joint, the induction coil is moved along the weld line continuously.

Carbon fibers are electrically conductive and able to function as a susceptor, a material that absorbs electromagnetic energy and converts it to heat. Therefore, carbon fibers (generally continuous carbon fibers) in thermoplastic composites can be heated when an alternating electromagnetic field is applied during induction heating. Eddy currents are generated from the electromagnetic field, and the current flows through carbon fibers that form a closed loop and produces heat in the carbon fibers. The heat is conducted to the surrounding thermoplastic matrix and raises its temperature to above melting temperate. It is noted that woven carbon fiber thermoplastic composites can be heated effectively, while unidirectional carbon fiber thermoplastic composites do not have efficient heating because of limited contact among carbon fibers and a limited number of closed loops. For thermoplastic composites that are reinforced by other nonconductive fibers (glass fibers, SiC fibers, etc.), a susceptor such as metal mesh is placed between the thermoplastic composites to be joined. When an alternating electromagnetic field is applied, the susceptor generates eddy currents and heat for fusion joining the thermoplastic composites. The added susceptor, however, stays in the joined composites, and this can have an adverse impact on the joint strength because of stress concentration and reduced corrosion resistance. Figure 4.52(a,b) show the setup for induction welding thermoplastic composites with and without a susceptor, respectively.

Induction heating of the thermoplastic composite is limited by the penetration depth of the electromagnetic field, commonly known as skin depth or penetration depth. The skin depth, δ, is dependent on electrical conductivity and magnetic permeability of the workpiece, and the electromagnetic field frequency, as shown in the following equation:

$$\delta = \frac{1}{\sqrt{\pi f \sigma \mu}} \qquad (4.15)$$

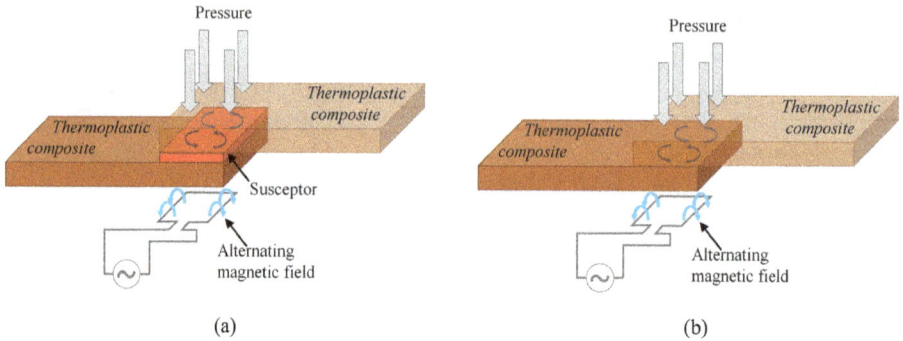

Fig. 4.52: Schematic of induction welding setup for joining thermoplastic composites (a) with susceptor and (b) without susceptor.

where f is the electromagnetic field frequency (unit: Hz), σ is the electrical conductivity of the workpiece (unit: S/m), and μ is the magnetic permeability of the workpiece (unit: H/m).

4.9.3 Ultrasonic welding

Ultrasonic welding is a welding method that uses very high-frequency vibrations to generate frictional heat at the interface between two thermoplastic composites to be joined (Fig. 4.53). The vibration is applied through a sonotrode or horn, and the vibration frequency is typically between 20 and 40 kHz. The vibration induces heat generation at the interface; therefore, it is not necessary to have any other materials, such as a susceptor, at the interface for welding.

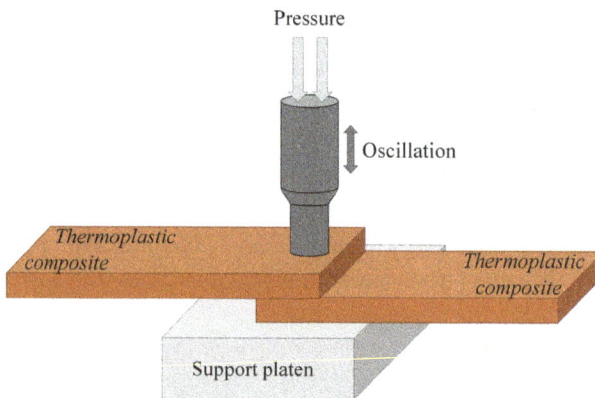

Fig. 4.53: Schematic of the ultrasonic welding process for joining thermoplastic composites.

There are two heating mechanisms during ultrasonic welding, surface friction and viscoelastic friction. Initially, surface friction occurring to the interface creates heat to increase the temperature of the thermoplastic composites. When the thermoplastic matrix reaches its glass transition temperature, viscoelastic friction or hysteresis friction becomes the dominant heating mechanism. The thermoplastic undergoes loading–unloading cycles induced by the vibration, and heat is generated from the viscoelastic friction due to the viscoelastic characteristic of the thermoplastic. Figure 4.54 shows an equilibrium loading–unloading path (at a very low strain rate) and the hysteresis loop between the loading and unloading curve, indicating that the energy dissipates as viscoelastic heat. It is to be noted that for a nonequilibrium path (at a high strain rate), the loading–unloading curve shape changes. However, the hysteresis loop still exists and heat is generated from each loading–unloading cycle. The ultrasonic vibration induces a tremendous number of such loading–unloading cycles in a short period of time because of the high frequency. For example, a frequency of 20 kHz leads to 20,000 loading–unloading cycles in one second; therefore, a significant amount of heat can be generated in a short time for efficiently joining thermoplastic composites.

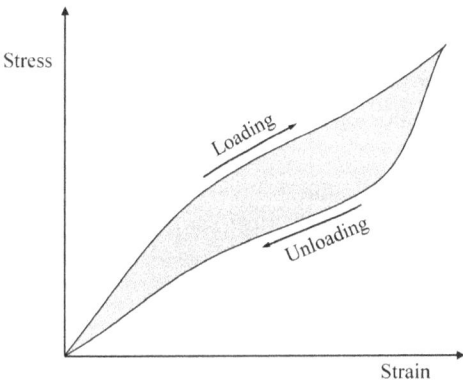

Fig. 4.54: The viscoelastic nature of thermoplastics leading to energy loss during a loading and unloading cycle. The shaded area between the loading curve and unloading curve is the energy loss resulting in viscoelastic heat generated in the thermoplastic for joining thermoplastic composites.

The viscoelastic friction heating rate, $\dot{Q}_v$, is proportional to the frequency of the vibration (ω), the loss modulus of the thermoplastic (E), and the square of the cyclic strain (ε^2) imposed in the thermoplastic that is determined by the amplitude of the vibration. The equation below describes the viscoelastic friction heating rate in relation to those parameters:

$$\dot{Q}_v = \omega E \varepsilon^2 / 2 \tag{4.16}$$

Ultrasonic welding is suited for lap joining thermoplastic composites. For example, uni-tapes are ultrasonically welded together after a uni-tape is laid up with the previous uni-tape for handling and consequent molding. Preforms with a desired layup sequence are spot welded for easy handling before being thermostamped in a set of matched molds (see Section 4.4.4). This welding process is similar to spot welding tailored blanks in the sheet metal stamping process.

4.9.4 Laser welding

Laser welding involves using laser to heat the material at the interface of two workpieces to be joined and fusion bonding them. It is also known as laser transmission welding or transmission laser welding. This welding method joins one material that is laser transparent or semitransparent with another material that absorbs the laser. The laser transparent or semitransparent material allows the laser beam to pass with minimal ionizing reaction and, therefore, with minimal heat generated. The laser semitransparent material includes amorphous thermoplastics that have high transmittance (~90%) in the near-infrared spectral range, such as PMMA, PC, and PS. On the other hand, the laser absorbing material, such as carbon fibers or carbon black in the thermoplastic composite, absorbs the laser and converts the energy to heat. The two materials are stacked, and a laser beam passes through the laser transparent material and irradiates the surface of the laser absorbing material. The heat generated by the laser absorbing material melts the surface at the interface and joining is achieved with the help of a clamping pressure. Figure 4.55 illustrates the setup for laser welding of thermoplastic composites.

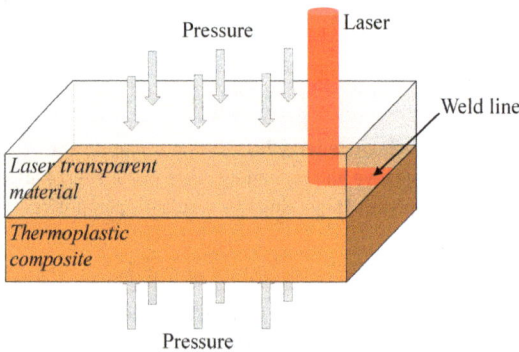

Fig. 4.55: A typical setup of the laser welding process consisting of a laser transparent material and a thermoplastic composite that absorbs the laser and converts the energy to heat for fusion bonding.

4.9.5 Friction welding

Friction welding includes a number of welding techniques that use the heat generated from friction to join thermoplastic composite workpieces. The thermoplastic composite does not undergo melting; therefore, it is a solid-state welding process. A typical friction welding method is friction stir welding, a joining method that uses a rotating tool to stir thermoplastic composites at the interface and generate frictional heat to bond the materials. The process was initially developed for metallic materials, such as aluminum, that are difficult to join using other methods. It has been adopted for joining thermoplastics and thermoplastic composites because of several advantages including localized heating (less heat-affected zone), no surface pretreatment required, no additional material required, low energy requirement, capability of joining different materials, reduced residual stress and workpiece warpage, and so on.

In the friction stir welding, thermoplastic composite workpieces are placed side by side. A metal tool with a head pin is used to generate frictional heat through rotation as shown in Fig. 4.56. The welding involves plunging the rotational tool into the workpieces at the joint line and generating heat through friction. The tool shoulder pushes against the top surface of the workpieces and applies pressure to the thermoplastic composite that is stirred up by the head pin. The continuous rotation of the tool and the translation of the tool along the joint line results in bonding between the thermoplastic composite workpieces. This method is most suited for short fiber-reinforced or particulate-filled thermoplastic composites. Friction welding can also be used to join thermoplastic composites with dissimilar materials, for example, glass fiber-reinforced PPS composites and aluminum alloys.

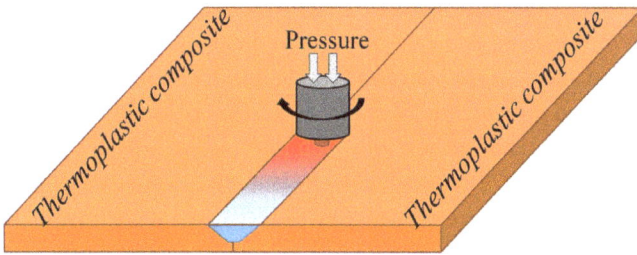

Fig. 4.56: Friction stir welding of thermoplastic composites.

Spin welding, or rotational welding, is another frictional welding method. It uses the rotational motion of one thermoplastic composite workpiece along circular mating surfaces. One thermoplastic composite workpiece rotates while being pushed against another workpiece. Friction heat is generated at the interface of the workpieces. When the temperature at the interface reaches close to the melting point of the thermoplastic matrix, diffusion, penetration and entanglement of polymer chains occur at the interface of the workpieces. After the interface is adequately cooled, fusion bonding is achieved.

4.9.6 Other joining methods

The following sections describe other fusion bonding methods and conventional adhesive bonding and mechanical fastening methods for joining thermoplastic composites.

1. **Adhesive bonding** is a traditional method for joining thermoplastic composites. The bonding surface is cleaned and treated before applying adhesives such as epoxy. Surface energy of the thermoplastic matrix plays a critical role in the bonding strength. A thermoplastic with higher surface energy generally favors adhesive bonding. A thermoplastic with low surface energy generally results in low bonding strength. Typical thermoplastics with a low surface energy include PTFE, PVDF, PP, and PE (see Section 2.4). Several treatment methods including surface roughening, chemical treatment, and plasma treatment are used to modify the morphology or chemistry of the surface of the thermoplastic composite workpieces to improve their bonding with adhesives. For example, surface roughening not only increases the surface area but also exposes fibers for better bonding.

2. **Mechanical fastening** is another joining method that has been traditionally used to join thermoplastic composites. This method involves drilling holes in thermoplastic composite workpieces and using mechanical fasteners such as bolts, screws, and rivets to join the workpieces together. This method can join thermoplastic composites with other materials such as metals. However, drilling results in cutting of the fibers and can introduce delamination in laminated thermoplastic composites. The stress concentration at the location of the hole can also cause initiation and propagation of cracks and weaken the composite.

3. **Microwave welding** is a joining method in which microwave energy is used to generate heat at the weld line. Microwave energy is absorbed by a susceptor that is placed between thermoplastic composite workpieces. The molecules with dipoles, or polar molecules, in the electromagnetic absorbent material reorient to realign with alternating electromagnetic fields and heat is generated via internal friction among molecules. The heat is conducted to the thermoplastic composite workpiece and raises the temperature to the melting temperature of the thermoplastic matrix. Pressure is also applied until the interface is cooled down below the glass transition temperature of the thermoplastic matrix to achieve joining.

4. **Infrared welding** uses an infrared heating source to achieve fusion bonding of thermoplastic composites. Fig. 4.57 shows a schematic of infrared welding. Infrared

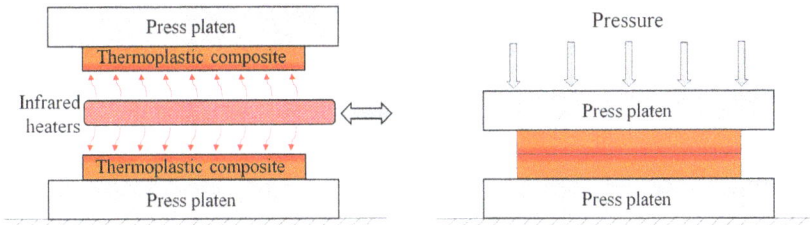

Fig. 4.57: Schematics of the infrared welding process: (a) thermoplastic composites being heated by infrared heaters and (b) the composites being joined under pressure.

heaters are placed between two thermoplastic composite workpieces and their surfaces are heated to above their melting temperature. The heaters are then removed, and the workpieces are pressed and joined together under pressure. Adequate cooling is required at the interface before pressure is removed.

5. **Hot plate welding** or **conduction welding** is another joining method that takes advantage of the reheating and reshaping capability of thermoplastic composites. Heat is conducted from a hot plate to two thermoplastic composite workpieces. When the thermoplastic matrix at the interface is melted, the hot plate is removed and the two workpieces are pushed toward each other and make a contact. Pressure is applied to bond the workpieces until the workpiece is cooled down. Fig. 4.58 shows a schematic of the hot plate welding process for bonding thermoplastic composites. Fig. 4.58a illustrates thermoplastic composite workpieces heated by a hot plate; Fig. 4.58b shows the thermoplastic composite workpieces being joined after the hot plate is removed.

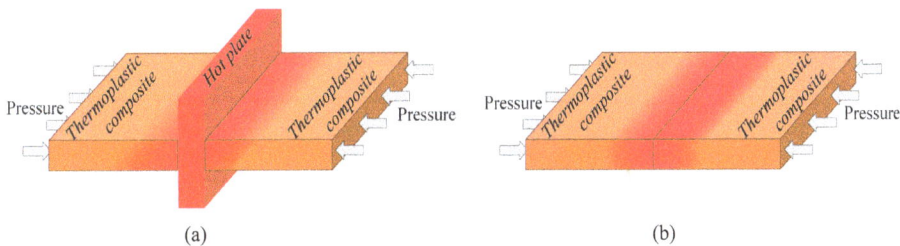

Fig. 4.58: Schematic of the hot plate welding process for joining thermoplastic composites: (a) thermoplastic composite workpieces heated by a hot plate; (b) thermoplastic composite workpieces being joined under pressure after the hot plate is removed.

Case study 4.2: Joining thermoplastic composites with metals
With the increasing use of thermoplastic composites in automotive industries, assembling components made of thermoplastic composites with other materials, such
as metals, has become more prevalent. Although conventional joining methods,
such as adhesive bonding and mechanical fastening, are still being used for assembling thermoplastic composites, more attention has been drawn toward taking advantage of their capability of being reshaped through application of heat and using
fusion bonding methods to join them with other materials.

Flexible Production Cell for Hybrid Joining, or FlexHyJoin, is a technique developed to join steel brackets onto thermoplastic composites through fusion bonding.
Those metal brackets are intended for assembling with other structural components,
for example, car roof panels. The technique demonstrates the capability of different fusion joining methods, such as induction welding and laser welding, in bonding thermoplastic composites with metals. The thermoplastic composite used in the technique
demonstration is a lightweight high-performance roof stiffener made of continuous
glass fiber-reinforced nylon 6 composite, a typical thermoplastic composite material
used in automobiles. Induction welding and laser welding are used to join central and
side brackets to the thermoplastic composite, respectively. Fig. 4.59a shows the thermoplastic composite roof stiffener with joined metal brackets. Fig. 4.59b shows the
composite component assembled onto a car roof cutout using the metal brackets.

Both induction welding and laser welding involve local melting of the thermoplastic matrix in the composite stiffener. Pressure is applied to force the molten

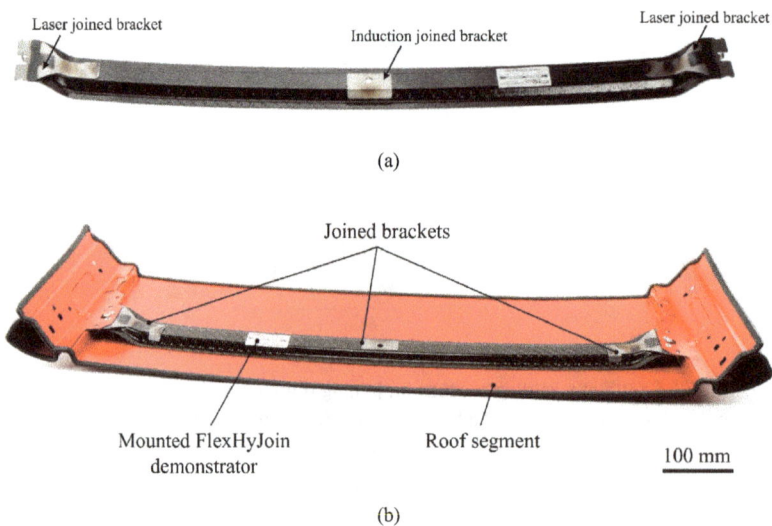

(a)

(b)

Fig. 4.59: (a) Induction welding and laser welding used for bonding metallic brackets to a roof
stiffener made of continuous glass fiber-reinforced nylon 6 composite; and (b) the roof stiffener
assembled to a car roof segment (reprinted from Reference [28] with permission).

thermoplastic to make a full contact with the metal bracket surface. Laser texturing is also used to create undercuts on the surface of the steel bracket. The thermoplastic melt is forced into the undercuts to achieve mechanical interlocking and improved bonding. Joining between the thermoplastic composite and the metal is realized after both materials are adequately cooled down. The joined materials have shown great bonding strength and their lap shear strength can reach up to 15 MPa [27]. The FlexHyJoin technique has also demonstrated that efficient joining can be achieved through full automation for bonding thermoplastic composites with metals.

4.10 Summary

- Processing of thermoplastic composites is more challenging than that of thermoset composites, because of the significantly higher viscosity of the thermoplastic matrix even when it is heated above its processing temperature.
- A pre-impregnation step and a consolidation step are normally required in processing thermoplastic composites due to the high viscosity of the thermoplastic.
- Pre-impregnation processes for producing continuous fiber thermoplastic composite preforms include melt impregnation, powder impregnation, commingling, film stacking, and solvent impregnation. The preform is consequently molded to achieve consolidation.
- Various processes for molding continuous fiber-reinforced thermoplastic composites, including pultrusion, double belt molding, continuous compression molding, thermostamping, roll forming, filament winding, tape placement, compression molding, autoclave molding, diaphragm forming, and bladder molding, have been developed. Those processes are able to manufacture thermoplastic composite products with different geometry complexity at different costs.
- Discontinuous fiber-reinforced thermoplastic composites mainly include random fiber mat thermoplastic, long fiber thermoplastic, and short fiber thermoplastic composite.
- Processing methods for discontinuous fiber thermoplastic composites include injection molding, compression molding, extrusion, overmolding, and so on.
- The overmolding process allows the integration of discontinuous fiber thermoplastic composites with other materials, including continuous fiber thermoplastic composites and metallic materials, at a relatively high production rate. Different combinations of materials can be overmolded together to produce multifunctional and cost-effective hybrid materials.
- Fusion bonding is a method that can join thermoplastic composites by locally heating and melting the thermoplastic matrix at the bond line. Based on the energy source for heat generation, fusion bonding can be categorized into resistance

welding, induction welding, ultrasonic welding, laser welding, friction welding, microwave welding, infrared welding, and conduction welding.
- Fusion bonding of thermoplastic composites involves diffusion, penetration, and entanglement of polymer chains as well as fiber migration at the interface, all of which improve the bonding between the thermoplastic composites.

References

[1]   Ajinjeru C, Kishore V, Chen X, Lindahl J, Sudbury Z, Hassen A, et al. The influence of rheology on melt processing conditions of amorphous thermoplastics for big area additive manufacturing (BAAM). Solid Freeform Fabrication. 2016;2016:754–61.
[2]   Silva VA, Folgueras LdC, Cândido GM, Paula ALd, Rezende MC, Costa ML. Nanostructured composites based on carbon nanotubes and epoxy resin for use as radar absorbing materials. Materials Research. 2013;16(6):1299–308.
[3]   Lewis P. Chapter 2-Sample Examination and Analysis. Forensic Polymer Engineering (Second Edition). Woodhead Publishing: 2016. 33–69.
[4]   Samanta B, Kumar P, Nanda D, Sahu R. Dielectric properties of Epoxy-Al composites for embedded capacitor applications. Results in Physics. 2019;14:102384.
[5]   Abdennadher A, Vincent M, Budtova T. Rheological properties of molten flax-and Tencel®-polypropylene composites: Influence of fiber morphology and concentration. Journal of Rheology. 2016;60(1):191–201.
[6]   Ho K, Shamsuddin SR, Riaz S, Lamorinere S, Tran M, Javaid A, et al. Wet impregnation as route to unidirectional carbon fibre reinforced thermoplastic composites manufacturing. Plastics, Rubber and Composites. 2011;40(2):100–7.
[7]   Wiegand N, Mäder E. Commingled yarn spinning for thermoplastic/glass fiber composites. Fibers. 2017;5(3):26.
[8]   Chiu S-H, Chen J-Y, Lee J-H. Fiber recognition and distribution analysis of PET/rayon composite yarn cross sections using image processing techniques. Textile Research Journal. 1999;69(6):417–22.
[9]   Kravaev P, Stolyarov O, Seide G, Gries T. Influence of process parameters on filament distribution and blending quality in commingled yarns used for thermoplastic composites. Journal of Thermoplastic Composite Materials. 2014;27(3):350–63.
[10] Holmes M. Large quantity composite part production makes further progress. Reinforced Plastics. 2020;64(2):84–91.
[11] Bhagat H. Linear Polyphenylene Sulfide (PPS) for Thermoplastic Composites. Ticona Engineering Polymers. 2008.
[12] Barile M, Lecce L, Iannone M, Pappadà S, Roberti P. Thermoplastic composites for aerospace applications. Revolutionizing Aircraft Materials and Processes. Springer: 2020. 87–114.
[13] Gardiner G. Aerospace-grade compression molding. High Performance Composites. 2010;63 (2010):34–40.
[14] Burkhart A, Cramer D, editors. Feasibility of continuous-fiber reinforced thermoplastic tailored blanks for automotive applications. SPE automotive composites conference & exposition; 2005.
[15] Dykes R, Mander S, Bhattacharyya D. Roll forming continuous fibre-reinforced thermoplastic sheets: experimental analysis. Composites Part A: Applied Science and manufacturing. 2000;31(12):1395–407.

[16] Sonmez FO, Hahn HT, Akbulut M. Analysis of process-induced residual stresses in tape placement. Journal of Thermoplastic Composite Materials. 2002;15(6):525–44.

[17] Ning H, Vaidya U, Janowski GM, Husman G. Design, manufacture and analysis of a thermoplastic composite frame structure for mass transit. Composite Structures. 2007;80(1): 105–16.

[18] Barhoumi N, Maazouz A, Jaziri M, Abdelhedi R. Polyamide from lactams by reactive rotational molding via anionic ring-opening polymerization: Optimization of processing parameters. Express Polymer Letters. 2013;7(1).

[19] Kazemi M, Shanmugam L, Lu D, Wang X, Wang B, Yang J. Mechanical properties and failure modes of hybrid fiber reinforced polymer composites with a novel liquid thermoplastic resin, Elium®. Composites Part A: Applied Science and Manufacturing. 2019;125:105523.

[20] https://www.compositesworld.com/articles/wind-blade-manufacturing-part-ii-are-thermoplastic-composites-the-future. accessed in Feb 2021.

[21] Thattaiparthasarathy KB, Pillay S, Ning H, Vaidya U. Process simulation, design and manufacturing of a long fiber thermoplastic composite for mass transit application. Composites Part A: Applied Science and Manufacturing. 2008;39(9):1512–21.

[22] Sherman LM. https://www.ptonline.com/articles/the-new-lightweights-injection-molded-hybrid-composites-spur-automotive-innovation. Plastics Technology. Accessed in April 2021.

[23] Gardiner G. Metal + Composite = Less weight, more room. Composites World. 2016.

[24] Award for multi-material system for car seats. Reinforced Plastics. 2015;59(2):79.

[25] Gardiner G. CAMISMA's car seat back: Hybrid composite for high volume. Composites Technology. 2014:34–40.

[26] Chesser P, Post B, Roschli A, Carnal C, Lind R, Borish M, et al. Extrusion control for high quality printing on Big Area Additive Manufacturing (BAAM) systems. Additive Manufacturing. 2019;28:445–55.

[27] Weidmann S, Hümbert M, Mitschang P. Suitability of thickness change as process control parameter for induction welding of steel/TP-FRPC joints. Advanced Manufacturing: Polymer & Composites Science. 2019;5(2):55–68.

[28] https://www.flexhyjoin.eu/, accessed in Jan 2021.

Chapter 5
Additive manufacturing of thermoplastic composites

5.1 Introduction

Additive manufacturing (AM), often called 3D printing or rapid prototyping, is a group of processes that produce three-dimensional parts by depositing, fusing, joining, or solidifying materials layer by layer. It was initially developed in the 1980s. However, not until the 2010s did it gain tremendous interest from scientists and engineers. Since then, the AM technology has evolved dramatically from a mere method of prototyping to sophisticated and versatile solutions in many fields. Novel AM materials, innovative AM methods, and new AM-related software and hardware have been extensively and continuously developed. With the capability of combining with more innovative materials developed from the material science and engineering field, AM has become a process that can print high-performance components along with other benefits such as low investment in equipment and space. It has been considered to be the preferred prototyping method in new product development for geometry verification, concept demonstration, and so on. AM has become so versatile that it is employed in almost every science, technology, engineering, and medical field for various purposes, including part prototyping, design verification, tooling, customized part production, and producing fully functional components.

The versatility of AM technology has allowed the use of various materials in the printing process. These materials include thermoplastics, thermoplastic composites, thermosets, thermoset composites, concrete, metals, ceramics, food, tissues, medicines, and so on. Among all the materials, neat thermoplastics such as ABS and PLA are the most commonly used because of low cost of printers required, wide variety of commercially available feedstock materials, low cost of the feedstock material, and ease of printability. However, the AM parts printed using neat thermoplastics generally have limited mechanical properties. Such parts are often conceptual prototypes rather than functional components. One main approach to improve the mechanical properties of the neat thermoplastics is to add reinforcements, especially fibers. The addition of discontinuous fibers offers improved specific modulus and strength without a considerable compromise in printability. With the development of continuous fiber-reinforced thermoplastic composites for AM in 2014, it is feasible to print components with significantly enhanced strength and modulus.

AM of thermoplastic composites is a unique process with characteristics that other manufacturing processes of thermoplastic composites do not have. It differs from the conventional processing methods described in Chapter 4 in several aspects, including manufacturing approach, flexibility, production rate, cost, and so

https://doi.org/10.1515/9781501519055-005

on. Therefore, AM of thermoplastic composites is described in a separate chapter as a unique process. AM possesses the following advantages that differentiate it from the conventional processing methods:

1. Capable of creating very complex geometry
2. No tooling required
3. Flexibility in design iterations
4. Capable of functionality integration
5. Ease of integrating multiple materials
6. Waste reduction

There are several steps typically involved in AM of thermoplastic composite parts. They are explained below.

1. **Computer-aided design** of the geometry to be printed. The geometry of the component to be printed is first designed using computer-aided design (CAD) software such as PTC Creo, Solidworks, Autodesk Inventor, and so on. Object scanners can also be used to obtain the geometry of the object to be printed. For a structural component to be printed, finite element analysis is often carried out for optimizing dimensions and/or fiber orientations to meet design criteria before printing.

2. **Conversion** of files. The CAD file from the previous step is first converted to STL file, the most common format for AM. STL is also known as Stereolithography, Surface Tessellation Language, Standard Tessellation Language, or Standard Triangle Language. During the file conversion, the surface of the object is tessellated into connected triangles. The STL file carries the surface information of the object. A higher number of triangles denote a higher resolution for geometric features such as arcs, circles, fillets, etc. This conversion can be done through the CAD software as previously mentioned.

3. **Slicing** of the geometry. Slicing of the geometry in an STL format is a critical step in AM. It demonstrates the working principle of AM. The STL file is imported to a printer equipped with an algorithm that can carry out slicing of the geometry. Different printer manufacturers may develop different algorithms and allow a different amount of control on slicing parameters, such as the thickness of a sliced layer.

4. **Setting of printing parameters**. Various printing parameters, including nozzle temperature, printing speed, printing path and orientation, printing bed temperature, material selection, and support design (if needed), have to be established before printing. When continuous fiber thermoplastic composites are used in printing, the printing path is important as it determines the fiber orientation and the material anisotropy. If multiple materials are involved in the printing, the material type has to be specifically assigned for certain layers. For open-source printers, those settings can often be adjusted. However, closed-source printers do not allow much freedom in adjusting certain printing parameters.

5. **Printing of the geometry**. The geometry is printed, layer by layer, according to the layer thickness, printing path, and material setting that are pre-determined from the previous steps. When the printing is finished, the component is removed from the printing bed for post-processing, if needed.
6. **Post-processing**. Post-processing refers to the operation conducted on the printed component to achieve desired performance, surface finish, and so on. The post-processing operation includes removal of support structures, trimming, sanding, polishing, or painting, for better surface appearance or annealing for a higher degree of crystallinity.

A variety of printing methods have been developed for printing thermoplastic composites. Those methods include fusion deposition modeling (FDM), fused pellet fabrication (FPF), selective laser sintering (SLS), liquid deposition modeling (LDM), and laminated object manufacturing (LOM). Different thermoplastic composites developed for those processes include discontinuous fiber, continuous fiber, particulate, and nanomaterial reinforced thermoplastics. Fibers include carbon fiber, aramid fiber, glass fiber, basalt fiber, cellulose fiber, and liquid crystal polymer fiber. Particulates include glass microsphere, graphite, iron particles, copper particles, and so on. Nanomaterials include nanoclay, nanotube, nanosilica, and nanographene. The thermoplastic used in AM includes ABS, nylon 6, nylon 11, nylon 12, PBT, PC, PE, PEEK, PLA, PMMA, PP, PPS, and TPU.

Tab. 5.1 summarizes the common AM methods for printing thermoplastic composites, and their advantages and disadvantages. The following sections describe each method in detail.

5.2 Fusion deposition modeling

Fusion deposition modeling (FDM) or fused deposition modeling is a typical extrusion-based AM process that is predominantly used in printing thermoplastics and thermoplastic composites. The FDM process was initially developed in the late 1980s for printing neat thermoplastics. It was adopted to print discontinuous fiber-reinforced thermoplastic composites in the 1990s and continuous fiber-reinforced thermoplastic composites in the 2010s. FDM is also called fused filament fabrication (FFF) because its feedstock material is in a filament form.

In the FDM process, the thermoplastic composite filament feedstock is guided through a heated nozzle. The thermoplastic in the filament is heated above its processing temperature in the nozzle and the composite is extruded from the nozzle at a controlled rate. When the nozzle moves according to a pre-determined path, the thermoplastic composite melt is deposited on a printing bed (for the first layer) or the previously deposited layer. The deposited thermoplastic composite cools and solidifies before the next layer is printed.

Tab. 5.1: Common additive manufacturing methods for printing thermoplastic composites.

AM method	Feedstock	Reinforcements	Advantages	Disadvantages
Fusion deposition modeling (FDM)	Filament	– Discontinuous fiber – Particulate – Nanomaterial – Continuous fiber	– Great mechanical properties – Able to integrate different materials	– Printing of continuous fiber composite is still under development – High filament quality required
Fused pellet fabrication (FPF)	Pellet	– Discontinuous fiber – Particulate – Nanomaterial	– Cost-effective – High printing rate	– Relatively low mechanical property
Selective laser sintering (SLS)	Powder	– Discontinuous fiber – Particulate – Nanomaterial	– No support required	– Limited feedstock options – Powder form required – Low mechanical property
Laminated object manufacturing (LOM)	Fabric	– Continuous fiber	– Great mechanical properties	– Waste material generated – Relatively high cost

Most thermoplastic composites used in FDM are mainly reinforced by discontinuous fibers, particulates, or nanomaterials. Those reinforcements have a length ranging from nanometers to hundreds of microns. They can be readily blended into the thermoplastic matrix by different approaches such as extrusion (see Section 5.2.1). Printing such thermoplastic composites is relatively easy as their flow characteristics are comparable to those of neat thermoplastics although the viscosity of the composite is higher. However, composites with such reinforcements have relatively low mechanical properties because of the low aspect ratio of the reinforcements. The general relationship of processability and properties with the reinforcement length/aspect ratio in the AM process follows the trend as illustrated in Fig. 3.3. In other words, the printability of thermoplastic composites increases with decreasing aspect ratio of discontinuous fibers, particulates, or nanomaterials in the composite, while its properties decrease with decreasing aspect ratio of the reinforcement.

Continuous fiber-reinforced thermoplastic composite feedstock materials are developed to address the low property issue existing in the discontinuous fiber, particulate, or nanomaterial reinforced thermoplastic composites. The first continuous fiber-reinforced thermoplastic composites for AM processes and their printers were introduced into the market by Markforged in January 2014. The printer, as well as its newer generation printers, is equipped with dual nozzles or dual extrusion units, one of which is for printing continuous fiber-reinforced thermoplastic composites, including carbon fiber, glass fiber, and Kevlar® fiber-reinforced nylon 6 composites. The printers are able to integrate the continuous fiber-reinforced thermoplastic composites in a neat thermoplastic (the first-generation printer) or a discontinuous carbon fiber-reinforced nylon 6 composite (a newer generation printer).

Figure 5.1 shows a schematic of the FDM process that prints continuous fiber-reinforced thermoplastic composites. The filament is guided through Teflon tubes. Rollers are used to feed the filament into the nozzle continuously during printing. When the continuous fiber-composite filament is extruded after being heated to its printing temperature (the hot end nozzle temperature at which thermoplastic composites are extruded and deposited), it is squeezed by the nozzle shoulder against the last layer, which is already solidified. The schematic shows the small gap between the nozzle shoulder and the last layer. The gap is much smaller than the filament diameter. For example, the filament has an original diameter of approximately 0.35 mm (Fig. 5.3); however, the gap is only about 0.1–0.125 mm. The difference between the filament diameter and the gap creates a pressure that forces the pliable filament to spread and thin down. More importantly, the pressure significantly promotes the adhesion of the filament with the previously deposited composite, which is critical to achieve adequate bonding among adjacent filaments as well as between adjacent layers. The printer is

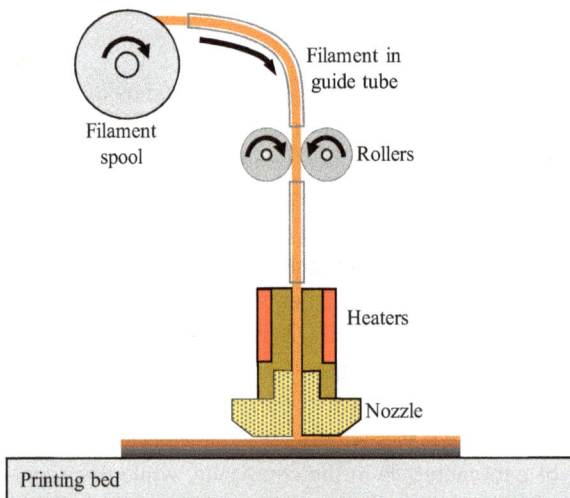

Fig. 5.1: Fusion deposition modeling process for printing continuous fiber-reinforced thermoplastic composites.

equipped with a cutter (not shown in the schematic) for cutting the continuous fiber filament after the printing of each layer is completed.

5.2.1 Filament-based feedstock

The form of the thermoplastic composite feedstock varies in different AM processes (Tab. 5.1). Filaments are the feedstock material for FDM printers. There are two types of filaments used in the FDM process, based on the continuity of the reinforcements in the thermoplastic composite feedstock, namely, discontinuous reinforcement thermoplastic composites and continuous fiber-reinforced thermoplastic composites. Discontinuous reinforcement thermoplastic composites include discontinuous fiber, particulate, or nanomaterial reinforced thermoplastic composites. It is to be noted that discontinues fiber-reinforced thermoplastic composite filaments are often called chopped fiber thermoplastic composite filaments. The filaments have diameters ranging from hundreds of micrometers (for continuous thermoplastic composite filaments) to a couple of millimeters (for discontinuous reinforcement thermoplastic composite filaments). The filament is wound on a spool and can be continuously fed into the printer during printing. It continuously moves from the filament spool to the nozzle of the printer before it is heated to its printing temperature for depositing.

The methods used to produce these two types of filaments differ. The feedstock made of discontinuous reinforcement thermoplastic composite is generally produced via the extrusion process. The schematic in Fig. 5.2a illustrates a twin screw extruder used in producing the discontinuous reinforcement thermoplastic composite filament. Thermoplastic pellets and reinforcements are first fed into the twin screw extruder. It is to be noted that a significant difference in size and density between the pellet and the fiber/particulate/nanomaterial can result in segregation and separation in the feeder. To avoid that issue, the pellets and the reinforcements are often fed through different ports with controlled feeding rates that result in a desired weight percentage of the reinforcement in the filament. Thermoplastics in a powder form can also be used to pre-mix with the reinforcement to avoid segregation. After the thermoplastic and the reinforcement are fed into the extruder, the extruder heats the thermoplastic to its processing temperature and compounds the reinforcements into the thermoplastic melt through a large shear stress resulted from the co-rotating or counter-rotating of twin screws in the extruder. The large shear stress is used to achieve homogeneous mixing of the reinforcements, especially nanomaterials that intend to agglomerate, in the thermoplastic. The composite is then extruded through a die that has an aperture with a desired diameter to form a filament. The composite filament is finally cooled through circulating air or water bath and

(a)

(b)

(c)

500

(d)

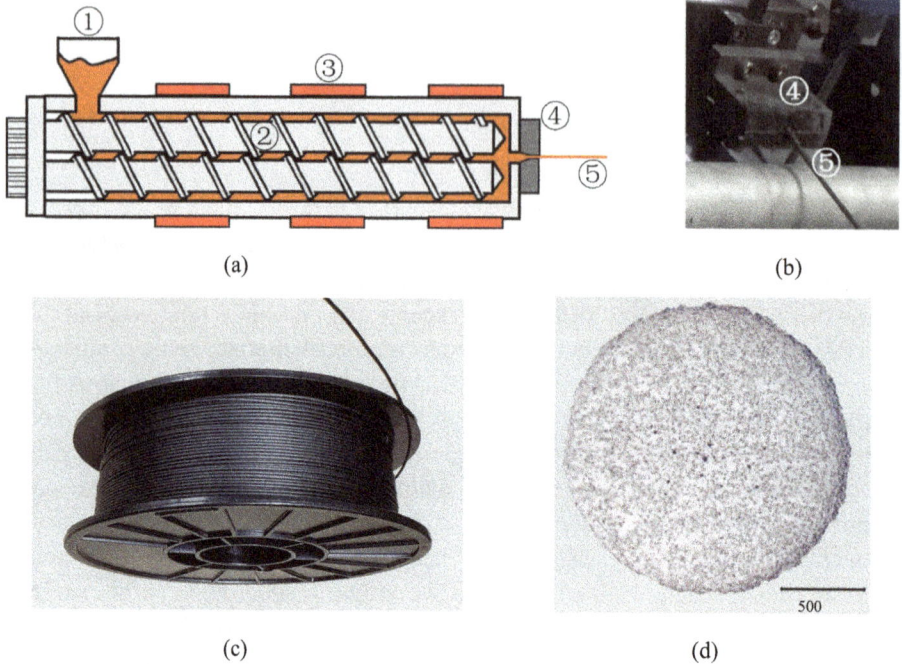

Fig. 5.2: Processing of discontinuous reinforcement thermoplastic composite filament via twin screw extrusion; (a) a schematic of the twin screw extrusion process (① – feeder; ② – screw; ③ – heater; ④ – extrusion die; and ⑤ – extruded filament.); (b) a thermoplastic composite filament extruded out of a die mounted on a twin screw extrusion extruder; (c) a spool of discontinuous carbon fiber-reinforced nylon 6 composite filament produced through the twin screw extrusion process; (d) the cross-sectional microstructure of the discontinuous carbon fiber-reinforced nylon 6 composite filament.

taken up on a spool. Figure 5.2b shows a filament made of discontinuous glass fiber-reinforced nylon 6 composites being extruded out of a die mounted on a twin screw extruder. Figure 5.2c shows a spool of a discontinuous carbon fiber (short carbon fiber)-reinforced nylon composite filament and Fig. 5.2d shows its cross-sectional microstructure. The microstructure demonstrates uniform distribution of discontinuous carbon fibers in the thermoplastic matrix.

The filament, made of continuous -reinforced thermoplastic composite, is essentially a unidirectional fiber reinforced thermoplastic composite in which all the fibers are aligned in the filament-length direction. The filament is one type of thermoplastic composite preform, similar to uni-tapes, but with a circular cross section. The filament is normally produced via the melt impregnation process illustrated in Section 4.3.1 that results in fully impregnated fibers. Fig. 5.3(a,b) show the cross-sectional microstructure of continuous carbon and glass fiber-reinforced nylon 6 composite filament, respectively. Both filaments show great fiber impregnation and minimal voids.

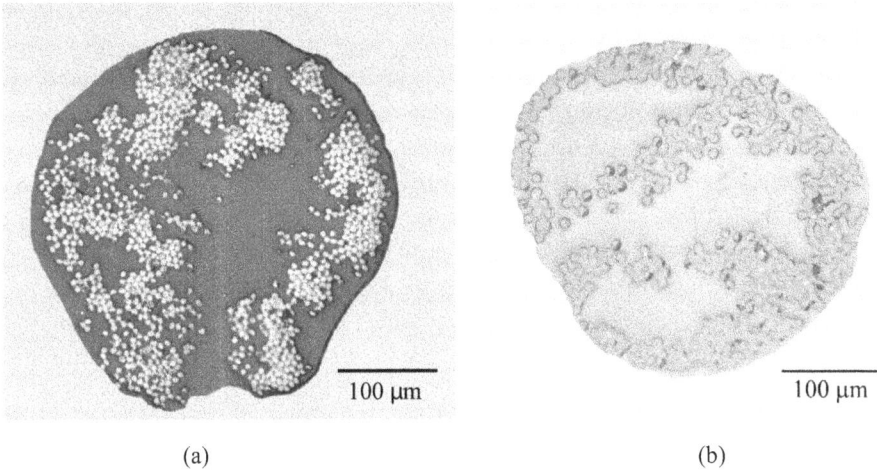

Fig. 5.3: (a) The cross-sectional microstructure of a continuous carbon fiber-reinforced thermoplastic composite filament; (a) the cross-sectional microstructure of a continuous glass fiber-reinforced thermoplastic composite filament.

The quality of the filament is a crucial factor that determines the quality of printed parts. As mentioned in Section 4.3, the melt impregnation process is a pre-impregnation process that produces preforms with complete fiber impregnation. Unlike the conventional molding processes, such as compression molding or auto-clave molding, in which high consolidation pressure and long consolidation time are involved, both consolidation pressure and time are limited for printing the fila-ment. The pressure resulting from the squeezing of the pliable filament between the nozzle and the previous layer (Fig. 5.1) is the main pressure for consolidation. The consolidation time for the filament is limited to its brief contact with the noz-zle shoulder after it comes out of the nozzle because of the continuous movement of the nozzle. Minimal fiber impregnation occurs in the filament during printing because of the limited consolidation pressure and short consolidation time. Com-plete fiber impregnation has to be done at the filament level; therefore, it is essen-tial to have fully impregnated filaments for printing high-quality composite parts.

It is possible to blend other materials into thermoplastics and integrate multiple functions with the feedstock material through the melt impregnation method. For example, electrically conductive materials such as metallic powders and carbona-ceous materials are blended with thermoplastics to form thermoplastic composites that possess both load-bearing capability and high electrical conductivity; glass mi-crospheres are added to thermoplastics to produce filaments for printing structures with good buoyancy and low thermal conductivity.

5.2.2 Mechanical behaviors

The properties of a material or a component can significantly vary with its processing methods because different structures can result from different processing methods. Typical processing parameters, including temperature, pressure, cooling rate, etc., determine the final structure of the manufactured components, and therefore, their mechanical properties. AM processes, including FDM, differ from conventional processes in various aspects, such as the way a component is produced, the amount of consolidation pressure involved, the cooling rate, and so on. AM components are built layer by layer while bulk materials are formed into components in one molding step. Additionally, the consolidation pressure in AM processes is significantly less than that in the conventional molding processes and the flow in the AM processes is very limited. Those distinctly different processing conditions in AM processes result in a composite component that is different from the one processed via conventional processes. Therefore, the material properties and behaviors of the AM thermoplastic composite can be distinctly different.

5.2.2.1 Tensile behavior

Tensile test is used to study the relationship between the tensile stress and tensile strain of FDM composites and their failure mechanisms. Elastic modulus (also called tensile modulus), tensile strength, and strain-to-failure are the basic material properties obtained from the test. There is a significant difference in the tensile properties of printed discontinuous reinforcement thermoplastic composites and printed continuous fiber thermoplastic composites. Overall, the discontinuous reinforcement thermoplastic composite, including discontinuous fiber, particulate, or nanomaterial reinforced thermoplastic composites, shows relatively low tensile properties. Fig. 5.4 compares the tensile properties among neat thermoplastics and their discontinuous fiber-reinforced composites that are printed using the FDM process. Elastic modulus can be considerably improved by adding discontinuous reinforcements. However, the increase in tensile strength is marginal compared to the modulus increase. One main reason for the limited strength increase is the presence of voids introduced by the printing process. These defects cause high stress concentration and weaken the composite.

Short fiber-reinforced thermoplastic composites printed via the FDM process show various failure modes such as matrix fracture, fiber pullout, and fiber fracture. Such failures modes are also observed in short fiber-reinforced composites that are molded using conventional processes such as injection molding and compression molding. Fig. 5.5 shows the fracture surface of a printed composite consisting of short carbon fibers and ABS matrix [6]. The fiber in the composite has an average length of 150 μm and a diameter of about 7 μm. There are obvious fiber pullout, matrix fracture, and fiber fractures. A large number of voids are also present. Those

(a)

(b)

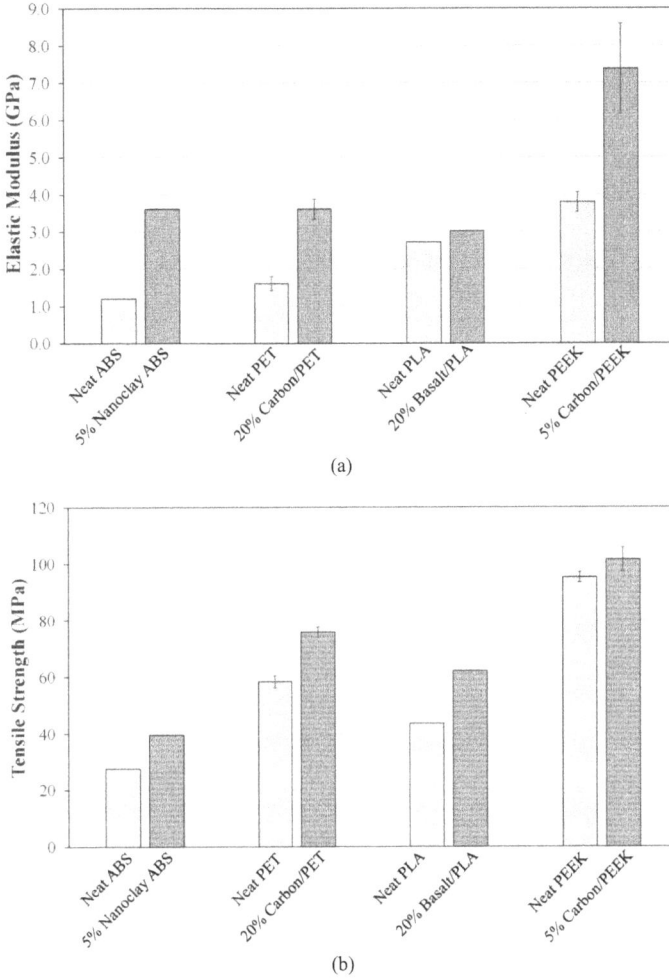

Fig. 5.4: (a) Elastic modulus and (b) tensile strength of neat thermoplastics and their composites printed by FDM process (adapted from References [1–5]).

voids are inter-voids, the voids that exist between filaments, and inner-void, the voids that are inside the filament.

The fibers used in the continuous fiber-reinforced thermoplastic composite filament are unidirectional and aligned in the filament-length direction. As previously mentioned, the unidirectional fiber-reinforced composite possesses superior mechanical properties in the fiber direction (see Section 3.2.1). Fig. 5.6 shows the tensile properties of some printed continuous fiber thermoplastic composites. The tensile strength and elastic modulus of printed unidirectional carbon fiber-reinforced nylon 6 composite can reach around 800 MPa and 68 GPa, respectively (note: those properties are still lower than the properties of uni-tapes as listed in Tab. 3.2 because of the

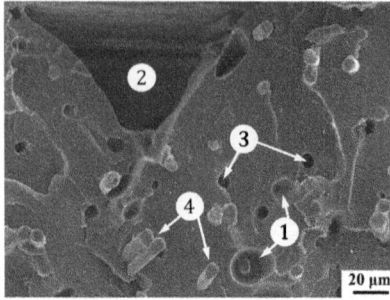

Fig. 5.5: Fracture surface showing fiber pullout and matrix fracture as well as the voids in a carbon fiber ABS composite printed by FDM process (①- inner voids; ②- inter voids; ③- cavities from fiber pullout; ④- fibers that are pulled out) (adapted from Reference [6]).

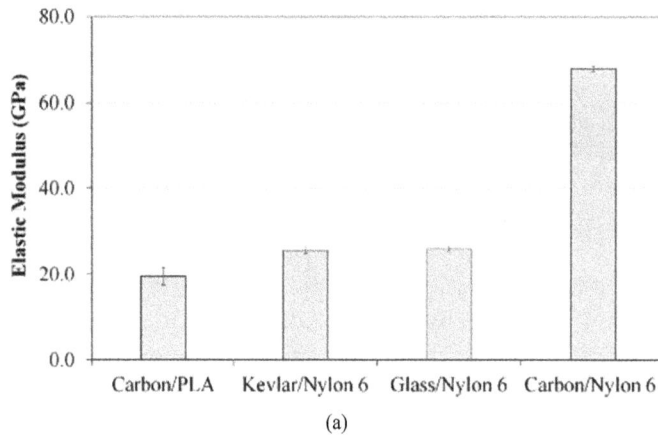

(a)

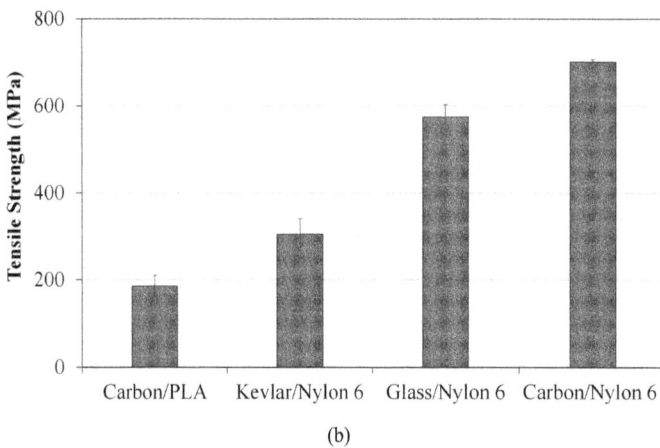

(b)

Fig. 5.6: (a) Elastic modulus and (b) tensile strength of printed continuous fiber-reinforced thermoplastic composites (adapted from References [7–9]).

limited fiber content in the filament). Therefore, integration of the unidirectional fiber-reinforced thermoplastic composite in a printed component can result in significantly improved mechanical properties compared to the one made of only neat thermoplastics or discontinuous fiber-reinforced thermoplastic composites.

The properties of the printed unidirectional fiber-reinforced thermoplastic composite can vary significantly even when the same type of fiber and matrix are used. The difference in properties is attributed to factors such as percentage of the continuous fiber-reinforced composite filament, fiber content in the filament, build orientation, infill percentage, number of walls, layer thickness, printing speed, nozzle temperature, printing chamber temperature, printing bed temperature, and so on. Because of the difference in the abovementioned factors, it is common to see a large variation in the mechanical properties of printed continuous fiber thermoplastic composites, even with the same type of fiber and matrix. Nevertheless, the part printed using the continuous fiber-reinforced thermoplastic composite can possess properties several times higher than the ones printed using only neat thermoplastics or discontinuous reinforcement thermoplastic composites.

Printed continuous fiber thermoplastic composites normally have a rough surface (Fig. 5.19); therefore, discontinuous fiber thermoplastic composites are generally printed onto the surface of the continuous fiber composite using dual nozzles. The printing path is designed such that the discontinuous fiber composite surrounds the continuous fiber thermoplastic composite to provide a better surface finish (Fig. 5.19).

5.2.2.2 Impact behavior

The impact strength of the FDM thermoplastic composite is an indication of its capability of absorbing impact energy. There are several factors that determine its impact properties. Fiber content is one of most important material variables. As the fiber content increases, its impact strength increases. Other processing related variables, such as build orientation, printing speed, layer thickness, nozzle temperature, printing chamber temperature, and printing bed temperature, can also influence the impact performance of the FDM thermoplastic composite.

Figure 5.7 compares the impact strength among various thermoplastics and continuous fiber thermoplastic composites printed via FDM. Neat thermoplastics generally have low impact strength. With continuous fibers added, their impact strengths are significantly improved. Glass fiber has more advantage in enhancing the impact strength of thermoplastics than the other fibers, such as carbon fiber and Kevlar® fiber. It is to be noted that the degradation of Kevlar® fibers during printing can happen when the Kevlar® fiber-composite filament goes through the nozzle at a temperature that is normally around 40 °C higher than the melting temperature of the thermoplastic matrix. Fig. 2.51 shows the temperature effect on the tensile strength of Kevlar® fibers. Their tensile strength suffers a drastic decrease when the Kevlar® fiber is exposed to a temperature of 250 °C or above. In addition, the melt impregnation process that is used for producing Kevlar® fiber thermoplastic composite filament can also cause the degradation of the Kevlar® fiber properties. Therefore, it is necessary to select a thermoplastic matrix with a low melting temperature to combine with Kevlar® fiber to minimize the degradation in its mechanical properties, including impact properties.

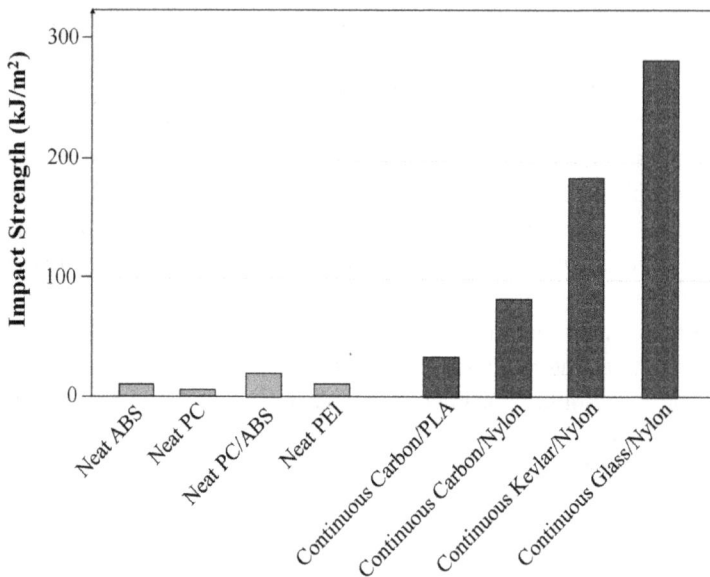

Fig. 5.7: Impact strength of neat thermoplastics and continuous fiber thermoplastic composites printed using the fusion deposition modeling process (adapted from Reference [10]).

The printed discontinuous and continuous fiber-reinforced thermoplastic composites show different failure mechanisms during impact. The failure modes are generally matrix fracture and fiber pullout for discontinuous fiber thermoplastic composites, as the fiber length is lower than the critical fiber length (see Section 3.3.2.2). On the other hand, the printed continuous fiber thermoplastic composite mainly shows fiber fracture, delamination, fiber splitting, and matrix fracture, during impact. The delamination occurs between the printed layers while the fiber splitting happens between filaments in the same layer.

5.3 Fused pellet fabrication

Filaments are commonly used as the feedstock in the AM of thermoplastics and thermoplastic composites. The filament is a material form for both discontinuous and continuous fiber thermoplastic composite. The feed rate of the filament-based AM process is largely limited to the size of the filament diameter and the filament feed rate. The filament is also required to have a tight tolerance for proper feeding and printing. The need for a higher feed rate and a lower raw material cost has led to the development of FPF, another extrusion-based AM process, especially for printing large scale parts.

FPF process is also called fusion pellet fabrication or fused particle fabrication process. The working principle of the FPF process is similar to that of the FDM process, namely, extruding thermoplastic composite melt through a nozzle and depositing it. However, the FPF process uses pellets as its feedstock material, instead of filaments. The FPF printer is equipped with an extruder that converts thermoplastic composite pellets into viscous melts for printing. Thermoplastic composite pellets are fed into the extruder and heated to the processing temperature of the thermoplastic composite before being extruded out of the nozzle for depositing.

In contrast to the FDM process or other AM processes that only allow printing small build volumes at low deposition rates, the FPF process is capable of printing components with significantly larger build volumes at a relatively high deposition rate. The process has been successfully used to print large structures, such as molds, in a relatively short time (Fig. 4.48). Additionally, the FPF process has other advantages compared to the FDM process. The FPF process uses pelletized feedstock material, which is easier to produce and handle. The issues caused by low filament quality, such as inconsistency of the filament diameter and inadequate rigidity of the filament, etc. encountered in the FDM process, can be circumvented in the FPF process. However, the FPF process is suited for printing discontinuous fiber thermoplastic composites, which results in limited mechanical properties, while the FDM process is able to print continuous fiber thermoplastic composites.

5.4 Selective laser sintering

Another method of printing thermoplastic composites is SLS, a powder-bed-based AM method. It uses laser as the energy source to sinter thermoplastic composite powders, layer by layer, into a part. Common thermoplastic composites used for the SLS process include particulate, nanomaterial, and discontinuous fiber-reinforced thermoplastic composites.

Fig. 5.8 shows a schematic of the SLS process. A laser beam is used to sinter the powders on the surface of a powder bed. When the powder is sintered, the bottom support of the powder bed is lowered and a roller or rake levels the powder. The previous sintered powder is covered with a layer of powders to be sintered by the laser beam. The steps are repeated until the part is built. The SLS process does not need a support structure, as the powder bed provides adequate support for the printed part.

Sintering occurs below the melting temperature of the thermoplastic matrix and does not involve any liquidation of the composite particles (note: if melting of the particle is involved, the process is called selective laser melting, another powder-bed-based AM process). It is desirable to delay the recrystallization of

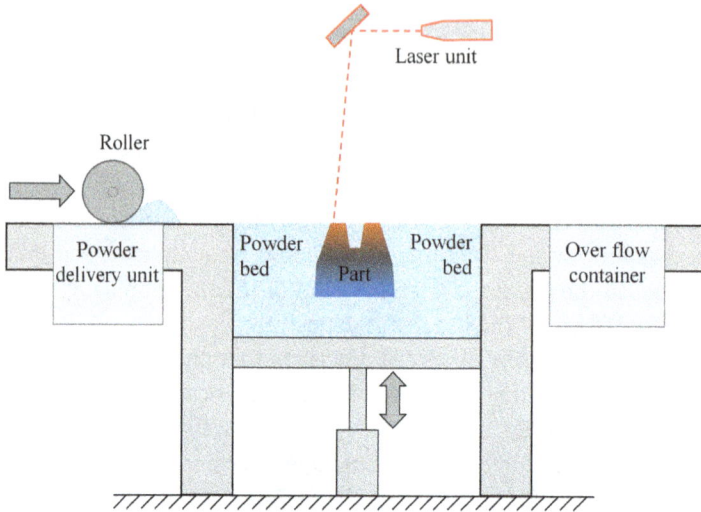

Fig. 5.8: Selective laser sintering process for printing thermoplastic composites.

the thermoplastic during sintering to maintain the material in a metastable status, for better coalescence among particles in the same layer as well as adequate adhesion with the previous layer. Therefore, onset temperatures of melting and recrystallization of the thermoplastic matrix in the composite particles are used to determine the processing window in the SLS process or sintering window. Fig. 5.9 illustrates the DSC curve of a nylon 12 matrix composite and the sintering window that is between the onset temperatures of melting and recrystallization of the nylon 12 matrix.

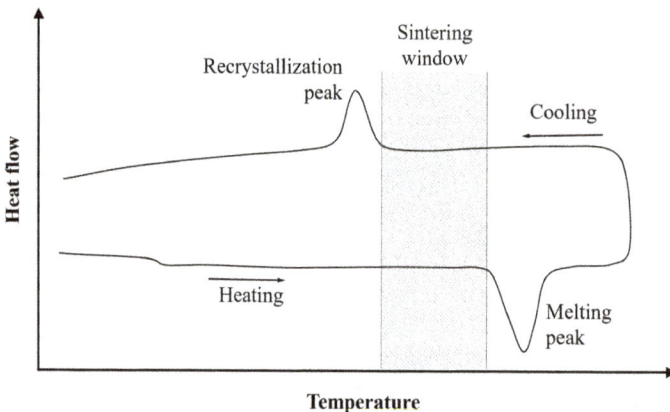

Fig. 5.9: The differential scanning calorimetry curve of a nylon 12 matrix composite showing the sintering window, the temperature difference between the onset temperatures of melting and recrystallization of the nylon 12 matrix.

Because the raw material for SLS is in a powder form and the reinforcement material, such as fibers, has very limited aspect ratio, the increase in mechanical properties by adding particles, nanomaterials, or discontinuous fibers to thermoplastics are limited when compared to neat thermoplastics. The increase in elastic modulus is normally more pronounced than that of the strength. For example, the tensile strength is increased from 45 to 80 MPa and elastic modulus from 1.3 to 5.8 GPa, respectively, when 30 wt% carbon fibers with a length of 150–250 μm are added to nylon 12 [11].

The printing resolution for the SLS process is mainly determined by the particle size in the powder feedstock. Smaller particles result in printed composites with smooth surfaces. However, the printed composite possesses lower mechanical properties because of the low aspect ratio of the reinforcement, such as fibers in the smaller particles. For example, the fibers in the feedstock have a microscale length, in the range of 150–250 μm, which results in a fiber aspect ratio of less than 50. Although some of the nanomaterials have extremely high modulus and strength (see Section 2.3.3), their low aspect ratio and moderate bonding with thermoplastic matrices result in limited strength of sintered thermoplastic composites.

5.4.1 Powder-based feedstock

The feedstock material for the SLS process is limited to discontinuous fiber, particulate, or nanomaterial-reinforced thermoplastic composites, as it has to be in a powder form. A number of processes have been developed for producing the powder-based feedstock for laser sintering. The feedstock-making process determines the powder size and morphology, which affect the resolution of the laser sintering process and performance of the sintered composite.

One of the methods used to prepare thermoplastic composite powders is dissolution–precipitation process. The process is illustrated in Fig. 5.10. A thermoplastic (normally in a pellet form) is firstly dissolved in a solvent at an elevated temperature. Reinforcements, such as particles, nanomaterials, or discontinuous fiber, are mixed with the same solvent to form a suspension. After the thermoplastic is completely dissolved and the reinforcements are uniformly dispersed via stirring and/or ultrasonic oscillation, the solution (consisting of solvent and dissolved thermoplastic) and the suspension (consisting of solvent and dispersed fibers) are mixed. The temperature of the mixture is then lowered to induce precipitation of the thermoplastic. During the dissolution and precipitation at the elevated temperature, an inert gas such as nitrogen is normally used to protect the thermoplastic from oxidation. After the solvent is removed, powders consisting of the thermoplastic matrix and the reinforcements are cleaned and dried. Ball milling may be used afterwards to refine the powder size. The dissolution-precipitation method can produce powders with a relatively narrow size distribution. A high fiber content (up to 50 wt%) can also be achieved. However, due to the challenge in finding suitable solvents that can effectively dissolve the thermoplastic and be readily removed, its use is limited to only certain thermoplastics.

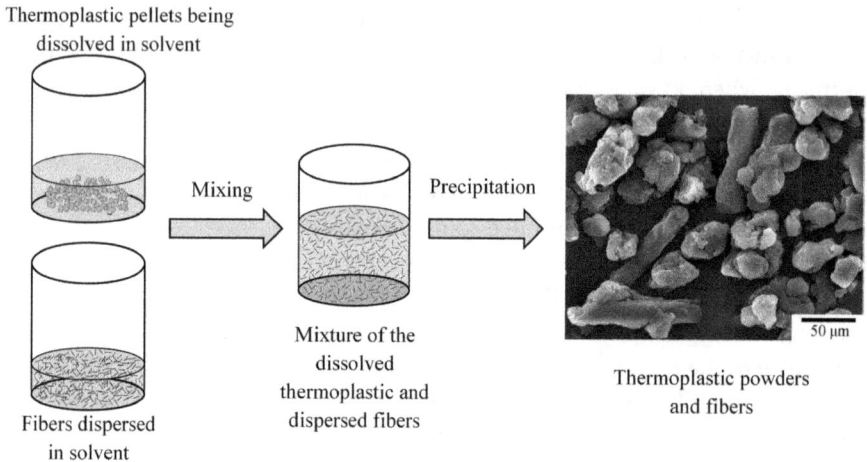

Fig. 5.10: The dissolution-precipitation method used for preparing thermoplastic composite powder feedstock for selective laser sintering (partially reprinted from Reference [12] with permission).

Nylon 12 is the most common thermoplastic used to produce powder feedstock by the dissolution-precipitation method. It dominates the SLS market because of its wide sintering window (Fig. 5.9). Different reinforcements, including carbon nanofibers, carbon nanotube, nanographene, carbon fibers, nanosilica, and metallic powder, are added to produce nylon 12-based thermoplastic composite powder feedstock for the SLS process. Other thermoplastics that have been made into powders include PP, nylon 11, PEK, PAEK, and PEEK.

Another method for preparing thermoplastic composite powders is cryogenic milling. Thermoplastic composite pellets, from compounding or melt impregnation, are normally used as the initial raw material for milling. Cryogenic milling of thermoplastic composites takes place in a cryogenic pulverizer at a temperature below −50 °C. The low temperature helps alter the thermoplastic, which is commonly ductile and deforms plastically during milling at higher temperatures, to a relatively brittle material that fractures and gets pulverized into small particles effectively. Sieving can be used to selectively remove the powders with undesired sizes. However, powders prepared from the cryogenic milling method generally have a large size distribution and their shape is irregular.

Besides the dissolution-precipitation and cryogenic milling processes, there are other powder-making methods, such as spray drying and melt emulsification. However, these methods have been mainly used in preparing neat thermoplastic powders and reinforcements are rarely involved in those powder-making methods.

5.4.2 Zero shear viscosity

Sintering of thermoplastic composite powders relies on the surface tension of the heated and viscous thermoplastic, the driving force for sintering. Frenkel's theory [13] defines the factors that affect the sintering at its initial stage. It describes the growth of the radius of a neck, x, that is formed between two droplets with equal initial radius as a function of sintering time, t:

$$\frac{x^2}{R} = \frac{3\gamma}{2\mu_0}t \tag{5.1}$$

where x is the radius of the sintering neck, R is the original thermoplastic particle radius (Fig. 5.11), γ is the surface tension of the viscous thermoplastic, μ_0 is the zero shear viscosity of the thermoplastic composite, and t is the sintering time. Zero shear viscosity is defined as the viscosity of a material when there is no shear stress in processes such as sintering. Sintering is more effective when the zero shear viscosity is low and the surface tension is high.

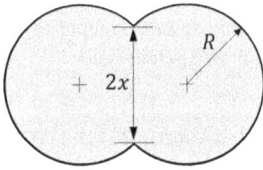

Fig. 5.11: Sintering of two equal-sized thermoplastic particles with initial particle radius, R, and neck radius, x.

The zero shear viscosity of a thermoplastic composite is affected by its reinforcement content and sintering or printing temperature. It increases with increasing reinforcement content and decreasing printing temperature. Figure 5.12 shows the change in zero shear viscosity of a carbon fiber-reinforced PEEK composite with fiber content and printing temperature. It demonstrates the same trend, as seen in the viscosity of thermoplastic composite, in relation to reinforcement content and temperature (Fig. 4.1).

5.5 Laminated object manufacturing

LOM is another AM process that can be used to print continuous fiber-reinforced thermoplastic composites. It was initially developed for thermoset-based continuous fiber prepregs. The same concept is adopted for printing thermoplastic composites after several modifications are made, for example, increasing the capacity of the heating unit from curing thermosets to melting thermoplastics.

The LOM process has the similar principle as the automated tape layup process in which thermoplastic composite preforms are stacked and consolidated, layer by layer.

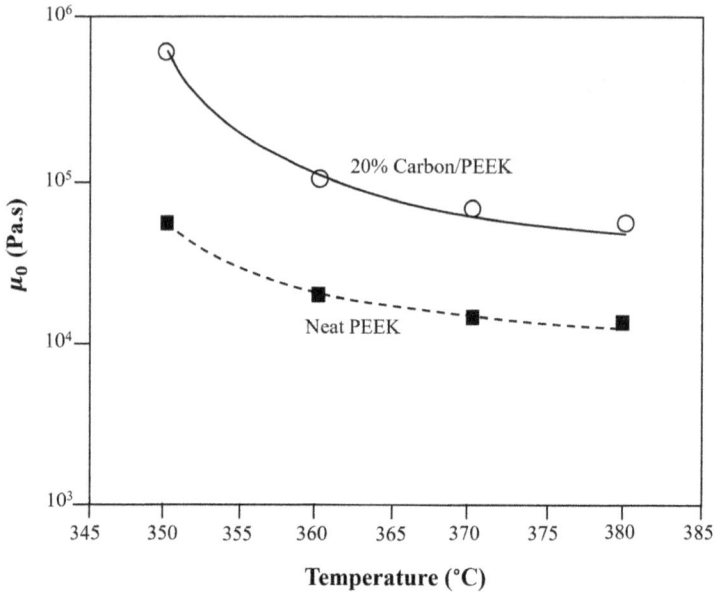

Fig. 5.12: The change of zero shear viscosity of PEEK and carbon fiber-reinforced PEEK composite with fiber content and printing temperature (reprinted from Reference [14] with permission).

Fig. 5.13 illustrates the setup of the LOM process. Continuous fiber thermoplastic composite preforms, for example, woven fabric-reinforced thermoplastic composite preforms, move through a forming station where heating, consolidation, and cutting of the composite preform take place. The preform is first heated by a laser unit and consolidated by a heated roller. A laser or a blade cuts along the periphery of the consolidated preform area. The platform is moved down with the consolidated layers and fresh preform is rolled into position for consolidation. The preform width is normally greater than the part width in order for the take-up roll to pull the new preform to the printing station.

Commercial available LOM printing processes include the selective lamination composite object manufacturing (SLCOM) process developed by EnvisionTEC and the composite-based AM (CBAM) process developed by Impossible Objects. In the SLCOM process, woven fabric-composite preforms with a variety of thermoplastic matrices such as PEEK, PEI, PPS, PP, PET, PC, PES, PEKK, and nylons are used and a setup similar to Fig. 5.13 is used for printing parts. The CBAM process is another derivative of the LOM process. Inkjet technique is used to selectively deposit an aqueous-based binder/glue on a fiber mat (feedstock material). A layer of thermoplastic powder is then deposited onto the fiber mat. The powder adheres only to the fiber mat where the aqueous binder is applied. After excess powder is removed, the fiber mats are stacked, heated, and compressed for consolidation. Any excess fibers and material are mechanically or chemically removed to obtain the final part. The mechanical property

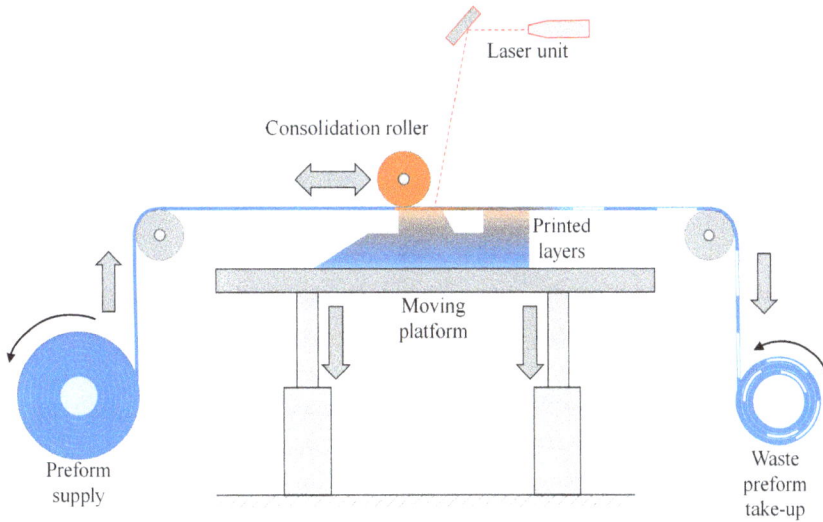

Fig. 5.13: Laminated object manufacturing process for additive manufacturing of thermoplastic composites.

of the part produced by this technique is relatively low because of the use of discontinuous fibers.

5.6 Other additive manufacturing processes

Besides the abovementioned AM processes, there are other AM processes used for printing thermoplastic composites. Those processes include liquid deposition modeling and selective laser melting.

1. **Liquid deposition modeling (LDM)** is an AM process that uses a solution as the printing material to produce parts. It is also referred to as direct ink writing. The solution normally has a high viscosity and is in a gel or paste state. After the solution is extruded from the printer nozzle, its solvent vaporizes rapidly and the material solidifies; the geometrical shape can, therefore, be retained after printing. It is mainly used for printing thermosets and thermoset composites. It allows for a high loading of reinforcements because of the low viscosity of the solution. The process is sometimes used in printing thermoplastic composites, mainly PLA matrix composites. PLA is dissolved in a highly volatile solvent, such as dichloromethane, and reinforcements are mixed and dispersed in the solution for printing.

2. **Selective laser melting (SLM)** is sometimes referred to as laser beam melting. It is a process that is similar to SLS. Thermoplastic or thermoplastic composite powders are fused together by laser. Instead of sintering the powders in the SLS process that maintains the material in a solid form, the laser beam melting

process uses high-power laser to melt and fuse thermoplastic powders or thermoplastic composite powders together. After one powder layer is melted and fused together, the powder bed lowers down and a roller or rake levels the powders for the next melting cycle. Since complete melting of the thermoplastic composite powder is involved in the SLM process, the printed part possesses better fusion and enhanced mechanical properties compared to the SLS part. However, geometry control is much more challenging in SLM because of the geometry instability induced by its higher processing temperature (more thermal shrinkage) and the transitions from solid to liquid and back to solid.

5.7 Effect of material variables

AM of thermoplastic composites relies on thermal transitions to achieve bonding of the thermoplastic matrix in the same layer and between adjacent printed layers. A typical example is melting of the thermoplastic matrix in extrusion-based AM processes for realizing the adhesion among filaments and layers. Therefore, thermal properties of the thermoplastic composite, such as melting temperature and glass transition temperature, play a critical role in affecting the AM process and performance of printed parts. In addition, other material variables, such as rheological and optical properties, also affect the printing. The effects of the thermal, rheological, and optical properties of the composite on AM are described below.

5.7.1 Thermal properties

The thermal property of a thermoplastic composite refers to the property that describes its response to heat. The thermal property includes melting temperature, recrystallization temperature, coefficient of thermal expansion (or shrinkage), thermal conductivity, and heat capacity. The following sections describe the effects of those major thermal properties of the composite on AM.

1. **Melting temperature** is one of the most important thermal properties that affect the AM process. Most of the AM processes for thermoplastic composites, such as FDM and FPF, require the material to be in a molten state. The melting temperature of the thermoplastic matrix basically determines the temperature for printing the thermoplastic composite, or printing temperature. Printing temperature is referred to as the hot end nozzle temperature at which thermoplastic composites are extruded and deposited for extrusion-based AM processes. The printing temperature results in a thermoplastic composite melt with an adequately low viscosity, while it continues to keep the material from degradation. For the semicrystalline thermoplastic matrix composite, the printing temperature is normally about 40 °C higher than the melting temperature of the thermoplastic matrix. On the other

hand, amorphous thermoplastics and their composites do not possess a melting temperature and the printing temperature is often determined to be considerably higher than their glass transition temperature. For example, ABS, an amorphous thermoplastic, has a glass transition temperature of 100 °C and its composite is normally printed at a temperature between 220 and 255 °C.

The printing temperature of the thermoplastic composite influences the geometric stability of the printed part. Thermal shrinkage occurs when the deposited thermoplastic composite is cooled down from the printing temperature. The thermal shrinkage is proportional to the temperature difference ΔT (Eq. (5.2)), that is, the difference between the printing temperature and the room temperature. A higher printing temperature decreases the viscosity of the composite, which benefits printing by enhancing adhesion among the deposited layers and reducing the porosities. However, a higher printing temperature can result in larger thermal shrinkage and higher thermal stress in the printed part, which can lead to severe warpage. Therefore, a trade-off has to be considered between the thermal shrinkage and the printability of a thermoplastic composite when selecting the temperature for printing a thermoplastic composite.

2. **Recrystallization temperature.** Recrystallization takes place in the semicrystalline thermoplastic matrix when its composite is cooled down from the molten state at an adequately low cooling rate. Some disordered polymer chains are transitioned into an ordered status during recrystallization. Because it is an exothermal process, heat is released, which can benefit bonding between the deposited layers. However, the disorder-to-order transition induces a relatively large volume change (shrinkage) (Fig. 2.6). The shrinkage can cause thermal stress in the composite. For amorphous thermoplastics and their composites, no crystallization occurs and the volume change is from thermal shrinkage only (Fig. 2.6).

Recrystallization temperature is also used as a reference to determine the sintering temperature for the SLS process. Sinter temperature should be above the onset temperature of recrystallization but below the onset temperature of melting of the thermoplastic matrix. When thermoplastic composite particles are heated above the recrystallization temperature of the thermoplastic matrix, diffusion, penetration, and entanglement of the polymer chains in the thermoplastic are enhanced for better adhesion among the particles.

3. **Coefficient of thermal expansion or shrinkage (CTE)** is sometimes known as coefficient of linear thermal expansion. It defines the change in dimension of a material with temperature (Eq. (5.2)). When a thermoplastic composite cools from the printing temperature, its dimension decreases and thermal stress is induced. It is another thermal property of the thermoplastic composite that considerably influences the AM process and the quality of the printed part:

$$\Delta L = \alpha L \Delta T \qquad (5.2)$$

where ΔL is change in dimension (length, width, or thickness), α is the coefficient of thermal expansion or shrinkage (unit: μm/m K), L is the original dimension (length, width, or thickness), and ΔT is the change in temperature.

The CTE of the thermoplastic matrix can vary at its glass transition temperature and melting temperature (Fig. 2.6). The CTE becomes greater when the temperature is raised to glass transition temperature and melting temperature. Addition of reinforcements generally lowers the CTE of thermoplastics, and therefore reduces the stress caused by thermal shrinkage. For example, the addition of short carbon fibers to thermoplastics can effectively reduce their thermal shrinkage and, therefore, residual stresses. However, continuous fiber thermoplastic composite filaments have significantly different coefficients of thermal expansion or shrinkage in the fiber direction and the transverse direction, which can result in severe geometric instability (see Section 5.8.5).

4. **Thermal conductivity** is defined as the intrinsic ability of a material to transfer heat. The thermal conductivity of thermoplastics is relatively low. Reinforcements in the thermoplastic composite used for AM have a wide range of thermal conductivity due to the variety of materials used in the reinforcement. Carbon-based reinforcements, including carbon fibers, graphite, graphene, and nanotubes, have a very high thermal conductivity and, therefore, can enhance the thermal conductivity of thermoplastic composite feedstock materials and their printed parts. For example, carbon fibers are added to thermoplastics to increase their coefficient of thermal conductivity, which promotes the heating of the filament and enhances the printing rate. Other reinforcements, such as glass fiber and cellulose fiber, have a relatively low thermal conductivity. Some other reinforcements, such as glass microspheres, have an extremely low thermal conductivity and can lead to a significantly low thermal conductivity of their thermoplastic composites.

The thermoplastic composite with a higher thermal conductivity heats up faster because heat can be conducted from the nozzle to the filament at a faster rate (Eq. (5.3)). Faster heating benefits the printing by shortening the heat time on the feedstock and improving the printing rate. On the other hand, the composite with a higher thermal conductivity undergoes faster cooling after it leaves the nozzle and makes contact with the printing bed or the previously deposited material:

$$\frac{Q}{t} = \frac{kA\Delta T}{d} \tag{5.3}$$

where Q is heat energy, t is time, k is coefficient of thermal conductivity of the thermoplastic composite (unit: W/m K), ΔT is change in temperature, A is the contact area, and d is the size of the filament.

5. **Specific heat capacity** (c) of a material is the amount of heat required to raise one degree or the amount of heat released to reduce one degree in one gram of

material. The thermoplastic composite material with a higher specific heat capacity implies that more heat (Q) is required to raise the composite to the same temperature compared to the composite with a less specific heat capacity, according to Eq. (5.4). Therefore, the composite with a high specific heat capacity needs a longer heating time to reach the printing temperature.

$$Q = mc\Delta T \qquad\qquad (5.4)$$

where Q is heat energy (unit: joule), m is mass, c is specific heat capacity (unit: joule/g °C), and ΔT is change in temperature.

The heat capacity of the composite also affects cooling of the composite after printing. When the composite has a higher specific heat capacity, there is more heat retained in the composite. As a result, the heat retained in a material with a great heat capacity can induce improved adhesion for the material in the same layer and between adjacent layers.

5.7.2 Rheological properties

In the AM processes, the matrix in the thermoplastic composite feedstock normally goes through a liquid state (except for the SLS process) and material flow or deformation is involved during printing. The properties that determine how the thermoplastic composite melt flows or deforms under an applied stress are called rheological properties. The rheological property of a thermoplastic composite material at the liquid state plays a critical role in its printing and mechanical properties.

Melt viscosity is one of the most important rheological properties in affecting the printing of thermoplastic composites. It determines the printability of a thermoplastic composite, the bonding strength between adjacent layers, the void content in the printed composite, etc. A common issue during the printing of thermoplastic composites is their high viscosity due to the high viscosity of the thermoplastic matrix, even when it is melted. When reinforcements, such as fibers, nanomaterials, and particulates, are added, the viscosity can be further increased (Fig. 4.1). A high viscosity may cause nozzle blockage, poor bonding between the adjacent layers, increased intravoid, and so on. Different approaches have been used to effectively reduce the viscosity of thermoplastic composite melts. Those approaches include increasing nozzle temperature, heating deposited layers, and heating printing chamber.

5.7.3 Optical properties

Optical properties are important in printing processes that involve lights, for example, the selective laser sintering or melting process that a light source, namely laser, is used for heating. When the photons in the laser interact with the thermoplastic composite particles, reflection, refraction, absorption, and transmission can occur. It is desirable to have the laser sintering feedstock with the least transmission and reflection but highest absorption. The addition of highly absorbing reinforcements, such as carbon fibers and carbon black, in the thermoplastic matrix can significantly enhance the absorption of the laser energy for more efficient heating and enhanced printing.

5.8 Challenges in AM of thermoplastic composites

A large amount of research and development work has been carried out on AM to render it as a versatile processing method that provides the capabilities that conventional processes do not have. However, AM of thermoplastic composites, especially continuous fiber-reinforced thermoplastic composites, is still under development and there are limitations that keep AM from realizing its full potential. Several challenges that remain in AM are listed below. These challenges offer opportunities for maturing the AM processes and furthering their use in more applications.
1. High void content
2. Inadequate bonding
3. Limited mechanical properties
4. Low production rate
5. Residual stress and warpage

5.8.1 High void content

Consolidation of thermoplastic composite preforms is the second step in manufacturing thermoplastic composite parts (see Section 4.2). A high consolidation pressure significantly improves the properties of the thermoplastic composite. Unlike conventional processes such as injection molding and compression molding that can apply a large amount of consolidation pressure, AM processes can only exert a limited force or pressure for consolidation. AM processes mainly rely on gravity, momentum from extrusion, pressure from squeezing the filament by the nozzle shoulder (for continuous fiber thermoplastic composites), surface tension (for powder-bed-based processes), etc. In addition, the time window for bonding with the adjacent layers is short. Because of the low consolidation pressure and short consolidation time involved in AM processes, it is common to have a larger amount of voids in the printed composite compared to the composite molded from conventional processes. The void content can be more than 12% in FDM thermoplastic composites that are printed using commercially available filaments on commercially available printers.

There are two types of voids in the FDM thermoplastic composites, inter-voids and inner-voids. Inter-voids exist between filaments in the printed thermoplastic composite and the inter-void is oriented in the printing direction (Fig. 6.5). In the FDM process, a thermoplastic composite melt is extruded from the nozzle that has a circular aperture, the extruded filament has a rounded or oval shape (because of gravity) when deposited. The deposition of the melt results in diamond-shaped voids in the printed component as shown in Fig. 5.14a. Shifting the extruded filament by half a diameter of the filament may change the shape from diamond to triangle (Fig. 5.14b). The same issue remains for the nozzle with other aperture shapes.

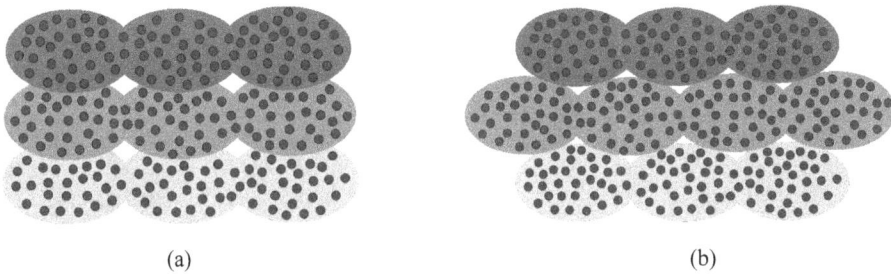

(a)   (b)

Fig. 5.14: Deposition of thermoplastic composite filaments resulting in inter-voids in (a) diamond shape and (b) triangular shape.

Besides the inter-void, the inner-void is another type of void in printed thermoplastic composites. It originates from the porosities inside each deposited filament. The inner-void is generally at a micro scale and much smaller than the inter-void. The inner-void content is also less than the inter-void content. Figure 5.5 shows both the inner and inter voids in a FDM carbon fiber ABS composite.

A large amount of voids can also be present in SLS or SLM parts. The void content can be more than 30% in SLS thermoplastic composites. The main reason is also due to the limited force or pressure for consolidation. In addition, another reason for the high void content in SLS thermoplastic composite is ineffective packing because of the size and geometry difference in thermoplastic matrix particles and fibers. The gaps among the fibers and thermoplastic particles are one of the main sources for voids.

Voids in AM thermoplastic composites are detrimental to their mechanical performance. The high void content results in the low strength and fracture toughness of printed thermoplastic composites because of the high stress concentration around the voids. Cracks can easily initiate around voids, propagate, and weaken the printed composite. The inter-voids also have an adverse effect on the bonding strength between the adjacent layers, which diminishes the fracture toughness of printed composites.

5.8.2 Inadequate bonding

The part made by AM processes can be identified by its characteristic layered structure. The layered structure provides the capability of building complex geometry. However, this layered structure induces many interfaces among deposited, sintered, or bonded thermoplastic composite layers. The interface can be noticed on as-printed thermoplastic composites under microscopes or even with naked eye. Figure 5.15a shows the side surface of an as-printed discontinuous carbon fiber nylon 6 composite. A distinct layered structure is noticeable. Some short fibers are also visible on the surface. Figure 5.15b shows the layered structure on the fracture surface of the same short carbon fiber nylon 6 composite.

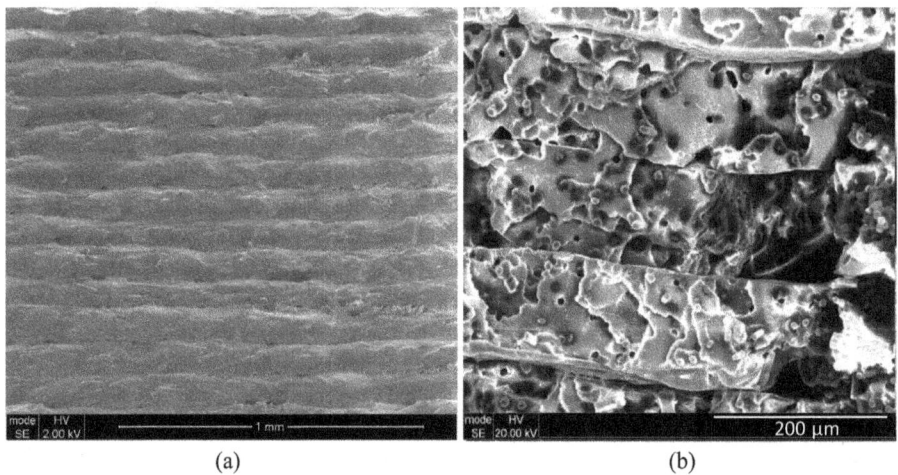

(a)   (b)

Fig. 5.15: (a) Surface morphology of an as-printed discontinuous carbon fiber-reinforced nylon 6 composite showing a distinct layered structure and (b) fractured surface of the composite showing the layered structure and the interface between adjacent layers.

The bonding at the interface of printed thermoplastic composites can be quantitatively evaluated using mechanical tests such as short beam test or lap shear test. These tests measure the interlaminar shear strength, which is a direct indication of the bonding strength at the interface between adjacent layers. Figure 5.16 compares the interlaminar shear strength of some printed thermoplastic composites and conventionally molded thermoplastic composites. All the composites have a nylon matrix. The printed composites have considerably less interlaminar shear strength compared to the conventionally molded composites. Figure 5.17 shows the delamination at the interfaces of FDM-printed fiber-reinforced nylon 6 composite resulting from short beam testing.

Compared to printed neat thermoplastics, thermoplastic composites show higher interlaminar shear strength. For example, short carbon fiber-reinforced composites have higher shear strength than printed neat nylon (Fig. 5.16). Typically, an

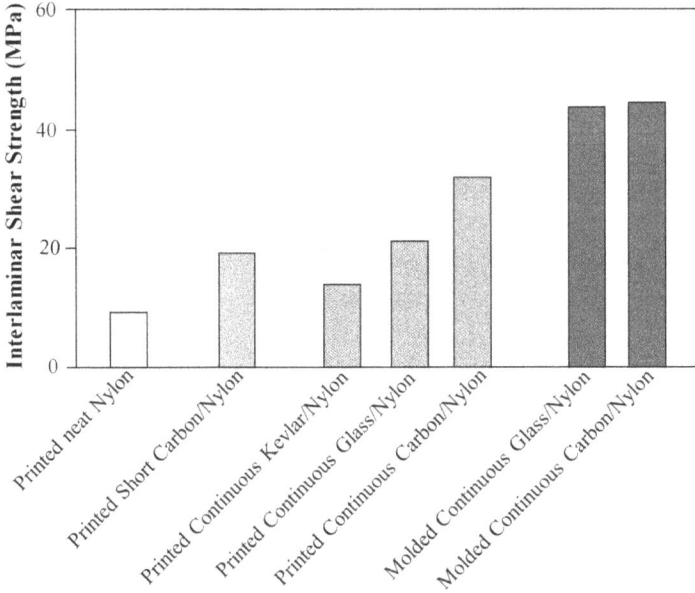

Fig. 5.16: Interlaminar shear strength of printed thermoplastic composites and compression molded composites (partially adapted from Reference [15]).

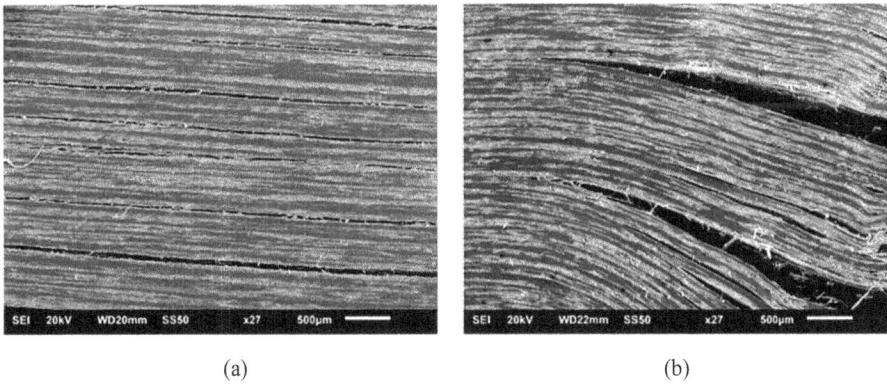

(a)   (b)

Fig. 5.17: (a) Continuous carbon fiber-reinforced nylon 6 composite and (b) continuous glass fiber-reinforced nylon 6 composite printed through FDM showing delamination resulting from short beam testing (reprinted from Reference [15] with permission).

unreinforced thermoplastic has less viscosity than its composites (Fig. 4.1), which enhances bonding between adjacent layers. However, the fibers in the composite can strengthen the interface by providing a bridging effect at the interface (Fig. 4.49).

Various methods have been used to address the inadequate bonding issue. These methods include increasing printing temperature, heating printing bed, heating printing chamber, locally heating previously deposited material, and applying low chamber pressure. The following sections describe each method.

1. **Increasing printing temperature** is one of the simplest but most effective methods to enhance the bonding strength at the interface between adjacent layers. A higher printing temperature reduces the viscosity of thermoplastic composites (Fig. 4.1), provided the temperature is below the degradation temperature of the thermoplastic matrix. The decrease in viscosity can improve the flow of the composite during printing, which can increase the contact area with the previous layer and, therefore, enhance the bonding. The higher printing temperature also allows for more material exchange across the interface, including diffusion and penetration of the polymer chains and fiber migration that can provide a bridging effect at the interface, to improve the bonding.

   A higher printing temperature, however, can have adverse effects on the quality of the printed composites. It can induce greater temperature gradients in the composite, which can cause more residual stress. As a result, severe warpage can occur to the printed composite. In addition, higher printing temperatures can lead to degradation of the thermoplastic matrix and decreased mechanical properties of the printed composite.

2. **Heating printing bed** is another method to enhance the bonding among printed layers. The heat conducted from the printing bed is able to maintain the previously deposited material at an elevated temperature for an extended time, which enhances its interaction with the newly deposited material, such as diffusion and penetration of the polymer chains between the materials at the interface for better bonding. Figure 5.18 compares the fractured surface of the printed short carbon fiber-reinforced nylon 6 composite after tensile testing. Figure 5.18a shows an SEM image of the fracture surface of a composite printed without a heated bed. Obvious individual filaments and print layers are present. In contrast, the composite printed with a heated printing bed shows better bonding between the adjacent layers (Fig. 5.18b). The filament contour and the layer structure are not obvious anymore. As a result, the bonding strength at the interface of the printed composite is enhanced.

3. **Locally heating previously deposited material** is another method to maintain the previously deposited composite material at an elevated temperature for a longer time. The material can be locally heated by noncontact heating methods, such as infrared radiation, laser heating, and microwave heating, to above the recrystallization temperature but below the melting of the thermoplastic. When the composite melt layer is deposited onto the heated surface, its viscosity can be maintained at a low level for a longer time because of a lower temperature gradient and a lower rate of heat dissipation to the previously printed material (Eq. (5.3)). Therefore, the deposited composite has a longer effective interaction time with the previous layer, which enhances diffusion, penetration, and entanglement of molecular chains at the interface as well as fiber migration.

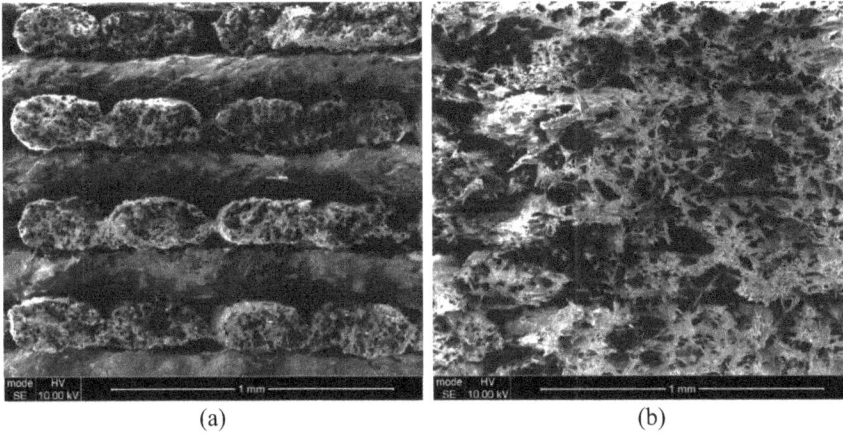

(a)   (b)

Fig. 5.18: The fracture surface of short carbon fiber-reinforced nylon composites printed (a) without a heated printing bed and (b) with a heated printing bed.

Better bonding strength can, therefore, be achieved at the interface. Heated printing chamber has the same principle in enhancing the bonding strength of printed composites.

4. **Applying low printing pressure**. Low pressure can be applied to reduce voids and improve the bonding strength between layers during the printing of thermoplastic composites. The printing takes place in a low pressure setting such as in a vacuum bag, a vacuum chamber, or an autoclave. The low pressure can reduce the inter-void content at the interface and improve the interfacial bonding. The low oxygen content in the low pressure condition also minimizes the thermal oxidation of certain thermoplastic matrices (such as nylons) that are susceptible to oxidation. In addition, the low pressure condition reduces the heat loss from convection, resulting in a higher temperature for enhanced diffusion and better bonding.

5.8.3 Limited mechanical property

The mechanical properties of additively manufactured thermoplastic composites are determined by several major factors such as fiber content, void content, and bonding among print layers. Some of their mechanical properties are shown in Figs. 5.4, 5.6, 5.7, 5.16. In general, additively manufactured thermoplastic composites possess limited mechanical properties. The major contributing factors are limited fiber content, high void content (see Section 5.8.1), and inadequate bonding among the layers (see Section 5.8.2).

Fiber content is a major variable that determines the mechanical properties for AM thermoplastic composites. The mechanical properties, such as modulus and strength, increase with fiber content (to a certain extent). Conventional discontinuous and continuous fiber thermoplastic composites have fiber volume percentages up to around 45 and 60 vol%, respectively. However, the fiber content in the thermoplastic composite feedstock (filament, pellets, and powder) for AM processes is still considerably less than those fiber contents. The main reason for that is that a balance between the fiber content and printability needs to be considered when fibers are added to the thermoplastic during the manufacture of the feedstock material for AM processes. With increasing fiber content, the viscosity of the thermoplastic composite increases (Fig. 4.1). The high viscosity can cause issues such as blockage in the nozzle, poor bonding among printed layers, and high void content. For the SLS process, the zero shear viscosity of the powder feedstock increases with fiber content, which can result in ineffective sintering (Eq. (5.1)).

5.8.4 Low production rate

AM, in general, has a low production rate due to its layer-by-layer construction nature. The production rate for printing thermoplastic composites is limited to the material deposition rate in extrusion-based AM processes such as the FDM process or the sintering rate in the SLS process. In comparison to the high production rate in conventional molding processes, such as injection molding (seconds per part), the production rate in printing thermoplastic composites is low (generally hours per part, depending on the part volume), which is one of the major hurdles for any of the AM processes to become a mass production process.

The production rate in the extrusion-based printing process, such as FDM and FPF, is closely related to printing resolution. Printing resolution or the resolution of a FDM printer is defined by its accuracy along three axes, namely, x, y, and z, especially the z axis. A smaller thickness of the deposited layer indicates a higher resolution of the printer. The production rate can be increased by increasing the flow volume. However, the printing resolution is compromised because of the larger thickness of each deposition layer. Since the thermoplastic composite melt from the nozzle is a flow process, its volumetric flow rate, q, can be calculated as the product of the cross section of the nozzle, A, and the printing speed, v:

$$q = Av$$

For a nozzle aperture with a rounded cross section that has a diameter of d, the volumetric flow rate is expressed by the following equation:

$$q_l = \pi(d/2)^2 v$$

where q_l is the deposition rate at a low resolution.

If the resolution of the printer needs to be improved by n times, that is, the diameter of the nozzle has to be reduced to (d/n), the volumetric flow rate is expressed by the equation below, assuming the print speed remains the same.

$$q_h = \pi[(d/n)/2]^2 v$$

where q_h is the deposition rate at a high resolution.

The ratio between the volumetric flow rates at different resolutions is defined as follows:

$$\frac{q_l}{q_h} = \frac{\pi(d/2)^2 v}{\pi[(d/n)/2]^2 v} = \frac{1}{n^2} \qquad (5.5)$$

Equation (5.5) indicates that the deposition rate (printing rate) is inversely proportional to the square of increase in printing resolution. When the resolution increases by n, the deposition rate decreases by n^2. The same relationship stands for nozzle apertures with different cross sections as long as the aperture shape changes in the same proportion. In order to balance the production rate and the printing resolution, thermoplastic composites can be deposited at a high flow rate to areas that do not require a high resolution. For areas that require a high resolution, for example outer surfaces, the material can be deposited through a smaller nozzle aperture. This approach is able to maintain a good balance between a high production rate and a good printing resolution [16].

5.8.5 Residual stress and warpage

Residual stress is the stress that remains in manufactured parts, such as AM parts, when there is no externally applied force. It is normally caused by thermal cycles such as heating and cooling during manufacturing and, therefore, it is also called thermal stress or residual thermal stress. AM of thermoplastic composites uses heat to achieve adhesion between adjacent layers and the composite undergoes a rapid transition from one state to another, for example, from a solid state to a molten state and back to a solid state, within seconds, during the fusion deposition modeling process. Residual stress is generally developed when the thermoplastic composite solidifies. The residual stress can cause undesired geometric change, such as warpage. The warpage in a printed part can be so severe that it can debond from the printing bed, which normally results in the lifting of the unfinished part and, eventually, unsuccessful printing.

Residual stress exists in the printed thermoplastic composite at different levels, that is, constituent level, filament level, print layer level, and part level. The residual stress at the constituent level is mainly caused by the difference in the coefficient of thermal shrinkage between the fiber and the thermoplastic matrix. The CTE between the fiber and the matrix can be significantly different. When the composite filament is heated and extruded through the nozzle, its temperature can generally reach more than 200 °C or even 300 °C, for advanced engineering thermoplastics. When the fiber and the matrix cool down after the composite is deposited, both of them undergo thermal shrinkage. The thermoplastic matrix normally has a significantly higher coefficient of thermal shrinkage than the fiber and, therefore, a much higher volume decrease. In addition, more shrinkage can be induced by the rearrangement of polymer chains into an ordered structure (or recrystallization) in semicrystalline thermoplastics (Fig. 2.6). Due to the difference in volume reduction between the fiber and the thermoplastic matrix, residual compressive stress is developed in the fiber and residual tensile stress in the matrix.

The residual stress at the filament level is due to the intended shrinkage of a deposited filament and constraints applied by previously printed filaments that are already solidified. When a filament is extruded out of the nozzle and deposited, it makes contact with the previously printed filaments and bonds with those filaments at the contact area. When it cools down, the filament undergoes change from a molten state to a solid state, and shrinks. However, the shrinkage is constrained at the contact area by the previously printed filaments. Therefore, residual stress is developed in the filament. A higher temperature difference between the deposited filament and the previously printed filaments induces a higher residual stress.

The residual stress at the print layer level is caused by the difference in the shrinkage between adjacent layers. After the top layer is deposited and bonded with the previous layer, the top layer shrinks during cooling. Although the previous layer also undergoes thermal shrinkage due to cooling, the amount of shrinkage of the previous layer is less than that of the top layer due to the temperature difference. Therefore, residual stresses are developed in both layers. The top layer develops tension stress and the previous layer, compressive stress.

The residual stress can also be developed at the part level. There exist temperature gradients and difference in cooling rate, not only due to the thickness of the part but also from the center to the edge. The bottom layer generally has the lowest temperature and the top layer has the highest temperature. The center of the part has a higher temperature than the edge, or the outer section, of the part. The cooling rate can also vary at different sections of the part due to the difference in contact condition, air circulation, etc. Both the temperature gradient and difference in cooling rate induce residual stresses at the part level. The residual stress at the part level can be minimized by reducing the temperature gradient in the printed part through approaches such as reducing printing temperature, heating printing bed, and heating printing chambers. The same approaches can be used to reduce the residual stress at other levels as well as enhance the bonding between adjacent layers, and improve the fracture toughness of the printed part.

Continuous fiber thermoplastic composite filaments have significantly pronounced material anisotropy. It is essentially a unidirectional fiber thermoplastic composite that has all the fibers in the filament direction. The fiber direction has minimal shrinkage because of the low coefficient of thermal shrinkage of the fibers in that direction. On the other hand, there is significant shrinkage in the transverse direction. When one layer is printed using the unidirectional fiber thermoplastic composite filament, the shrinkage in the transverse direction is constrained by the printing bed (if it is the first layer) or the previous layer and residual stress is developed. The residual stress, accumulated from printing multiple layers in the same printing direction, can overcome the bonding between the part and the printing bed. It can cause debonding between the part and the printing bed, and eventually severe bending of the unidirectional fiber composite. When debonding happens, printing has to be stopped. Figure 5.19a shows a continuous Kevlar® fiber-reinforced nylon composite (fibers aligned in the width direction) debonding from the printing bed during printing (top) and the composite after being removed from the printing bed (bottom). Figure 5.19b shows another continuous Kevlar® fiber-reinforced nylon composite (fibers aligned in the length direction) debonding from the printing bed during printing (top) and the composite after being removed from the printing bed (bottom). The wall material is a short carbon fiber nylon composite (in black color) that provides a good surface finish to the printed continuous fiber composite.

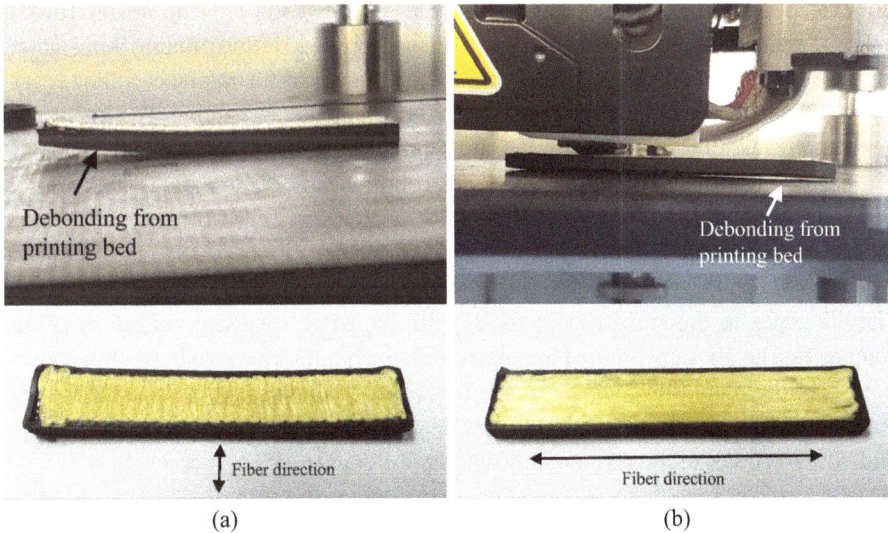

(a)   (b)

Fig. 5.19: (a) A continuous Kevlar® fiber nylon composite with all fibers in its width direction debonding from the printing bed (top) and the composite after being removed from the printing bed (bottom) and (b) a continuous Kevlar® fiber nylon composite with all fibers in its length direction debonding from the printing bed (top) and the composite after being removed from the printing bed (bottom).

When printing continuous fiber thermoplastic composites, it is necessary to reduce the shrinkage and warpage by properly designing the printing direction (the fiber direction) for each layer to avoid unsuccessful printing. Although it is not possible to eliminate the shrinkage at a specific layer, caused by the high material anisotropy, the shrinkage of the layer can be constrained by printing the next layer with an orientation perpendicular to that layer. For example, the layers can be printed in alternate 0° and 90° directions, identical to the cross-ply layup in conventionally processed thermoplastic composites (see Section 7.3.4). The shrinkage of the 0° fiber layer in the 90° direction is constrained by the 90° fiber layer, and vice versa.

Discontinuous fiber thermoplastic composites tend to have preferred fiber orientation, resulting from processing when material flow is involved. These processes include injection molding, compression molding, and extrusion. AM processes such as FDM and FPF are extrusion-based processes and deposition of a thermoplastic composite melt through the nozzle is essentially an extrusion process. Flow of the thermoplastic composite melt through the printer nozzle induces fibers oriented in the printing direction. In spite of the small fiber aspect ratios, the oriented fibers can result in certain material anisotropy in the printed composite. Fibers can also develop an orientation in powder-bed-based AM processes. The feedstock material for SLS or selective laser melting processes initially has random fiber orientation. There is limited flow involved in the selective laser melting process even for the feedstock that is heated above the melting temperature. However, the sintered thermoplastic composite part may still develop oriented fibers from the raking or rolling motion that supplies the feedstock during printing. After sintering is completed on one layer, the part is lowered down and the raking or rolling is carried out to deposit the powder on the part for laser sintering or melting. The interaction between the rake/roller and the powder can align fibers in the raking/rolling direction.

The developed fiber orientation in the discontinuous fiber thermoplastic composite can induce material anisotropy in each layer, which can cause different amount of shrinkage in different directions. The shrinkage in the printing direction is the least while the direction perpendicular to the printing direction has the highest shrinkage. The difference in the shrinkage increases with increasing material anisotropy of the composite. The abovementioned approach for reducing the warpage in printed continuous fiber thermoplastic composites can be used to minimize the warpage for the discontinuous fiber thermoplastic composite by printing the composite with alternate 0° and 90° directions, or more commonly, alternate 45° and −45°.

5.9 Summary

– AM processes for thermoplastic composites include fusion deposition modeling, fused pellet fabrication, selective laser sintering, and laminated object manufacturing. The fusion deposition modeling process is predominantly used to print thermoplastics and thermoplastic composites.

- Continuous fiber thermoplastic composites possess excellent mechanical properties but low printability. Limited consolidation pressure, short consolidation time, and high viscosity of thermoplastic composite melts lead to difficulty in printing continuous fiber thermoplastic composites.
- Discontinues fiber thermoplastic composite filaments are produced using the extrusion process. Continues fiber thermoplastic composite filaments are produced by the melt impregnation process and have fully impregnated fibers.
- Selective laser sintering is a powder-bed-based AM method. It uses laser as the energy source to sinter thermoplastic composite powders layer by layer into a part. Sintering occurs between the onset temperatures of melting and recrystallization of the thermoplastic matrix and does not involve any liquidation of the material.
- Several major processing-related challenges exist in AM. Those challenges include high void content, inadequate bonding, limited mechanical properties, low production rate, and residual stress and warpage.
- Voids in fusion deposition modeling are categorized into inner-voids and inter-voids. The inner-void originates from the porosities inside each feedstock filament, while the inter-void comes from the gaps between filaments during printing. Both types of voids, especially the inter-void, contribute to the high void content in the printed thermoplastic composite.
- There exist many interfaces due to the layered structure of AM thermoplastic composites. Inadequate bonding can result at the interface. Increasing the printing temperature, heating printing bed, heating printing chamber, locally heating previously deposited material, and applying low chamber pressure, can be used to improve bonding at the interface.
- The production rate in AM is not comparable to the rate achieved in conventional processing methods. The low production rate of AM is considered the largest hurdle for it to become a mass production process.
- There is a trade-off between printing rate and printing resolution. The printing rate is inversely proportional to the square of increase in the printing resolution.
- Continuous fiber thermoplastic composite filaments have significantly pronounced material anisotropy. It is necessary to reduce the shrinkage by properly designing the printing direction (the fiber direction) for each layer to avoid unsuccessful printing.
- Oriented fibers in the printing direction caused by the directional flow in the nozzle induce material anisotropy in discontinuous fiber thermoplastic composites. The shrinkage in the printing direction is less than in the other directions, while the direction perpendicular to the printing direction has the most shrinkage. The difference in shrinkage increases with increasing material anisotropy of the composite.
- Printing thermoplastic composites in alternate printing directions that are perpendicular to each other, for example, 0° and 90°, or 45° and −45°, can minimize the warpage of the printed part.

References

[1]   Kováčová M, Kozakovičová J, Procházka M, Janigová I, Vysopal M, Černičková I, et al. Novel hybrid PETG composites for 3D printing. Applied Sciences. 2020;10(9):3062.

[2]   Han X, Yang D, Yang C, Spintzyk S, Scheideler L, Li P, et al. Carbon fiber reinforced PEEK composites based on 3D-printing technology for orthopedic and dental applications. Journal of Clinical Medicine. 2019;8(2):240.

[3]   Yu S, Hwang YH, Hwang JY, Hong SH. Analytical study on the 3D-printed structure and mechanical properties of basalt fiber-reinforced PLA composites using X-ray microscopy. Composites Science and Technology. 2019;175:18–27.

[4]   Gupta A, Fidan I, Hasanov S, Nasirov A. Processing, mechanical characterization, and micrography of 3D-printed short carbon fiber reinforced polycarbonate polymer matrix composite material. International Journal of Advanced Manufacturing Technology. 2020; 107 (7–8): 3185–205.

[5]   Weng Z, Wang J, Senthil T, Wu L. Mechanical and thermal properties of ABS/montmorillonite nanocomposites for fused deposition modeling 3D printing. Materials & Design. 2016;102: 276–83.

[6]   Ning F, Cong W, Qiu J, Wei J, Wang S. Additive manufacturing of carbon fiber reinforced thermoplastic composites using fused deposition modeling. Composites Part B: Engineering. 2015;80:369–78.

[7]   Chacón J, Caminero M, Núñez P, García-Plaza E, García-Moreno I, Reverte J. Additive manufacturing of continuous fibre reinforced thermoplastic composites using fused deposition modelling: Effect of process parameters on mechanical properties. Composites Science and Technology. 2019;181:107688.

[8]   Justo J, Távara L, García-Guzmán L, París F. Characterization of 3D printed long fibre reinforced composites. Composite Structures. 2018;185:537–48.

[9]   Matsuzaki R, Ueda M, Namiki M, Jeong T-K, Asahara H, Horiguchi K, et al. Three-dimensional printing of continuous-fiber composites by in-nozzle impregnation. Scientific Reports. 2016;6:23058.

[10]   Caminero M, Chacón J, García-Moreno I, Rodríguez G. Impact damage resistance of 3D printed continuous fibre reinforced thermoplastic composites using fused deposition modelling. Composites Part B: Engineering. 2018;148:93–103.

[11]   Jing W, Hui C, Qiong W, Hongbo L, Zhanjun L. Surface modification of carbon fibers and the selective laser sintering of modified carbon fiber/nylon 12 composite powder. Materials & Design. 2017;116:253–60.

[12]   Yan C, Hao L, Xu L, Shi Y. Preparation, characterisation and processing of carbon fibre/ polyamide-12 composites for selective laser sintering. Composites Science and Technology. 2011;71(16):1834–41.

[13]   Frenkel J. Viscous flow of crystalline bodies under the action of surface tension. Journal of Physics. 1945;9:385.

[14]   Yan M, Tian X, Peng G, Li D, Zhang X. High temperature rheological behavior and sintering kinetics of CF/PEEK composites during selective laser sintering. Composites Science and Technology. 2018;165:140–7.

[15]   Caminero M, Chacón J, García-Moreno I, Reverte J. Interlaminar bonding performance of 3D printed continuous fibre reinforced thermoplastic composites using fused deposition modelling. Polymer Testing. 2018;68:415–23.

[16]   Chesser P, Post B, Roschli A, Carnal C, Lind R, Borish M, et al. Extrusion control for high quality printing on Big Area Additive Manufacturing (BAAM) systems. Additive Manufacturing. 2019;28:445–55.

Chapter 6
Characterization of thermoplastic composites

6.1 Introduction

Characterization of thermoplastic composites refers to probing their composition, fundamental structure, properties, as well as of their constituents using various techniques. This characterization helps engineers and scientists gain better understanding of the interrelationship among the microstructure, processing, and the property of the thermoplastic composite. It can realize a more efficient design, cost-effective processing, and improved performance of the thermoplastic composite for various applications.

Many techniques are developed for characterizing polymer matrix composite materials to evaluate their chemical, physical, and mechanical properties. These methods include general ones that can be used to characterize both the thermoset matrix composites and thermoplastic matrix composites. For example, micro-CT is used to evaluate the void content for both polymer matrix composites. However, some characterization techniques are developed only for characterizing thermoplastic composites because of the structure differences in the matrix polymer at the molecular level between the thermoplastic composite and the thermoset composite. For example, X-ray diffraction (XRD) is used to study the crystallinity of the semi-crystalline thermoplastic composite.

A variety of characterization techniques have been used to study many material variables in thermoplastic composites, including fiber content, fiber orientation distribution (FOD), fiber length distribution (FLD), crystallinity of the matrix, rheology, thermal properties, and fiber/matrix interfacial bonding. Each variable plays a significant role in determining the performance of the thermoplastic composite. It is critical to characterize the variables and achieve an in-depth understanding of not only the material behavior of the thermoplastic composite as a whole, but also the structure and property of the individual constituents of the composite.

The sections below illustrate the typical techniques for characterizing several major material variables of thermoplastic composites.

6.2 Fiber content

Fibers are one of the main constituents of a thermoplastic composite. The fiber carries a majority of the load applied to the composite. The structural performance of the thermoplastic composite increases with increasing fiber content until the matrix is not enough to impregnate the fiber anymore. Therefore, fiber content is one of the most important variables that determine the mechanical properties of the thermoplastic composite.

https://doi.org/10.1515/9781501519055-006

Fiber content is also called fiber loading, fiber percentage, or fiber fraction, and it can be described as percent by volume or percent by weight. Fiber content as percent by volume, or fiber volume fraction, of a thermoplastic composite refers to the percentage of the fiber volume in the total volume of the composite. Similarly, fiber content as percent by weight, or fiber weight fraction, of a thermoplastic composite refers to the percentage of the fiber mass with respect to the total mass of the composite. Fiber volume fraction is mainly used in design and modeling of thermoplastic composites, while fiber weight fraction is used in processing and characterization. Fiber volume fraction can be converted to fiber weight fraction when densities of the fiber and the matrix are known. Equation (6.1) shows the conversion from the fiber weight fraction to the fiber volume fraction:

$$V_f = \frac{\rho_m}{\rho_f\left(\frac{1}{W_f} - 1\right) + \rho_m} \tag{6.1}$$

where V_f is the fiber volume fraction of the thermoplastic composite, W_f is its fiber weight fraction, ρ_m is the density of the thermoplastic matrix, and ρ_f is the density of the fiber.

Fiber volume fraction can also be converted to fiber weight fraction provided the densities of the fiber and the matrix are known. Equation (6.2) converts the fiber volume fraction to the fiber weight fraction:

$$W_f = \frac{\rho_f V_f}{\rho_f V_f + \rho_m (1 - V_f)} \tag{6.2}$$

The fiber weight fraction and the volume fraction for a thermoplastic composite (or any other composites) are often not equal unless the densities of its fiber and matrix are the same. The difference can be more than 20% when there is a large difference in the densities of the fiber and the matrix of the composite; for example, glass fiber-reinforced PP composites (see Example question 7.1). Therefore, it is always beneficial to specify if the fiber content is in weight fraction or volume fraction to avoid any confusion or errors. Commonly wt% or wt.% is used to specify fiber weight fraction while vol% or vol.% is used to specify fiber volume fraction.

Several methods have been developed to measure the fiber content in thermoplastic composites and thermoplastic composite preforms. These methods include burn-off, acid digestion, carbonization, and microscopy methods. The following sections describe those methods.

6.2.1 Burn-off method

Burn-off is a typical method to remove the thermoplastic matrix at high temperatures with an adequate supply of air. This method is mainly used for thermoplastic composites, reinforced by inorganic fibers (such as glass fiber and basalt fiber) that are not affected by the high temperature. Common temperatures used for the burn-off range from 500 to 600 °C. Sometimes, the burn-off method is called ignition method because the thermoplastic matrix is burned off in air by ignition. Thermoplastic composites, reinforced with carbon fibers, can also be characterized using this method. However, the burn-off temperature for the composite has to be kept at not more than 500 °C for a limited time to protect the carbon fiber from oxidation (Fig. 6.1). It is to be noted that the heat generated from the ignition of the thermoplastic matrix can raise the temperature to greater than 500 °C, resulting in thermal degradation of the carbon fibers and an underestimation of the fiber content.

It is necessary to remove the moisture from the thermoplastic composite samples before the burn-off, especially for the composite with thermoplastic matrices (such as nylons) that tend to absorb moisture. The mass of the thermoplastic composite before the burn-off and the mass of the fibers after the burn-off are recorded, and used to calculate the fiber weight fraction of the composite as follows:

$$W_f = m_f/m_c \tag{6.3}$$

where W_f is the fiber weight fraction of the thermoplastic composite, m_f is the mass of the fiber after burn-off, and m_c is the mass of the composite before burn-off.

6.2.2 Acid digestion method

Acid digestion is an approach to remove the matrix in the thermoplastic composite by dissolution. Typical acids used for dissolving the thermoplastic matrix include sulfuric acid. The acid breaks down the polymer chains into monomers or oligomers that dissolve in the acid. Fibers are the only residue, after separating the fibers from the solution, through filtration. The fibers are then rinsed in a weak base solution multiple times to remove any residual acid before being dried for mass measurement. The same equation, Eq. (6.3), is used to calculate the fiber weight fraction.

This method is mainly used for the carbon fiber-reinforced thermoplastic composite whose fiber content cannot be characterized by the burn-off method, for example, the carbon fiber composite with advanced engineering thermoplastics. Those thermoplastics have high degradation temperatures and cannot be totally burned off before oxidation of the carbon fibers starts. It is worth noting that some high-performance semicrystalline thermoplastics, for example, PPS and PEEK, do not dissolve in the acid easily at room temperature because of their high chemical resistance. For instance, PPS does not achieve complete dissolution below 200 °C in any solvent.

6.2.3 Carbonization method

The carbonization method involves carbonizing the thermoplastic in an inert atmosphere at an elevated temperature to determine the fiber weight fraction. This approach does not need to separate the fiber from the thermoplastic matrix. It is mainly used for carbon fiber-reinforced thermoplastic composites, especially the ones with advanced engineering thermoplastics that other methods have difficulty in removing. This method also avoids the use of hazardous chemicals, which are required in the acid digestion method.

Carbon fiber can be oxidized at elevated temperatures when exposed to oxygen. Carbon dioxide is generated and the mass of the carbon fiber reduces because of oxidation. Figure 6.1 shows the mass loss of the carbon fiber at different temperatures with a hold time of 30 min. There is a slight mass loss at temperatures lower than 500 °C. However, the mass loss dramatically increases above 500 °C.

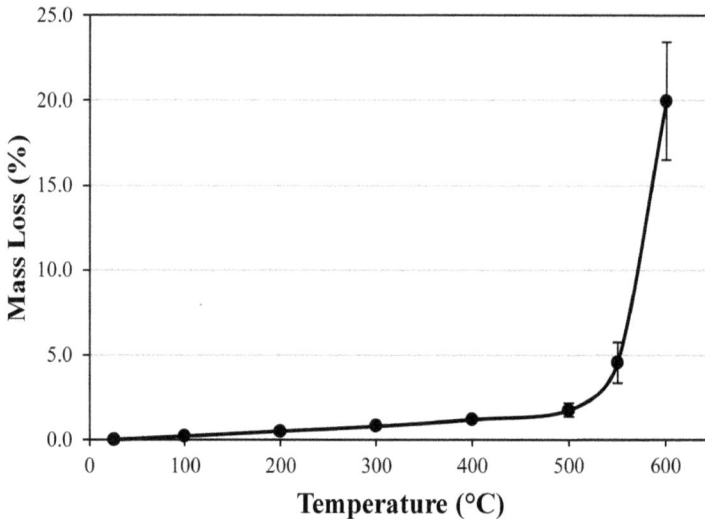

Fig. 6.1: The mass loss of carbon fibers at different temperatures after 30 min exposure in air (reprinted from Reference [1] with permission).

When there is no oxygen, carbon fibers show a minimal mass change at 600 °C, even with a hold time of one hour [2]. Based on this principle, a carbonization-in-nitrogen method is developed to determine the carbon fiber weight fraction in the thermoplastic composite. This method uses a reference neat thermoplastic as a calibration material to determine its carbonization ratio. The reference thermoplastic is the same as the matrix in the thermoplastic composite, to ensure accurate results. The reference thermoplastic and the composites are carbonized in a nitrogen-purging furnace at a carbonization temperature of up to 600 °C. Carbonization ratio (CR) is calculated in accordance with the following equation:

$$CR = \frac{m_r}{m_i} \qquad (6.4)$$

where CR is the carbonization ratio of the reference neat thermoplastic, m_r is the residue mass from the reference neat thermoplastic, and m_i is the mass of the reference neat thermoplastic before carbonization.

The mass of the thermoplastic matrix in the composite is calculated according to the following equation:

$$M_m = \frac{(M_i - M_r)}{(1 - CR)} \qquad (6.5)$$

where M_m is the mass of the matrix in the composite, M_i is the initial mass of the composite, and M_r is the residue mass of the composite.

The carbon fiber weight fraction of the thermoplastic composite is derived based on the following equations:

$$W_f = \frac{(M_i - M_m)}{M_i} \qquad (6.6)$$

or,

$$W_f = \frac{M_r - M_i \times CR}{M_i \times (1 - CR)} \qquad (6.7)$$

where W_f is the weight fraction of the carbon fiber. The weight fraction can be calculated when the initial mass of the composite, the residue mass of the composite, and the carbonization ratio of the reference neat thermoplastic are known.

6.2.4 Microscopy method

Another approach for characterizing the fiber content is by microscopy and image analysis. This approach can be used to directly measure the fiber volume fraction of a thermoplastic composite, typically a unidirectional fiber-reinforced thermoplastic composite. A cross section of the composite, perpendicular to the direction of fiber length, is polished and examined under an optical microscope. The cross-sectional areas of all the fibers and the composite are measured and the fiber volume fraction is calculated using the equation below. Image analysis software is normally used to assist in measuring the fiber cross-sectional area and the overall composite area:

$$V_f = \frac{A_f}{A_c}$$

where V_f is the fiber volume fraction, A_f is the cross-sectional area of the fibers, and A_c is the cross-sectional area of the composite. Besides the fiber volume fraction, the other information in the unidirectional fiber-reinforced thermoplastic composite, such as fiber impregnation, uniformity of fiber distribution, and void content, can also be obtained from the microscopy method.

6.3 Fiber length distribution

FLD is one of the most important variables that determine the mechanical properties of discontinuous fiber-reinforced thermoplastic composites. The FLD in a molded composite can be affected by several factors, including original fiber length, fiber content, processing conditions (pressures and temperatures), gate design, and screw design. As Fig. 3.3 indicates, the fiber length considerably influences the modulus, strength, and impact resistance of the discontinuous fiber-reinforced thermoplastic composite. Therefore, it is essential to characterize the FLD. Different methods are used to characterize the FLD and these methods include a combination of burn-off and microscopy method, and micro-CT method.

6.3.1 Burn-off and microscopy method

The combination of burn-off and microcopy techniques can be effectively used to investigate the FLD for glass or basalt fiber-reinforced thermoplastic composites. This approach starts with the burn-off method to remove the thermoplastic matrix and, consequently, uses microscopes to measure the lengths of the fibers collected from the burn-off residue. It is necessary to measure an adequately large number of fiber samples for accurate evaluation.

Figure 6.2a shows a micrograph of glass fibers after burn-off of a compression-molded long glass fiber PP composite part (Fig. 4.38). The LFT pellets used for molding the part have a length of 25 mm. The PP matrix in the composite is removed through burn-off for fiber length measurement under a microscope. The histogram in Fig. 6.2b illustrates the FLD of the composite from several hundreds of measurements. An average fiber length can also be obtained from this method.

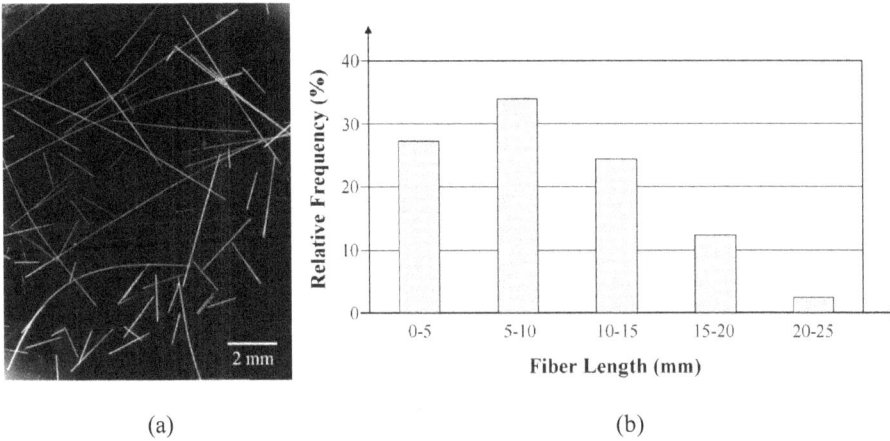

Fig. 6.2: (a) A microscopic image of glass fibers from burn-off of a compression molded long glass fiber PP composite for fiber length measurement; (b) fiber length distribution of the composite (reprinted from Reference [3] with permission).

6.3.2 Micro-CT method

Micro-CT, sometimes referred to as X-ray microtomography, high-resolution X-ray tomography, or micro-computed tomography, is a nondestructive technique that uses X-ray to create the three-dimensional geometry of an object. The object is scanned by X-ray and slice images are collected that are composed of voxels or volume elements. The slice images are then stacked to construct the three dimensional image of the object through tomographic reconstruction. This technique has been mainly used for medical uses. However, micro-CT has become a versatile technique that can be used for characterizing material variables in thermoplastic composites after its resolution is highly improved, which allows for imaging of micro-size fibers or voids in thermoplastic composites. It can characterize several material variables, including FLD, FOD, and void content.

X-ray is able to penetrate the thermoplastic composite and create a contrast among its constituents. The attenuation coefficient for a specific point with a coordinate (x, y, z), or $\mu(x, y, z)$, in the composite is defined by the following equation:

$$\mu(x, y, z) = K\rho \frac{Z^4}{E^3} \tag{6.8}$$

where K is a constant, ρ is the density of the material at a certain point, Z is the atomic number of the material at a certain point, and E is the energy of the incident photons.

According to the Beer–Lambert law, the intensity of synchrotron X-ray decreases with attenuation coefficient μ exponentially as indicated in Eq. (6.9).

$$I = I_0 e^{-\mu z} \tag{6.9}$$

where I_0 and I are the intensities of initial and final X-ray, μ is the linear attenuation coefficient of the material, and z is the length of the X-ray path.

For a thermoplastic composite material with multiple constituents, the intensity decrease can be expressed in Eq. 6.10 for synchrotron X-rays that are monochromatic:

$$I = I_0 \exp\left\{ \sum_{i=1}^{n} (-\mu_i z_i) \right\} \tag{6.10}$$

For polychromatic X-ray, its attenuation coefficient is dependent on X-ray energy, G, and the intensity is an integration over the range of the X-ray energy spectrum:

$$I = \int I_0 \exp\left\{ \sum_{i=1}^{n} [-\mu_i(E)z_i] \right\} dG \tag{6.11}$$

A material with a lower density and a lower atomic number exhibits lower X-ray attenuation compared to a material with a higher density and a higher atomic number. Materials with a larger difference in density and atomic number can be readily differentiated because of a larger intensity difference. Glass fiber-reinforced thermoplastic composites have a better contrast between the fiber and the matrix. On the other hand, micro-CT images of carbon fiber-reinforced thermoplastic composites do not have a distinct contrast between the fiber and the matrix. Synchrotron X-ray beam that is parallel, monochromatic, and more coherent is better suited for low contrast materials such as carbon fiber-reinforced thermoplastic composites. Polychromatic and divergent X-ray beams are suited for the glass fiber thermoplastic composite with a higher contrast between the fiber and the matrix.

The micro-CT method is mainly used to study discontinuous fiber thermoplastic composites. The discontinuous fiber thermoplastic composite has fibers with either random or preferred orientation from processing. In addition, different processing conditions can result in a wide range of fiber lengths in the composite. The micro-CT method provides an effective approach to quantitatively evaluate the FLD. The micro-CT image of a glass fiber-reinforced polypropylene composite with 20 wt% fiber content is shown in Fig. 6.3 [4]. Fibers are assigned random colors for better distinction.

6.4 Fiber orientation distribution

FOD refers to the distribution of the fiber angels in mainly discontinuous fiber-reinforced thermoplastic composites. The distribution of fiber orientation in the composite is governed by the same aforementioned factors, including original fiber length, fiber content, processing conditions (pressure and temperature), and

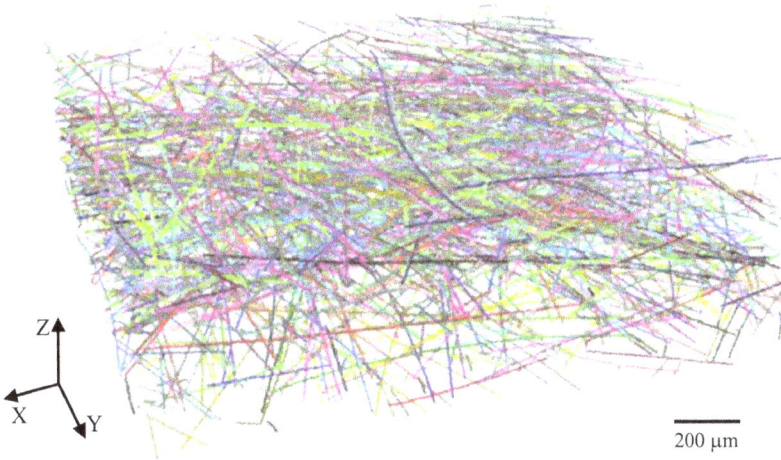

Fig. 6.3: Micro-CT image of a 20 wt% long glass fiber-reinforced polypropylene composite with a dimension of 1.8 × 1.8 × 0.6 mm (adapted from Reference [4]).

gate design (in injection molding), charge placement (in compression molding), and so on. The FOD can considerably affect the material properties of the composite. Below are descriptions of the methods used to characterize the FOD.

6.4.1 Micro-CT method

The micro-CT method is capable of quantitatively characterizing not only the FLD but also the FOD of thermoplastic composites. The method is detailed in Section 6.3.2.

6.4.2 Microscopy method

The microscopy method is another way to characterize the FOD. Thermoplastic composites are polished and evaluated under microscopes, such as optical microscope, confocal microscope, stereoscope, and SEM. Each image provides only a two-dimensional view of the fiber orientation. Polishing a series of sections from the same composite is required to provide a 3D construction of the fiber orientation. This is the same principle used in characterizing voids in thermoplastic composites (see Section 6.5.2). In addition, fracture surface of the thermoplastic composite can also be characterized on its FOD; however, the microscopy method is limited to only qualitative characterization.

6.5 Void content

Voids in composites are the empty spaces that are unoccupied by either the matrix
or fibers. A void is considered a major defect in composites, including thermoplastic
composites, as it acts as a stress concentration site and promotes crack initiation
and propagation. Therefore, voids are detrimental to material properties, especially
strength. The mechanical properties of the composite generally decrease with increasing void content. Void content is defined as the volume fraction of the total
voids in a composite. Because there is no mass or weight associated with the void,
void content only refers to the volume fraction of the void.

There are several methods available for characterizing the void content of thermoplastic composites. These methods include density method, microscopy method, and
micro-CT method. The most direct method is the density method that compares the
theoretical density of the composite without any voids and the actual density of the
composite, measured by the Archimedes method (or the water immersion method). A
void does not contribute to the mass of the composite but occupies space in the composite, which results in a density decrease in the composite. The density decreases
with increasing void content. In addition, microscopy and micro-CT methods are used
to characterize the void content. The microscopy method allows for direct observation
of voids in the composite, and measurement of the void content. The large density difference between the void and other constituents, that is, the fiber and the matrix, enables the micro-CT method to detect voids in the composite nondestructively. All the
characterization methods are summarized in Tab. 6.1 and are described in detail in the
following sections.

Tab. 6.1: Main methods for characterizing void content of thermoplastic composites.

Method	Information	Advantages	Disadvantages
Density	– Void content – Composite density	– Simple – Inexpensive – Quick	– Destructive – Densities of fiber and matrix required – Accuracy is dependent on input properties – No other void info provided
Microscopy	– Void content – Void location – Void size (2D) – Void shape (2D)	– Relatively simple – Relatively inexpensive – Relatively quick – Provides void info	– Destructive – Section- and location-dependent – Multiple analysis required

Tab. 6.1 (continued)

Method	Information	Advantages	Disadvantages
Micro-CT	– Void content – Void location – Void size (3D) – Void shape (3D)	– Highly accurate – Provides detailed void info	– Expensive – Image analysis required – Section- and location-dependent – Multiple analysis required

6.5.1 Density method

Voids in a thermoplastic composite contribute to its volume but not mass. Therefore, the density of the thermoplastic composite decreases with increase in void content. Density method is developed based on this principle. When the density of each constituent and the fiber weight fraction are known, the theoretical density of the composite free of voids can be calculated. Furthermore, the actual density of the composite can be measured by the Archimedes method. By comparing the difference between the theoretical density and the actual density, the void content in the composite can be determined. The derivation of the equation used to characterize the void content in a thermoplastic composite is described below. It is assumed that a thermoplastic composite is free of voids. The composite volume is the sum of the fiber volume and the matrix volume:

$$\frac{m_c}{\rho_{ct}} = \frac{m_f}{\rho_f} + \frac{m_m}{\rho_m}$$

where m_c, m_f, and m_m is the mass of the composite, the fiber, and the matrix, respectively; ρ_{ct}, ρ_f, and ρ_m is the density of the composite, the fiber, and the matrix, respectively.

The fiber weight fraction, W_f, and matrix weight fraction, W_m, can be measured using the burn-off, acid digestion, or carbonization methods defined in Section 6.2. Since $W_f = m_f/m_c$ and $W_m = m_m/m_c$,

$$\rho_{ct} = \frac{1}{\left(W_f/\rho_f + W_m/\rho_m \right)} \tag{6.12}$$

The actual density of the composite, ρ_{ca}, can be measured using the Archimedes method. It is always lower than the theoretical density because of the presence of voids in the composite. The void content, V_{void}, can be calculated using Eq. (6.13), based on the difference between the theoretical and actual densities:

$$V_{void} = \frac{(\rho_{ct} - \rho_{ca})}{\rho_{ct}} \tag{6.13}$$

The void content of the composite can also be calculated by subtracting the volumes of both the fiber and the matrix from the total volume of the composite, using the following equation:

$$V_{void} = 1 - (V_f + V_m) \tag{6.14}$$

where V_{void}, V_f, and V_m are the volume fraction of the void, the fiber, and the matrix, respectively. The volume fractions of both the reinforcement and the matrix of the thermoplastic composite can be determined using Eqs. (6.15) and (6.16), respectively:

$$V_f = \frac{\text{fiber volume}}{\text{composite volume}} = \frac{m_f/\rho_f}{m_c/\rho_{ca}} \tag{6.15}$$

$$V_m = \frac{\text{matrix volume}}{\text{composite volume}} = \frac{m_m/\rho_m}{m_c/\rho_{ca}} \tag{6.16}$$

Equation (6.14) can be written as follows:

$$V_{void} = 1 - \rho_{ca} \left(\frac{M_f}{\rho_f} + \frac{M_m}{\rho_m} \right) \tag{6.17}$$

where M_f and M_m are the mass fractions of the fiber and the matrix, respectively.

Equations (6.13) and (6.17) represent different approaches to calculate the void content of thermoplastic composites. However, Eqs. (6.13) and (6.17) can be converted to each other after derivation.

6.5.2 Microscopy method

Microscopy method is a versatile method that can be used to characterize several material variables of thermoplastic composites, including void content. After the thermoplastic composite is polished and microscopic images are taken, image analysis is carried out to obtain the void content and other information such as void size, void shape, and void location. Image analysis is generally carried out with the aid of software such as Image-Pro, ImageJ, and so on.

Figure 6.4 shows the optical microscopic image of a carbon fiber-reinforced PEEK composite, before and after image analysis. Figure 6.4a (before image processing) shows obvious voids (in dark color). Figure 6.4b shows the same image after a color threshold is applied to filter the image and highlight the porosity regions in red for void content measurement [5].

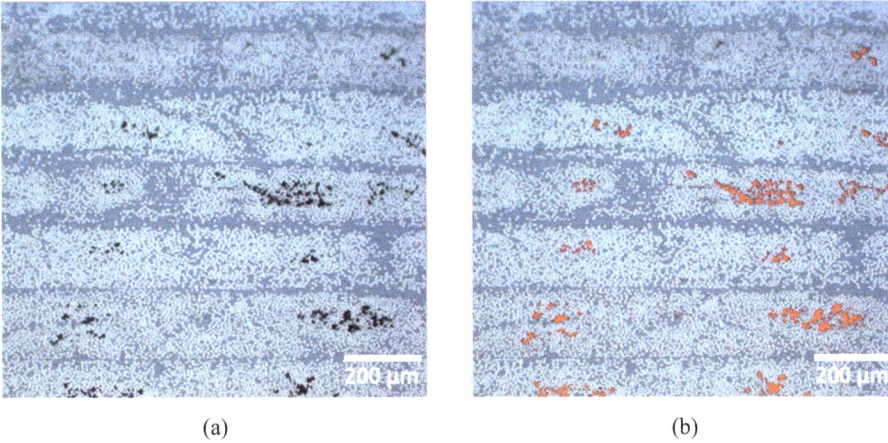

(a)   (b)

Fig. 6.4: (a) A microscopic image of a carbon fiber reinforced polyether ether ketone thermoplastic composite showing the voids in dark color and (b) image analysis carried out to measure the void content (reprinted from Reference [5] with permission).

The microscopy method has several limitations in characterizing the void content though. Firstly, the location where the sample is prepared from a thermoplastic composite plays an important role in the void content results. Because of possible variation in preforms and molding conditions (temperature and pressure) in the composite, voids may not be uniformly distributed within the composite. Multiple samples from different locations may be required to achieve an accurate mapping of the void content, size, and shape. Secondly, the sample needs to be polished to a sufficiently good finish. An inadequate finish considerably hampers image analysis and the consequent void content measurement. Thirdly, the method provides only a two-dimensional view of the void from one polish. Polishing a series of sections from the same composite can provide a 3D construction of the voids; however, it is time-consuming and tedious. Finally, each image only covers a small area and multiple images are required to get statistically significant void information.

6.5.3 Micro-CT method

The micro-CT technique is another method that can characterize the void content, size, shape, and location in thermoplastic composites. Voids in the thermoplastic composite have no density, which results in minimal X-ray attenuation and a great contrast with other constituents, such as the reinforcement and the matrix. The method has been described in Section 6.3.2. Three-dimensional images of the composite obtained from the micro-CT method provide a large amount of information on the voids and other constituents.

Case study 6.1: Characterization of 3D printed thermoplastic composites using micro-CT

Micro-CT has been used to characterize 3D printed thermoplastic composites. The technique is able to identify the constituents in the composite because of their density difference. Figure 6.5a shows the micro-CT image of a basalt fiber-reinforced PLA composite, printed through fusion deposition modeling. Different constituents such as fiber, matrix, and void are indicated with different colors in the image. Each constituent is also isolated from the composite and indicated in individual images (Fig. 6.5b–d). Figure 6.5d shows inter-voids in the printed composite and their orientation in the filament deposition direction. The volume percentage of each constituent can be calculated based on the three-dimensional images. This case study shows that micro-CT is a powerful and versatile tool in characterizing the constituents in thermoplastic composites, including the void, and can help us gain understanding of the fundamental interrelationship among the processing, structure, and properties of the thermoplastic composite.

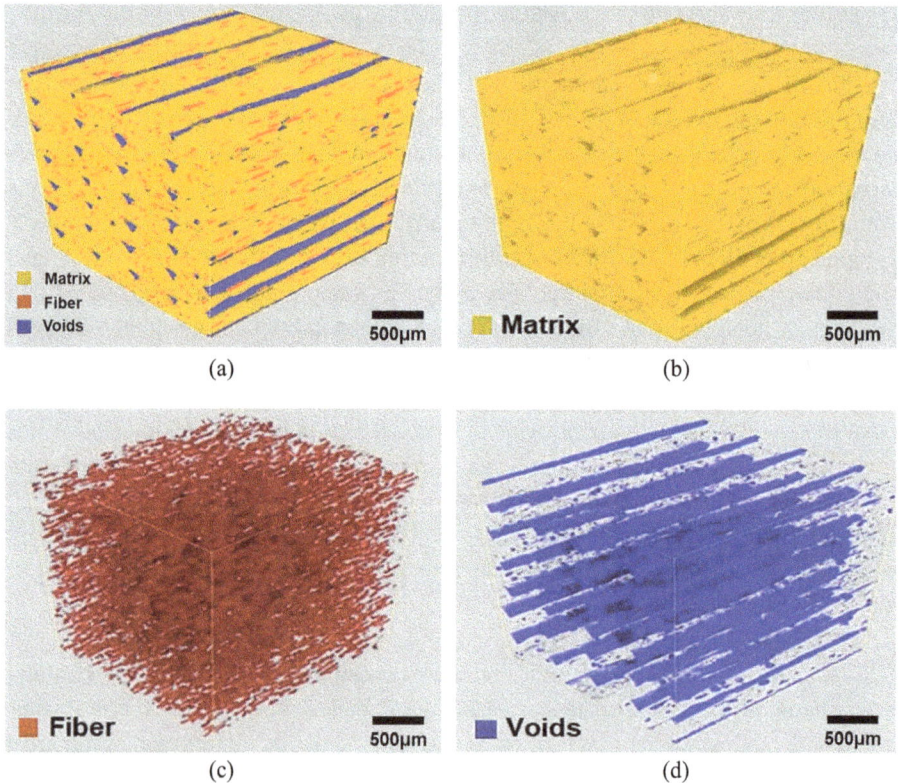

Fig. 6.5: (a) Micro-CT image of a 3D printed basalt fiber-reinforced PLA composite; (b) image of the PLA matrix only; (c) image of the basalt fibers only; and (d) image of the voids only (reprinted from Reference [6] with permission).

6.6 Thermal properties

Thermal property of a thermoplastic composite refers to its response to the application of heat. Because the thermoplastic matrix is the constituent most affected by heat, the thermal property of the composite is dominantly determined by the thermal properties of the thermoplastic matrix along with the contribution from the fiber. For example, the thermal properties of the thermoplastic matrix in a composite determine several attributes of the composite, including processing temperature, service temperature, and so on, while the fibers can play a role in affecting those attributes.

Characterization of the thermal property of thermoplastic composites is achieved by heating the composites and studying their thermal behavior or thermomechanical behavior. Typical thermal analysis methods include differential scanning calorimetry (DSC), dynamic mechanical analysis (DMA), and thermomechanical analysis (TMA). The sections below describe typical thermal properties of thermoplastic composites and the methods used to characterize those properties.

6.6.1 Glass transition temperature

Glass transition temperature is one important thermal property in thermoplastics and thermoplastic composites. It is defined as the temperature at which the thermoplastic polymer converts from its glassy state to rubber state, or vice versa (see Section 2.2.3). It is an indication of the thermal stability of the thermoplastic polymer. A thermoplastic polymer with a higher T_g possesses better high-temperature performance. However, T_g should not be considered as the service temperature of the thermoplastic composite, as the reinforcement that is added has a significant effect in improving its service temperature. This effect will be described in Section 6.6.2.

Typical tools for characterizing the glass transition temperature of thermoplastics or thermoplastic composites are DSC, TMA, and DMA.

1. **DSC** is a versatile tool capable of characterizing several important thermal properties of the thermoplastic matrix (or thermoplastic polymer fibers) in thermoplastic composites, including glass transition temperature (T_g), crystallization temperature (including both cold crystallization and recrystallization temperatures), melting temperature (for semicrystalline thermoplastics), and enthalpy (see Section 6.6). DSC requires only a small amount of the thermoplastic composite (in milligram scale) for analysis. The composite is sealed inside a set of aluminum crucibles. A reference sample, normally an empty aluminum crucible, is heated at the same time with the thermoplastic composite sample. The reference material has a known heat capacity. Heat capacity is defined as the amount of heat required to raise the temperature of 1 g of the material by 1 °C. The reference material does not undergo any physical or chemical change throughout

the analysis. Both the composite and the reference samples are maintained at the same temperature during heating or cooling, and the difference in the amount of heat that is required to increase the temperature is measured. The difference in the amount of heat is the heat required to raise the temperature of the thermoplastic composite sample.

The heat flow remains constant before the thermoplastic matrix reaches glass transition temperature. When the glass transition temperature is attained, its heat capacity increases and more heat is required to maintain the same temperature of the composite sample. The change in the heat flux is shown as a change in the slope of DSC and indicates the glass transition temperature. The change in heat capacity is shown as the heat flow. It is a function of temperature and indicates any change of the thermoplastic composite. It is to be noted that the glass transition temperature may not be noticeable from the DSC curve for some thermoplastic composites, especially those with high fiber contents and high DOC. The change in heat capacity in those composites is relatively small compared to the overall heat capacity of the thermoplastic composite, and the change in the heat flow required to maintain the same temperature is low.

The same principle is used to characterize the melting temperature (when crystalline regions become disordered and heat is released) and crystallization temperature (when disordered molecular chains form crystalline regions) of a thermoplastic composite. Heat flow, which is equal to the enthalpy change in the composite because of constant pressure used during analysis, is involved in the composite during those transitions. The peak for each transition can be used to calculate the enthalpy change during the transition by integrating the area under the peak. The enthalpy change is used to calculate the degree of crystallinity of the thermoplastic matrix (see Section 6.7.1).

2. **TMA** is a method used for measuring the change in the dimension of the thermoplastic composite with temperature through a probe that is placed on the composite surface. When the thermoplastic composite is in a glassy state, the dimension change caused from thermal expansion or thermal shrinkage moves the probe. The dimension change with temperature, or the coefficient of linear thermal expansion (CTE) of the material, is constant when there is no transition in the material. When the thermoplastic matrix is heated to its glass transition temperature, a second-order transition occurs and its CTE increases (Fig. 2.6). The transition is indicated by the slope change in a TMA curve, which can be used to determine the glass transition temperature of the thermoplastic matrix. A penetration method can also be used to determine the glass transition temperature. When an adequately high load is applied on the composite sample surface by the probe, the probe can penetrate the surface of the composite when the thermoplastic matrix reaches T_g and softens. The sudden change of the dimension in a TMA curve because of the penetration can be used to determine the glass transition temperature of the thermoplastic matrix.

3. **DMA** is another common method used to determine the glass transition temperature. It is considered the most sensitive method in characterizing the glass transition in thermoplastics and thermoplastic composites. DMA is a thermal analysis method for studying the viscoelastic behavior of materials, including thermoplastics and thermoplastic composites. Thermoplastics have long molecular chains and possess viscoelastic properties. Therefore, their composites are viscoelastic materials that exhibit both elastic and viscous characteristics when stress is applied. A perfectly elastic solid responds immediately when a stress is applied and instantaneously returns to its original state when the stress is removed. On the other hand, a perfectly viscous fluid has a response to stress but its strain lags stress by 90°. Because thermoplastics or their composites are viscoelastic, they exhibit a behavior between a perfectly elastic solid and a purely viscous liquid, exhibiting a certain lag in strain with respect to an applied stress.

   DMA applies an oscillatory stress that induces a response in a thermoplastic composite. Its storage modulus decreases when the temperature of the thermoplastic matrix in the composite reaches glass transition. The onset of the storage modulus drop is normally considered as the glass transition temperature. However, it should be noted that the peak in tan δ and the peak in loss modulus are also used to define the glass transition temperature.

6.6.2 Heat deflection temperature

Heat deflection temperature (HDT) of a thermoplastic composite is defined as the temperature at which the composite undergoes a specified deflection at a specified load. It is sometimes referred to as heat distortion temperature or deflection temperature under load. HDT is often used as an indication of short-term heat resistance. It considerably differs from glass transition temperature for thermoplastic composites, although it is closely related to the glass transition temperature for neat thermoplastics. Reinforcements added to the thermoplastic significantly increase its HDT and the HDT increases with the reinforcement amount to a certain extent. On the other hand, the glass transition temperature of a thermoplastic remains the same after reinforcements are added because the glass transition temperature describes the transition of only the thermoplastic portion from a glassy state to a rubbery state.

HDT is commonly measured through flexural testing. A thermoplastic composite sample with a specific dimension is supported at both ends and loaded at its center. The whole flexural setup is placed in an oil bath in which the oil temperature is controlled at a low heating rate, for example, 2°C/min. When the deflection reaches a certain value, for instance, 0.25 mm, at a certain temperature, that temperature is defined as HDT. Figure 6.6 shows the setup for measuring HDT. HDTs for common thermoplastics and their composites with different fiber contents are compared in Tab. 6.2. All the thermoplastics and their composites listed in the table are injection molded. The temperatures are obtained from two different loads that result in a flexural stress of 1.82 and 0.455 MPa, respectively. HDTs in the table show several general trends when fibers are added to thermoplastics: (1) a small amount of fibers added to thermoplastics can tremendously improve their HDTs; (2) HDT increases with increasing fiber content to a certain extent; (3) carbon fibers can improve HDT more effectively than glass fibers.

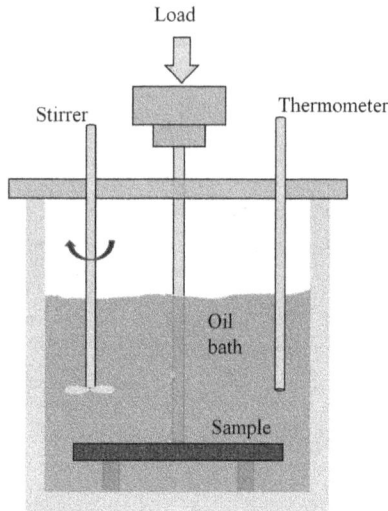

Fig. 6.6: A setup for characterizing the heat deflection temperature of thermoplastic composites.

Tab. 6.2: Heat deflection temperature of neat thermoplastics and their composites with different fiber contents (adapted from RTP product datasheets).

Thermoplastics or thermoplastic composites	Heat deflection temperature (°C)	
	1.82 MPa	0.455 MPa
Neat PP	54	107
10% glass/PP	113	149
20% glass/PP	129	152
30% glass/PP	141	157
40% glass/PP	141	157
Neat nylon 6	71	171
10% glass/nylon 6	182	199
20% glass/nylon 6	188	204

Tab. 6.2 (continued)

Thermoplastics or thermoplastic composites	Heat deflection temperature (°C)	
	1.82 MPa	0.455 MPa
30% glass/nylon 6	199	216
40% glass/nylon 6	215	–
Neat nylon 66	79	
10% glass/nylon 66	204	216
20% glass/nylon 66	227	246
30% glass/nylon 66	232	249
40% glass/nylon 66	232	249
10% carbon/nylon 66	232	238
20% carbon/nylon 66	238	–
30% carbon/nylon 66	249	–
40% carbon/nylon 66	252	–

6.6.3 Melting temperature

The crystalline regions in a semicrystalline thermoplastic contribute to its melting characteristic. The molecular chains in the crystalline regions transition from an ordered structure to a disordered structure during melting. Melting is an endothermal transition and heat is absorbed by the thermoplastic while it changes from a solid to liquid state. Since melting of the thermoplastic matrix is involved in processing, such as molding and additive manufacturing, of any semicrystalline thermoplastic composites, the melting temperature is often used as an important reference to determine the processing temperature.

DSC and DMA are also used to characterize the melting temperature of the thermoplastic composite. Those methods have been described in Section 6.6.1.

6.6.4 Crystallization temperature

The molecular chains in the thermoplastic matrix can undergo a transition from a disordered state to an ordered state during cooling or even heating. This transition is called crystallization and the temperature at which crystallization happens is called crystallization temperature. Crystallization in the thermoplastics or thermoplastic composites includes recrystallization, cold crystallization, and pre-melt crystallization.

When a thermoplastic composite is cooled from melting when all the molecular chains are not ordered, some of the chains will rearrange themselves and form crystalline regions. This transition is called recrystallization. This phenomenon can be detected using DSC by running a heat and cool cycle. The composite is first raised above the melting point of the thermoplastic matrix to remove crystallites. The thermoplastic composite melt is then cooled down at an adequately low rate to allow the transition of some molecular chains from a disordered structure to an ordered structure (Fig. 2.27). The recrystallization temperature can be determined from the recrystallization peak in the DSC curve.

Cold crystallization, on the other hand, can occur during heating of a thermoplastic composite if recrystallization of the matrix was previously suppressed by fast cooling in the earlier thermal cycle. Polymer chains in certain amorphous regions can rearrange themselves and form crystalline regions during heating. The transition is indicated by the exothermal peak on the DSC curve. Figure 6.7 shows the DSC curve of a carbon fiber-reinforced PEEK composite that was previously cooled at a very high rate. The onset temperature of cold crystallization is approximately 176 °C and the specific enthalpy from the cold crystallization is about 2.0 J/g [7].

The recrystallization, cold crystallization, and pre-melt crystallization release heat and show as exothermal peaks in the DSC curves. Although the released heat, especially from the cold crystallization and pre-melt crystallization, can be minimal, it can be detected and characterized using DSC. The heat released from the transition can be calculated by integration of the area under the exothermal peak.

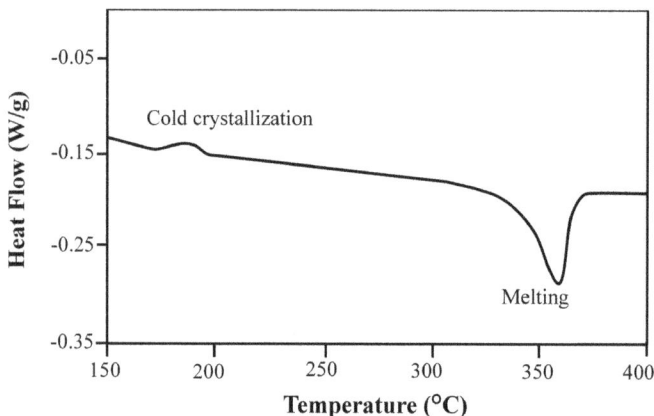

Fig. 6.7: A carbon fiber-reinforced PEEK composite showing a cold crystallization peak during heating (adapted from Reference [7]).

6.6.5 Thermal degradation temperature

As an organic material, thermoplastics can get degraded when they are heated to a temperature that is significantly higher than their melting temperature (for semi-crystalline thermoplastic matrix) or glass transition temperature (for amorphous thermoplastic matrix). During degradation, chain scissoring, oxidation, and generation of volatile products occur. The temperature at which these phenomena start to happen is called thermal degradation temperature, degradation temperature, or onset degradation temperature. Degradation of the thermoplastic composite normally happens during its processing when it is over-heated. The mechanical properties of the thermoplastic composite deteriorate due to the degradation. To avoid degradation, the processing temperature of the thermoplastic composite should stay below its thermal degradation temperature.

Thermal degradation temperature is commonly characterized by thermogravimetric analysis (TGA). TGA is a thermal analysis method for measuring the mass change of a material, including thermoplastic composites, as a function of temperature. When the thermoplastic composite is heated to its thermal degradation temperature, its mass decreases because volatiles are generated. TGA is able to detect the mass reduction with temperature and determine the thermal degradation temperature (Fig. 2.19).

A TGA apparatus typically consists of a sample holder, a microbalance, a chamber with atmospheric control, and a furnace. The microbalance measures the initial mass of the sample placed in the sample holder and continues to keep track of its mass change throughout the analysis. The analysis can be carried out in different atmospheres. Common gases used for the analysis are air and nitrogen. Nitrogen is commonly purged into the chamber for protecting the thermoplastic composite from oxidation. The mass of the composite decreases when the temperature reaches its degradation temperature, and decomposition of the thermoplastic matrix and generation of volatile products happen. The composite is often heated at a constant rate and its mass change with temperature is recorded and plotted. Gas chromatography-mass spectrometer can be connected to the TGA to evaluate the volatile products released from the composite.

6.7 Crystallinity of matrix

The semicrystalline thermoplastic matrix in a thermoplastic composite can have different degrees of crystallinity because of the varying thermal history during processing. Thermal history, such as cooling rate, considerably affects the nucleation rate and re-crystallization in the thermoplastic. The degree of crystallinity of the thermoplastic matrix can affect the mechanical properties, especially matrix-dominant properties, of its composites. The crystalline phase, or crystallite, in semicrystalline thermoplastic composites also helps improve their HDT and service temperature. In addition, the crystallite improves the chemical resistance of the thermoplastic composite.

Processing conditions affect the degree of crystallinity of the thermoplastic in the thermoplastic composite. During molding of a semicrystalline thermoplastic matrix composite, the thermoplastic matrix is firstly melted and the composite is molded, followed by a cooling process. If the cooling rate is adequately low, the thermoplastic undergoes a recrystallization process in which crystallites form and grow in the thermoplastic. Nucleation of the crystallites can occur in the thermoplastic and on the fiber surface. The crystallites can grow until they are blocked by neighboring crystallites (Fig. 6.8), or immobilized from the drop of temperature to below the recrystallization temperature.

Fibers can also affect the degree of crystallization of the thermoplastic matrix in thermoplastic composites. Reinforcements, including carbon fiber, basalt fiber, and nanographene, can act as nucleation sites (Fig. 6.8), which considerably promote nucleation of crystallites in the composite. However, further increase in the amount of the reinforcements beyond a certain limit can impede the crystallite formation, by hindering the migration and diffusion of the thermoplastic polymer chains.

Several methods can be used to characterize the crystallinity of the thermoplastic matrix and these methods are described below.

6.7.1 DSC method

As mentioned in Section 2.2.1, DSC can be used to determine the DOC of neat thermoplastics by measuring the enthalpy change (or heat flow) resulting from the order-to-disorder transition of its crystalline regions. Equation (2.1) is used to calculate the DOC based on the specific enthalpy measured from the endothermic peak during melting (see Section 2.2.1). The same method can be adopted to measure the DOC of the matrix in a thermoplastic composite. A thermoplastic composite is heated in DSC at a constant rate until the melting is completed. At the peak of the melting point, the thermoplastic shows endothermic heat flow and the enthalpy can be obtained by integrating the area under the peak. It is assumed that the fiber in the composite does not contribute to any enthalpy change. This is the case for inorganic fibers such as carbon fibers, glass fibers, and basalt fibers. The specific enthalpy of only the thermoplastic matrix is then calculated by dividing the specific enthalpy obtained from the melting peak in the DSC curve with the fiber weight fraction, $\Delta H_s/(1-W_f)$. ΔH_s is the specific enthalpy measured from DSC with respect to the total mass of the thermoplastic composite, W_f is the fiber weight fraction of the thermoplastic composite, $(1-W_f)$ is the matrix weight fraction of the thermoplastic composite.

The equation for calculating the DOC of the matrix in thermoplastic composites can be written as follows:

$$\text{DOC} = \frac{\Delta H_s}{(1 - W_f)\Delta H_f} \times 100\% \tag{6.18}$$

where ΔH_f is the specific enthalpy of the thermoplastic with an assumed 100% DOC. When there is no fiber or when $W_f = 0$, Eq. (6.18) becomes Eq. (2.1).

A modification to Eq. (6.18) is required when there is cold crystallization and pre-melt crystallization, if any, in the thermoplastic matrix. The cold crystallization and pre-melt crystallization normally occur in the thermoplastic matrix, whose re-crystallization was previously suppressed by fast cooling. The cold crystallization is caused by the crystallization of the segments of the neighboring chains in the amorphous regions when the composite is heated above the glass transition temperature but far below the melting temperature. Cold crystallization shows as an exothermal peak in the DSC curves (Fig. 2.25), which can be used to calculate the specific enthalpy of the cold crystallization (ΔH_c) in the composite, that is, enthalpy change (heat flow) divided by the total mass of the composite. Because the fiber in the composite does not contribute to ΔH_c, the specific enthalpy of cold crystallization of only the thermoplastic matrix is $\Delta H_c/(1 - W_f)$. In addition, a pre-melt crystallization can happen near but below the onset melting temperature. The process is caused by the rearrangement of the imperfect part of crystallites. The pre-melt crystallization peak can be used to calculate the specific enthalpy of the pre-melt crystallization or ΔH_p in the composite. The specific enthalpy of pre-melt crystallization of only the thermoplastic matrix is $\Delta H_p/(1 - W_f)$. Therefore, Eq. (6.18) is modified to Eq. (6.19) by considering the specific enthalpy of the thermoplastic matrix only from cold crystallization, $\Delta H_c/(1 - W_f)$, and the pre-melt crystallization, $\Delta H_p/(1 - W_f)$. ΔH_s has the opposite sign to ΔH_c or ΔH_p:

$$\text{DOC} = \frac{\Delta H_s + \Delta H_c + \Delta H_p}{(1 - W_f)\Delta H_f} \times 100\% \tag{6.19}$$

According to Eq. (6.19), when the specific enthalpy from the melting peak is equal to the sum of the specific enthalpy of both the cold crystallization and the pre-melt crystallization, the DOC of the thermoplastic matrix is zero and the thermoplastic is considered amorphous.

Table 6.3 lists the specific enthalpy of common neat thermoplastics with an assumed 100% degree of crystallinity. It is to be noted that some of the thermoplastics have a wide range of ΔH_f values due to several possible reasons, including small sample sets used, inconsistent crystallinity in calibration samples, and single point measurement. Different methods used in the characterization of the specific enthalpy include wide-angle X-ray scattering, density measurement, and Raman spectroscopy, can also contribute to the variation in the measurement of the specific enthalpy. For example, PP has been studied since 1957 and more than 20 different values have been reported, ranging from 138 to 221 J/g [8].

Tab. 6.3: Specific enthalpy, ΔH_f, of common thermoplastics with assumable 100% degree of crystallinity.

Thermoplastics	Specific enthalpy ΔH_f (J/g)
High-density polyethylene	293
Polypropylene	138–221
Polyoxymethylene	190
Nylon 12	95
Nylon 6	190–230
Nylon 66	185–226
Polyethylene terephthalate	140
Polybutylene terephthalate	142–145
Polyvinylidene fluoride	105
Polytetrafluoro ethylene	82
Polyvinyl chloride	176
Polyether ether ketone	130
Polyvinyl alcohol	156
Polyphenylene sulfide	50–150

6.7.2 Polarized light microscopy method

Crystallites exist in semicrystalline thermoplastic matrix composites. The crystallite forms when the semicrystalline thermoplastic undergoes transitions, typically re-crystallization, while cooling from a molten state to a solid state. Some polymer chains align themselves and form an ordered structure. Because of its spherical shape, normally, the ordered structure is also called spherulite. The crystallite size generally ranges from 1 to 100 μm. Crystallites can be directly observed under a po-larized light microscope (PLM) in a semicrystalline thermoplastic or its composite, provided that it is adequately thin, or the thickness of the thermoplastic or the com-posite is in the same scale as the crystallite size.

The polarized light microscope uses double refraction of the crystallite for imag-ing. A polarizer is placed between a light source and the semicrystalline thermoplastic composite. The polarizer allows only the plane-polarized light to go through and inter-act with the crystallite. When the crystallite has nonequivalent axes, that is, crystal structures other than cube structure that has equivalent axes ($a = b = c$), the crystallite possesses optical anisotropy and can refract the incident light in two rays, which is often known as birefringence or double refraction. The two light rays with different phases are then processed through an analyzer to generate an image.

Figure 6.8 shows the development of crystallites in a basalt fiber-reinforced poly(L-lactic acid) composite observed through a polarized light microscopy. Poly(L-lactic acid) is a semicrystalline thermoplastic of the PLA family. The images clearly show the nucleation of crystallites on the fiber and in the thermoplastic matrix (Fig. 6.8a) as well as the growth of the crystallites (Fig. 6.8b–d) at 130 °C. It indicates that the fiber plays a critical role in the nucleation of crystallites. The fiber surface functions as nucleation sites and the crystallite preferentially nucleates on the fiber surface. Impingement between adjacent crystallites often occurs when they grow to a certain size and contact with one another (Fig. 6.8d).

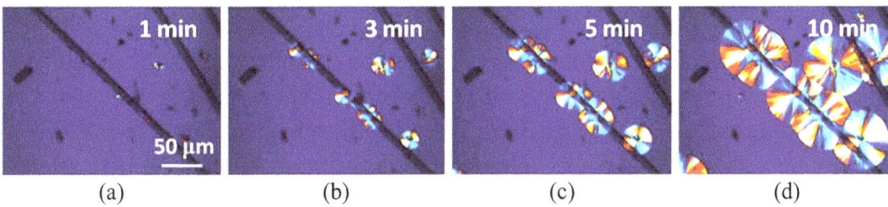

Fig. 6.8: (a) Nucleation of crystallites on the fiber and in the thermoplastic matrix of a basalt fiber-reinforced poly(L-lactic acid) composite at 130 °C; (b–c) growth of the crystallites with time; (d) impingement of crystallites after they grow to a certain size and contact with one another (reprinted from Reference [9] with permission).

6.7.3 X-Ray diffraction method

XRD is used for studying the crystallography of crystalline materials, typically metallic materials. The Bragg's law governs the diffraction behavior of the crystalline material. In the case of composites with semicrystalline thermoplastic matrices, the crystallites in the thermoplastic matrix diffract X-rays by following the Bragg's law defined in Eq. (6.20). The crystallite has a long-range order structure and generates sharp Bragg peaks in a diffraction pattern. The sketch of the semicrystalline thermoplastic matrix in Fig. 6.9a is the same as the one in Fig. 2.1a, and one of the crystallites is highlighted for illustrating how it diffracts X-rays (Fig. 6.9b):

$$n\lambda = 2d \sin \theta \tag{6.20}$$

where n is an integer, λ is the X-ray wavelength, d is the spacing between adjacent crystallite planes, and θ is the Bragg angle between the incident X-ray and the crystallite plane.

The degree of crystallinity of a semicrystalline thermoplastic can be characterized using the XRD technique. There is minimal temperature change in the thermoplastic

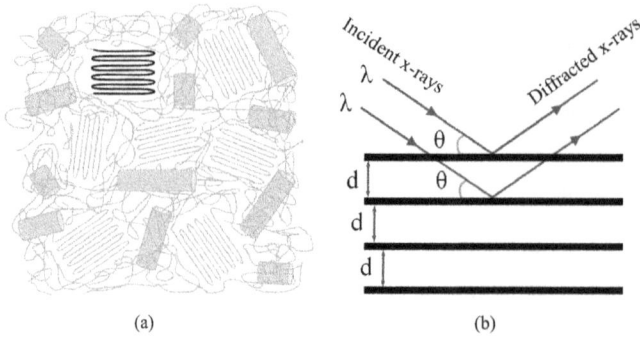

Fig. 6.9: (a) A crystallite highlighted in a semicrystalline thermoplastic matrix composite and (b) X-ray diffraction in the crystallite governed by the Bragg's law.

during characterization, therefore, XRD can avoid the crystallization (such as cold crystallization and pre-melt crystallization) induced by the temperature change in the thermal analysis, such as DSC or DMA. The XRD diffraction pattern consists of two components, sharp Bragg peaks generated from the crystalline phase with a long-range order in the thermoplastic and a broad halo caused by the short-range order structure in its amorphous phase. Figure 6.10 shows the diffraction pattern of a semicrystalline polypropylene. It is comprised of Bragg peaks and an amorphous halo. The areas under the Bragg peaks and under the amorphous halo can be used to calculate the DOC of the thermoplastic (Eq. (6.21)). Normally X-ray diffractometers are equipped with data analysis software and database, for example, the amorphous halo pattern, to carry out the calculation.

$$DOC = [A_c/(A_c + A_a)] \times 100\% \tag{6.21}$$

where A_c is the area under the diffraction peaks generated from the crystalline phase and A_a is the area under the halo from the amorphous phase.

The XRD technique can be used to determine the degree of crystallinity of the thermoplastic matrix in a thermoplastic composite. However, thermoplastic composites have reinforcements such as carbon fibers, glass fibers, and PP fibers that can also generate peaks and/or halos. These peaks or halos can overlap with the peaks generated from the crystalline phase of the thermoplastic matrix and the halo from its amorphous phase. Extensive analysis of the diffraction peaks and an understanding of the source of the peaks and halos are required to obtain an accurate degree of crystallinity of the thermoplastic. Although it is believed that the XRD method is capable of providing more accurate DOC results than DSC, extensive data analysis that is required can limit its practical use.

The XRD can also be used to determine the size of the crystallite, which can be calculated based on the XRD spectrum and the following Scherrer's equation:

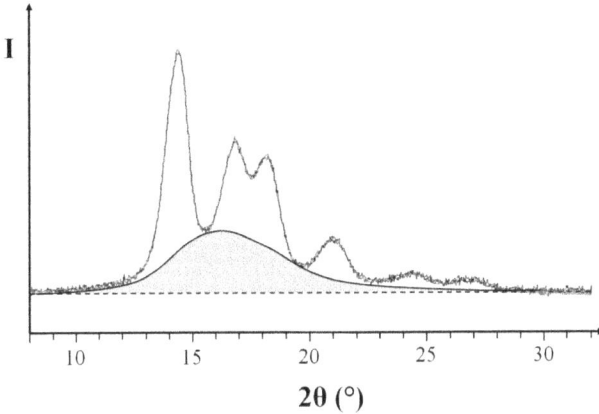

Fig. 6.10: A diffraction pattern of a semicrystalline polypropylene consisting of Bragg peaks and an amorphous halo (shaded area).

$$t = \frac{K\lambda}{\beta \cos\theta} \tag{6.22}$$

where K is the Scherrer constant that is dependent on the actual shape of the crystallite, λ is the X-ray wavelength, θ is the Bragg angle, and β is the full width at half the maximum intensity of the diffraction peak in radians. The Scherrer constant K is normally regarded as 1 for spherulitic crystallites. The full width at half the maximum intensity of the diffraction peak is inversely proportional to the crystallite size.

6.8 Rheology

Study of the flow behavior of a material, typically liquids, is called rheology. The rheology of the liquid, such as a discontinuous fiber thermoplastic composite melt, describes the relationship between its shear stress and shear rate. The flow response of the composite melt under load offers useful information for its processing, which in turn affects the property of the composite. For example, the viscosity change with shear rate in the discontinuous fiber thermoplastic composite melt has been used to determine the flow and mold filling in different processes, such as injection molding, extrusion, and compression molding, that involve different magnitudes of shear rate. Injection molding and extrusion processes use a large shear rate, while other processes, including compression molding, involve a lower shear rate. Therefore, an understanding of the rheological behavior in the discontinuous fiber thermoplastic composite melt at different shear rates provides a crucial basis for the processing and helps improve the property of the composite.

The rheological behavior of thermoplastic composite melts can be measured by different methods, such as rheometry and squeeze flow. These methods are described below.

6.8.1 Rheometry

Rheometry refers to the characterization technique used to study the flow behavior of a liquid or slurry, such as discontinuous fiber/particulate/nanomaterial-reinforced thermoplastic composites at their molten state when a load is applied. A rheometer is a common instrument used for the characterization. The rheometer typically measures the viscosity of the liquid or slurry as a function of the shear rate. In addition, it can determine the viscosity with time and temperature at a specific shear rate.

Thermoplastic and thermoplastic composite melts are generally shear-thinning fluids and their viscosity decreases as the shear rate increases. Their rheological behavior follows the Ostwald-de Waele equation or power-law equation:

$$\tau = K\dot{\gamma}^n \tag{6.23}$$

where τ is the shear stress, $\dot{\gamma}$ is the shear rate, n is the power-law index or flow behavior index, and K is the flow consistency index, which is equal to the viscosity at a shear rate of $1\ \mathrm{s}^{-1}$. For shear-thinning fluids such as thermoplastic or thermoplastic composite melts, the power-law index, n, is less than 1 ($n < 1$).

Since the apparent viscosity (μ) can be written as shear stress over shear rate, or $\mu = \tau/\dot{\gamma}$, τ can be replaced by $\mu\dot{\gamma}$ and the power-law equation can be written as follows:

$$\mu = K\dot{\gamma}^{n-1} \tag{6.24}$$

From Eq. (6.24), when the shear rate $\dot{\gamma}$ increases for shear-thinning fluids such as the thermoplastic or thermoplastic composite melts, the apparent viscosity, μ, decreases. This rheological behavior of the thermoplastics and their composites, reinforced with discontinues fibers/particulates/nanomaterials, is beneficial in high shear rate processes, for example, injection molding. The melt is injected into the mold cavity at a shear rate of more than $10^2\ \mathrm{s}^{-1}$. Their viscosity significantly reduces at the high shear rate, which facilitates the flow of the melt into the mold cavity and ensures complete fill of the mold cavity.

Rotational rheometer is a unit typically used to introduce stress to the thermoplastic composite melt and study its reaction to the stress. Figure 6.11 schematically shows a rheometer comprised of a stationary plate (bottom plate) and a rotating plate. The rheometer is equipped with heating units to heat the thermoplastic composite to a desired temperature. The rotating plate rotates and introduces shear stress to the thermoplastic composite melt. A typical rheometric curve shows the change of the complex viscosity with angular frequency at a specific temperature (Fig. 4.1), which indicates the same trend between the apparent viscosity and the shear rate according to the Cox–Merz rule.

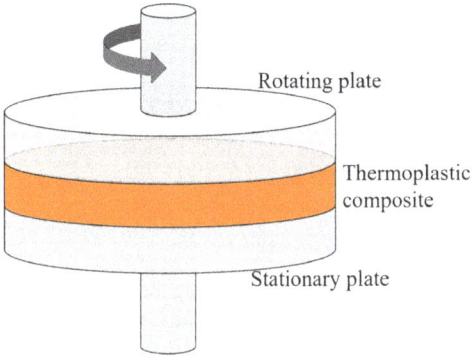

Fig. 6.11: Measurement of the flow behavior of thermoplastic composite melt in response to applied load in a rotational rheometer.

6.8.2 Squeeze flow

Squeeze flow is a method that uses two heated parallel plates to squeeze a disc-shaped thermoplastic composite placed between the two plates. Because the squeeze flow method simulates the actual flow that occurs during processes such as compression molding, it is increasingly used to characterize the rheological behavior of discontinuous fiber-reinforced thermoplastic composites, including GMT and LFT composites. It can reduce edge effects, nonhomogeneous flow fields, and preferential alignment of fibers that are normally experienced for composites tested in rheometry.

Figure 6.12 schematically shows the squeeze flow setup. The thermoplastic composite is placed between two parallel plates. It is heated to the temperature of interest, normally the processing temperature. The assembly is placed on a loading frame that is used for mechanical testing. One plate is maintained stationary while load is applied to one of the plates to squeeze the thermoplastic composite melt. The load is recorded in relation to the displacement. Different shear rates can be achieved by adjusting the loading rate of the test. Power-law parameters and corresponding viscosities can be obtained by calculating the velocity profile of the molten thermoplastic composite, converting the velocity profiles to strain rates, and obtaining the power-law parameters by the use of the experimental stress and calculated strain rates [10].

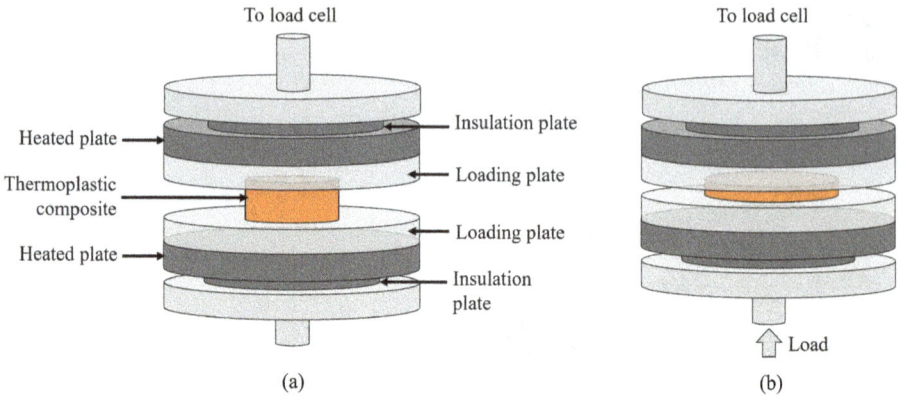

Fig. 6.12: Squeeze flow setup for studying the flow behavior of discontinuous reinforcement thermoplastic composite melt (a) before applying load and (b) after applying load (adapted from Reference [3]).

6.9 Fiber and matrix interface

The interface between the fiber and the matrix is a constituent that can significantly affect the mechanical property of thermoplastic composites because of its critical function of transferring load from the matrix to the fiber. It determines properties such as compression strength, fracture toughness, and fatigue performance. A number of methods can be used to characterize the fiber/matrix interface, qualitatively or quantitatively, and individual fibers are normally involved in the characterization.

6.9.1 Scanning electron microscopy

Scanning electron microscopy (SEM) can be used to qualitatively characterize the interface between the fiber and the matrix. It can directly observe the amount of thermoplastic matrix residue on the individual fibers from the fracture surface of the thermoplastic composites. If the fiber and the matrix have weak bonding, the surface of the fiber is generally clean and a minimal matrix residue remains on the fiber. On the other hand, a large amount of matrix residue remains on the fiber surface, resulting from a strong fiber/matrix interface. Figure 6.13 compares the fiber surface in a glass fiber-reinforced polypropylene composite, with and without maleic anhydride-grafted polypropylene (MAPP), a coupling agent commonly used to enhance the bonding between glass fibers and polypropylene. The clean fiber surface (Fig. 6.13a,b) indicates inadequate bonding between the fiber and the matrix when there is no coupling agent added. After the coupling agent is added, a large amount of residual matrix remains on the fiber surface (Fig. 6.13c,d), indicating a stronger fiber/matrix interface. This method is relatively quick and easy to use and provides useful information for the fiber/matrix interfacial bonding; however, it is qualitative and suited for comparative studies.

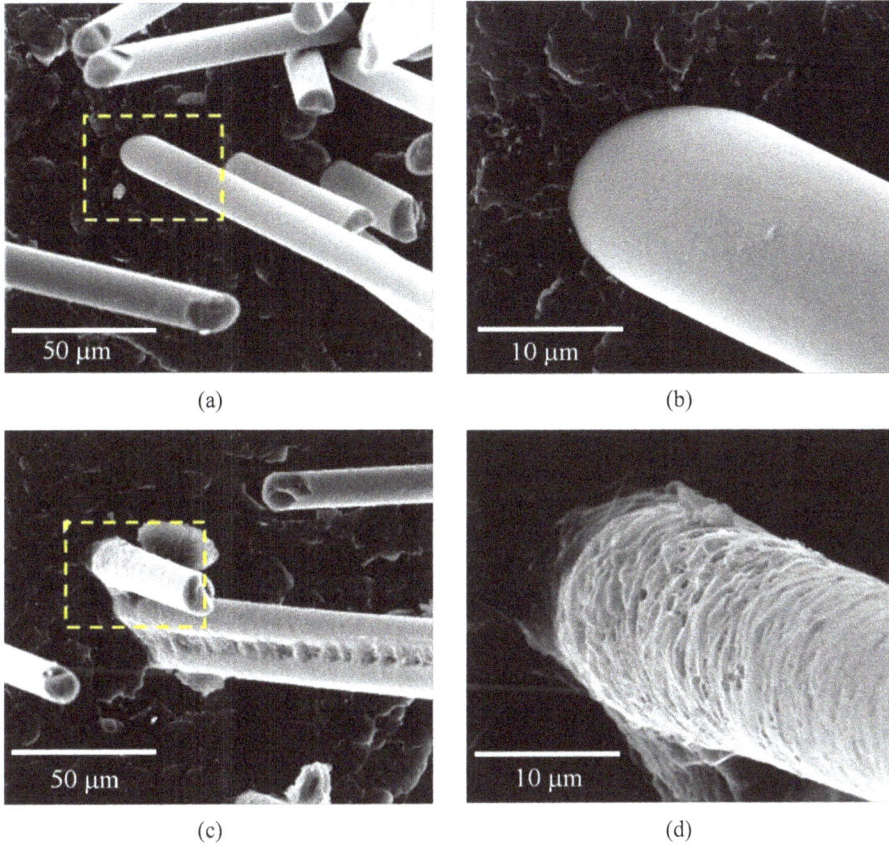

Fig. 6.13: (a) SEM image of the fracture surface of glass fiber-reinforced PP composite without the addition of coupling agent; (b) clean fiber surface from the highlighted region in (a); (c) SEM image of the fracture surface of glass fiber-reinforced PP composite with coupling agent added; (d) a large amount of residual PP matrix on the fiber surface from the highlighted region in (c) (reprinted from Reference [11] with permission).

6.9.2 Micromechanical test

Micromechanical tests refer to a variety of quantitative tests carried out at the microscale level to measure the fiber/matrix interfacial strength. Those micromechanical tests include fiber pushout test, fiber pullout test, microbond test, and fragmentation test. Microscopes are generally required for sample preparation, testing setup, observation of the test, etc. The fiber pushout test, fiber pullout test, and microbond test involve loading a single fiber in compression or tension and introducing separation between the fiber and the matrix. The fragmentation test involves loading a thermoplastic composite with only one single fiber embedded, and fracturing the single fiber. All the tests can quantitatively characterize the interfacial bonding between the fiber and the matrix as described below. It is noted that there are challenges existing in the sample preparation and testing which may hinder their use. Variation in fiber surface morphology and inconsistent sample preparation can also affect the testing result, and the test often needs to be repeated a large number of times to achieve statistical confidence.

1. **Fiber pushout test**, or single fiber pushout test, is a mechanical test that quantitatively measures the bonding strength at the fiber/matrix interface. A nanoindenter is normally used to apply the load. The nanoindenter has a tip that is small enough to apply the load on an individual fiber. However, the low load capacity of the nanoindenter generally requires a small thickness of the composite, in the micrometer range, to initiate separation between the fiber and the matrix. There is a strict requirement for preparing the sample. The sample is sectioned through microtome slicing or polishing. It is necessary to have the fiber perpendicular to the polished surface and ensure that the top and bottom surfaces are parallel. The sample is partially supported such that there is an adequate space for the fiber to be pushed out. Figure 6.14a illustrates the setup for the single fiber pushout test.

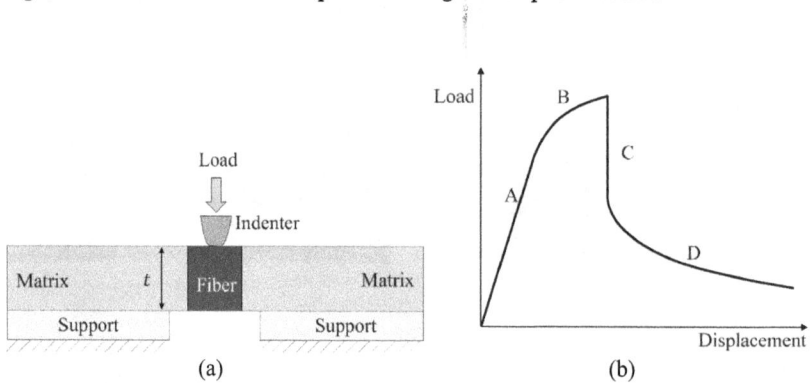

Fig. 6.14: (a) A single fiber pushout test setup; (b) a typical load-displacement curve from the single fiber pushout test.

There are several stages in the single fiber pushout test: (A) linear elastic loading of the fiber; (B) nonlinear section because of decrease in stiffness due to crack initiation and propagation at the interface; (C) sudden load drop caused by the debonding between the fiber and the matrix; and (D) frictional sliding of the fiber. These stages correspond to different sections on a typical load-displacement curve resulting from the single fiber pushout test (Fig. 6.14b).

The interfacial shear strength, τ, is calculated based on the peak load P, the diameter of the fiber d, and the fiber height, which is equal to the thickness of the composite at the test location, t. Below is the equation (Eq. 6.25) for calculating the interfacial shear strength based on these parameters:

$$\tau = \frac{P}{\pi dt} \tag{6.25}$$

2. **Fiber pullout test** involves application of a tension load on a single fiber that is half-embedded in a thermoplastic matrix. During the sample preparation, the fiber is sandwiched between two thermoplastic films, before being heated to above the melting temperature of the thermoplastic. A pressure is applied to ensure bonding between the fiber and the matrix. The sample is then tested in the setup shown in Fig. 6.15. This method is suited for comparing the bonding strength at different conditions. For example, it can be used to study the effect of different fiber sizing on the interfacial bonding strength.

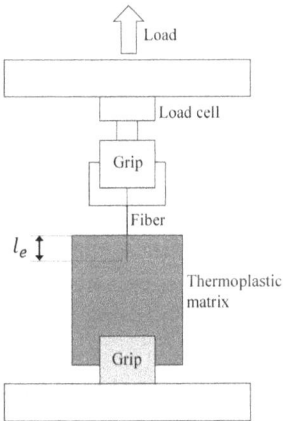

Fig. 6.15: Schematic of a single fiber pullout test.

The interfacial shear strength, τ, is calculated from the peak load that is obtained from the pullout test. The shear strength calculation is expressed by the following equation:

$$\tau = \frac{P}{\pi d l_e} \tag{6.26}$$

where P is the peak load, d is the fiber diameter, and l_e is the embedded length.
3. **Microbond test**, another micromechanical test, is similar to the fiber pullout test. A drop of a thermoplastic melt is placed on a single fiber to form a microdroplet on the fiber (Fig. 6.16a). The sample is pulled through knife edges to induce debonding between the fiber and the droplet (Fig. 6.16b). Equation

(6.26) from the single fiber pullout test is used to calculate the shear strength for the microbond test.

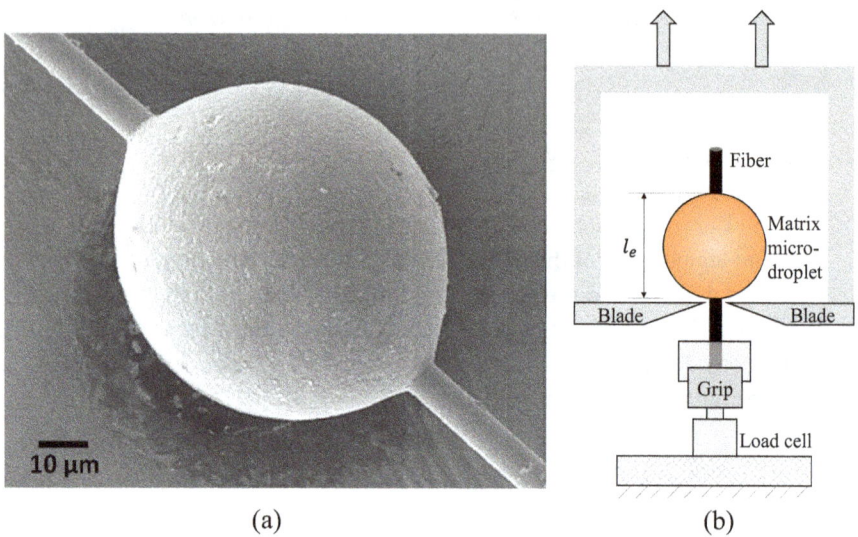

Fig. 6.16: (a) A microbond testing sample consisting of a carbon fiber and PPS matrix (reprinted from Reference [12] with permission) and (b) schematic of the microbond test.

4. **Fragmentation test** is used to determine the interfacial shear strength by fracturing a single fiber embedded in a thermoplastic matrix. A single fiber of interest is sandwiched between two thermoplastic films and compression molded into a composite by applying heat and pressure. A number of individual fibers are often compression molded with films at the same time to produce multiple samples. It should be ensured that the single fiber is straight during the sample preparation. Also, the fiber should be protected from damage, as surface defects can produce false testing results.

The fragmentation test involves loading the fiber embedded in the thermoplastic matrix in tension by pulling the ends of the sample (Fig. 6.17a). If the fiber length (L) is greater than the critical fiber length (L_c), the stress (σ_x) in the mid-section of the fiber can reach the fiber tensile strength (σ_f) (Figs. 6.17b and 3.12). Fracture can occur at any point within that fiber section. After the fiber fractures into two segments, if any fiber fragment has a length greater than L_c, the stress will again reach the fiber tensile strength and the fiber will fracture furthermore. For any fiber segment with a fiber length less than L_c, its stress will not reach the fiber tensile strength, as shown in the two fiber segments on the left in Fig. 6.17c. The fiber segments with a length great than L_c will continue to facture until all the segment have a length smaller than L_c, and the stress at any location of any fiber segment is less than σ_f (Fig. 6.17d). After that, sliding and separation between the fiber and the matrix occur. Figure 6.17e shows the development of fragmentation with applied load in a typical load–displacement curve.

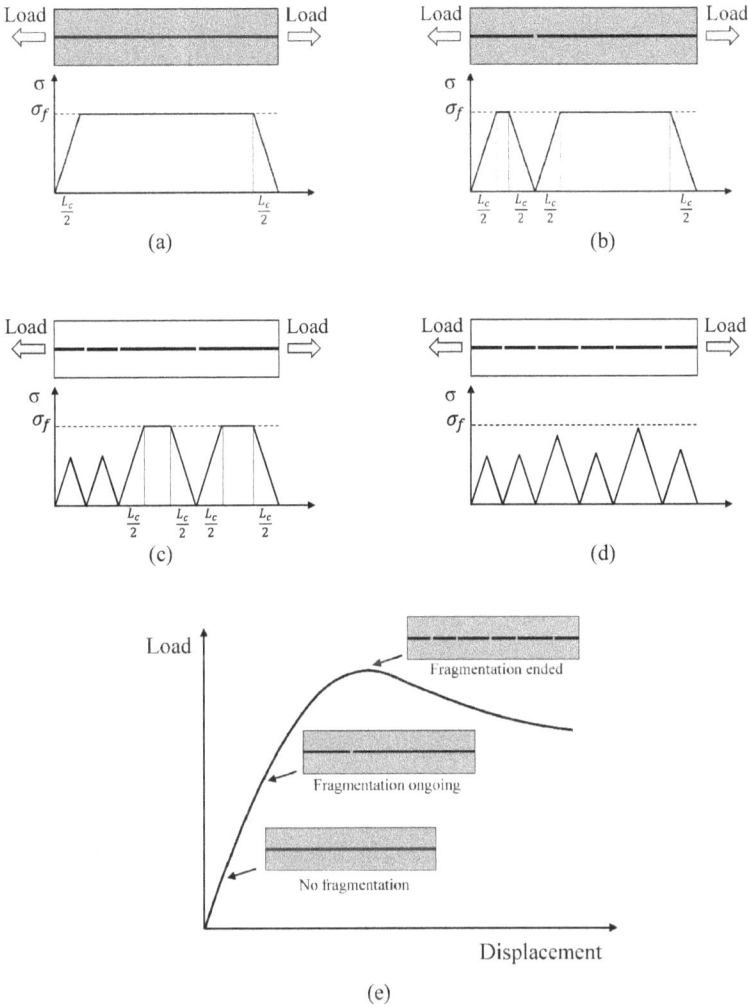

Fig. 6.17: Schematics of fragmentation test; (a) no fragmentation at low stress; (b) and (c) occurrence of fragmentation; (d) completion of fragmentation; and (e) a typical load–displacement curve showing development of fragmentation with applied load.

The essential goal of the fragmentation test is to find out the critical fiber length, based on the lengths of the fractured fibers. The fiber length of each fiber segment can be directly measured using optical microscopes if the matrix is transparent. It is more challenging to measure the fiber length if the thermoplastic matrix is not transparent. Acoustic emission technique, chemical dissolution, or polishing are normally used for evaluating the fiber length of each fiber segment.

The stress at a distance x from the fiber end, σ_x, is defined by the following equation:

$$\sigma_x = \frac{4\tau}{d}x \tag{6.27}$$

where τ is the interfacial bonding strength and d is the fiber diameter.

When x is half the critical fiber length, or $x = L_c/2$, the stress of the fiber σ_x reaches σ_f and Eq. (6.27) is identical to Eq. (3.6).

The stress σ_x in the fiber can increase up to the fiber tensile strength, σ_f, at a value of x_0 if the fiber segment length is adequately long (equal to or greater than L_c):

$$\sigma_f = \frac{4\tau}{d}x_0 \tag{6.28}$$

If any fiber fragment has a length greater than $2x_0$, fracture occurs to the fiber fragment until all the fiber segments have a length less than $2x_0$. The lengths of all fiber segments range between x_0 and $2x_0$. The average length of all the segments is calculated to be

$$\bar{l} = \frac{1}{2}(x_0 + 2x_0)$$

Because the critical fiber length L_c is equal to $2x_0$, $\bar{l} = \frac{3}{4}L_c$, or

$$L_c = \frac{4}{3}\bar{l} \tag{6.29}$$

After the average length $(\bar{l})$ is determined from the lengths of all fiber segments, the critical fiber length can be calculated according to Eq. (6.29).

6.10 Summary

– Characterization of thermoplastic composites refers to probing of the composition, fundamental structure, and property of the composite as well as of its constituents, using various techniques.
– Fiber content of thermoplastic composites can be characterized by a variety of methods, including burn-off, acid digestion, carbonization, and microscopy.

- Fiber length distribution can be characterized using a combination of burn-off and microscopy methods as well as micro-CT method.
- Fiber orientation distribution can be characterized using micro-CT and microscopy methods.
- Void content of a thermoplastic composite can be measured using density method, microscopy method, and micro-CT method.
- Degree of crystallinity of a thermoplastic composite can be calculated by
$$DOC = \frac{\Delta H_s + \Delta H_c + \Delta H_p}{(1 - W_f)\Delta H_f} \times 100\%.$$
- Fiber surface can function as nucleation sites and crystallites preferentially nucleate on the fiber surface.
- Impingement of crystallites often occurs between adjacent crystallites when they grow to a certain size and contact one another.
- X-ray diffraction by crystallites in thermoplastic composites follows the Bragg's law, $n\lambda = 2d \sin \theta$.
- Differential-scanning calorimetry is a versatile tool that can be used to characterize glass transition temperature, melting temperature, recrystallization temperature, and cold crystallization temperature of thermoplastic composites.
- DSC, TMA, and DMA are commonly used to determine the glass transition temperature of thermoplastic composites.
- HDT of a thermoplastic composite is defined as the temperature at which the composite undergoes a specified deflection at a specified load. It is similar to the glass transition temperature for neat thermoplastics. However, it is much higher than the glass transition temperature for thermoplastic composites because of the addition of fibers.
- TGA can be used to measure the thermal degradation temperature. Processing of thermoplastic composites should be carried out at temperatures below their thermal degradation temperatures to avoid deterioration of their mechanical properties.
- When the shear rate $\dot{\gamma}$ increases for shear-thinning fluids, such as thermoplastic or thermoplastic composite melts, their apparent viscosity decreases.
- The bonding strength at the fiber/matrix interface can be quantitatively characterized using micromechanical tests, such as fiber pushout test, fiber pullout test, microbond test, and fragmentation test.
- The critical fiber length can be determined using the fiber fragmentation test and is equal to $4\bar{l}/3$.

References

[1]   Wang Q, Ning H, Vaidya U, Pillay S, Nolen L-A. Fiber content measurement for carbon
      fiber–reinforced thermoplastic composites using carbonization-in-nitrogen method. Journal
      of Thermoplastic Composite Materials. 2018;31(1):79–90.
[2]   Wang Q, Ning H, Vaidya U, Pillay S, Nolen L-A. Development of a carbonization-in-nitrogen
      method for measuring the fiber content of carbon fiber reinforced thermoset composites.
      Composites Part A: Applied Science and Manufacturing. 2015;73:80–4.
[3]   Thattaiparthasarathy KB, Pillay S, Ning H, Vaidya U. Process simulation, design and
      manufacturing of a long fiber thermoplastic composite for mass transit application.
      Composites Part A: Applied Science and Manufacturing. 2008;39(9):1512–21.
[4]   Pinter P, Bertram B, Weidenmann KA, editors. A novel method for the determination of fibre
      length distributions from μCT-data. Conference on Industrial Computed Tomography (iCT);
      2016.
[5]   Saenz-Castillo D, Martín M, Calvo S, Rodriguez-Lence F, Güemes A. Effect of processing
      parameters and void content on mechanical properties and NDI of thermoplastic composites.
      Composites Part A: Applied Science and Manufacturing. 2019;121:308–20.
[6]   Yu S, Hwang YH, Hwang JY, Hong SH. Analytical study on the 3D-printed structure and
      mechanical properties of basalt fiber-reinforced PLA composites using X-ray microscopy.
      Composites Science and Technology. 2019;175:18–27.
[7]   Pistor CM, Güçeri SI. Crystallinity of on-line consolidated thermoplastic composites. Journal
      of Composite Materials. 1999;33(4):306–24.
[8]   Lanyi FJ, Wenzke N, Kaschta J, Schubert DW. On the determination of the enthalpy of fusion of
      α-crystalline isotactic polypropylene using differential scanning calorimetry, X-ray diffraction,
      and fourier-transform infrared spectroscopy: An old story revisited. Advanced Engineering
      Materials. 2020;22(9):1900796.
[9]   Pan H, Cao Z, Chen Y, Wang X, Jia S, Yang H, et al. Effect of molecular stereoregularity on the
      transcrystallization properties of poly (l-lactide)/basalt fiber composites. International
      Journal of Biological Macromolecules. 2019;137:238–46.
[10] Thattaiparthasarthy KB, Pillay S, Vaidya UK. Rheological characterization of long fiber
      thermoplastics–Effect of temperature, fiber length and weight fraction. Composites Part
      A: Applied Science and Manufacturing. 2009;40(10):1515–23.
[11] Luo G, Li W, Liang W, Liu G, Ma Y, Niu Y, et al. Coupling effects of glass fiber treatment and
      matrix modification on the interfacial microstructures and the enhanced mechanical
      properties of glass fiber/polypropylene composites. Composites Part B: Engineering.
      2017;111:190–9.
[12] Liu B, Liu Z, Wang X, Zhang G, Long S, Yang J. Interfacial shear strength of carbon fiber
      reinforced polyphenylene sulfide measured by the microbond test. Polymer Testing.
      2013;32(4):724–30.

Chapter 7
Micromechanics and macromechanics of thermoplastic composites

7.1 Introduction

A continuous fiber-reinforced thermoplastic composite structure is essentially constructed from basic constituent materials, namely, fiber and thermoplastic polymer. The fiber and the thermoplastic are firstly combined through different processing methods to produce preforms, for example, uni-tapes (see Section 4.3). Several layers of uni-tapes or other preforms are then stacked in different orientations and molded into a thermoplastic composite through various processes (see Section 4.4). Each uni-tape or preform layer is called lamina. The thermoplastic composite, comprised of two or more laminae, is called thermoplastic composite laminate or laminated thermoplastic composite. The thermoplastic composite laminate is then manufactured into a component for final assembly (Fig. 7.1). Therefore, laminated thermoplastic composites are a structural material with different levels of hierarchy.

Because of directionally dependent properties in the fiber and each lamina, as well as the thermoplastic composite laminate as whole, the mechanical behavior of the thermoplastic composite is much more complex than that of the isotropic materials that possess identical properties in all directions. Because of the different levels of hierarchy in the laminated thermoplastic composite, there are different continuum mechanics developed to study their mechanical behaviors at different levels, such as microscale and macroscale levels.

Micromechanics and macromechanics of thermoplastic composites study their mechanical behaviors, or mechanics, at the microscale and macroscale levels, respectively. It is indispensable to understand the micromechanics and macromechanics of a laminated thermoplastic composite to ensure that the stacked laminae with a specific thickness and orientation design achieve desired mechanical performance. Understanding of the mechanics of the laminate starts from the analysis of the thermoplastic composite preformed at the microscopic level, or micromechanics. For example, the material behavior of a uni-tape is analyzed based on the properties of its matrix and fiber and their volume fractions, through which the mechanical properties of the uni-tape are obtained. The properties from the micromechanics analysis of a lamina are then used for macromechanics analysis to predict the behavior of the laminated thermoplastic composite. This chapter discusses the micromechanics and macromechanics of thermoplastic composites.

https://doi.org/10.1515/9781501519055-007

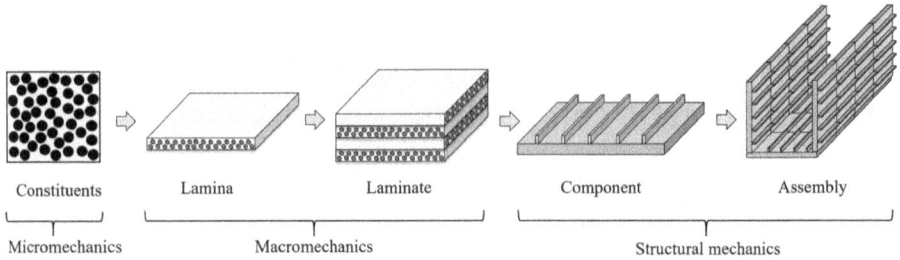

| Constituents | Lamina | Laminate | Component | Assembly |

Micromechanics | Macromechanics | Structural mechanics

Fig. 7.1: A schematic of hierarchical structure in thermoplastic composites.

7.2 Micromechanics of thermoplastic composite

Micromechanics refers to analysis of composites, including thermoplastic composites, at the level of their individual constituents, that is, the fiber and the matrix. Each individual fiber and the distance between adjacent fibers are at a micrometer scale, hence the name micromechanics. The essential objective of the micromechanics is to study interaction between the constituents at the microscale level and calculate the overall or average properties of the composite based on the attributes of constituent materials, including the properties and volume fraction of each constituent material, fiber aspect ratio, and fiber orientation.

Micromechanical models are the models developed through the micromechanics approach. Each model includes equations that are used to calculate the average or overall properties of the composite by integrating the attributes of each constituent in the composite. Micromechanical models have been developed for both discontinuous and continuous fiber-reinforced thermoplastic composites. The micromechanical models for continuous fiber-reinforced thermoplastic composites, such as unidirectional fiber thermoplastic composites, include Voigt model [1], Reuss model [2], concentric cylinder assemblage model [3], self-consistent model [4, 5], Mori–Tanaka model [6], and method of cells [7]. In addition to the continuous fiber-reinforced thermoplastic composite, discontinuous fiber- or particulate-reinforced thermoplastic composites can also be modeled using the micromechanical approach. The models for discontinuous fiber-thermoplastic composite include modified ROM, Halpin–Tsai equations, shear lag theory, Mori–Tanaka model, self-consistent model, bounding models, and any other model extensions that are based on those basic models. The models for particulate thermoplastic composites include modified ROM, Halpin–Tsai equations, Mori–Tanaka model, and Eshelby model. Figure 7.2 summarizes the micromechanics models for different types of thermoplastic composites.

The following general assumptions are made for the micromechanical models:
(a) The matrix and the fibers are linearly elastic, and the fibers are isotropic or transversely isotropic.

(b) The matrix and the fibers are perfectly bonded. When there is an external load applied, the matrix transfers the load to the fiber and there is no slipping between the fiber and the matrix or matrix microcracking during deformation.
(c) The fibers are uniform in diameter and shape and can be defined by a fiber aspect ratio.
(d) The matrix is free of voids.

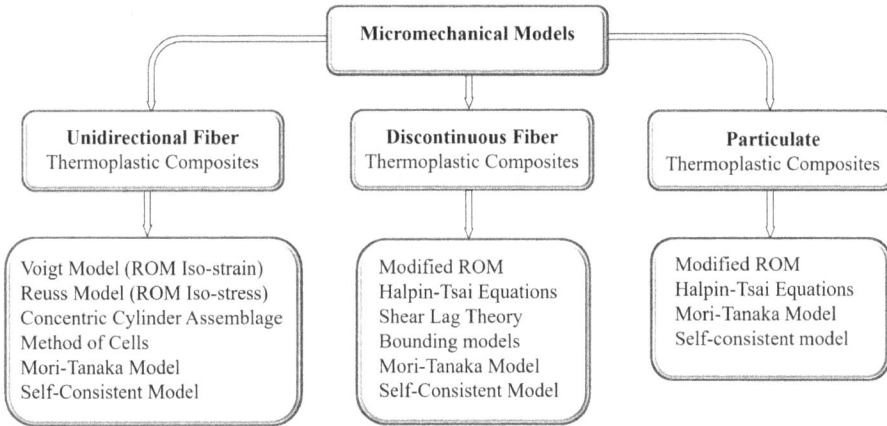

Fig. 7.2: Micromechanical models used for different types of thermoplastic composites.

The following sections describe micromechanical models commonly used in predicting the mechanical properties of thermoplastic composites.

7.2.1 Rule of mixture

Rule of mixture (ROM) is a mathematical method to calculate the properties of a mixture based on the volume fraction and property of each constituent in the mixture. It is one of the mostly used micromechanical models to predict a variety of properties including tensile modulus, shear modulus, and Poisson's ratio. It is used for different types of thermoplastic composites, including unidirectional fiber- and particulate-reinforced thermoplastic composites.

The properties of a unidirectional fiber thermoplastic composite preform, such as uni-tape, highly depend on the properties and volume fractions of its constituent materials. The ROM models are used to calculate the properties of the composite based on the properties and volume fractions of the constituent materials. The calculated properties of the uni-tape can be used as the input for macromechanical models (see Section 7.3). The following sections describe ROM models for unidirectional fiber-reinforced thermoplastic composites.

7.2.1.1 Elastic modulus

When a unidirectional fiber-reinforced thermoplastic composite is elongated by ΔL from its original length, L, under loading in the fiber direction (Fig. 7.3a), the fiber and the matrix have the same elongation, ΔL. Therefore, the composite, its fiber, and its matrix have the same strain:

$$\varepsilon_f = \varepsilon_m = \varepsilon_{cl} = \frac{\Delta L}{L} \tag{7.1}$$

where ε_f is the strain in the fiber, ε_m is the strain in the matrix, ε_{cl} is the strain in the composite in the longitudinal (or fiber) direction, and L is the original length. This condition is called iso-strain condition or action in parallel condition.

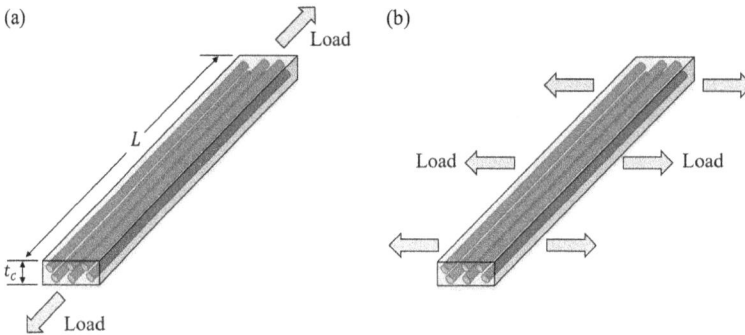

Fig. 7.3: A unidirectional fiber thermoplastic composite (thickness t_c and length L) applied with a load in (a) longitudinal direction and (b) transverse direction.

According to the stress–strain relationship of a material in the elastic region, the stress in the fiber, σ_f, is the product of its elastic modulus, E_f, and its strain, ε_f:

$$\sigma_f = E_f \varepsilon_f \tag{7.2}$$

Similarly, the stress in the matrix, σ_m, is the product of its elastic modulus, E_m, and its strain, ε_m. Therefore, it can be expressed in the following equation:

$$\sigma_m = E_m \varepsilon_m \tag{7.3}$$

After substituting ε_f and ε_m with ε_{cl} based on Eqs. (7.1)–(7.3), Eqs. (7.2) and (7.3) can be written as follows:

$$\sigma_f = E_f \varepsilon_{cl} \tag{7.4}$$

$$\sigma_m = E_m \varepsilon_{cl} \tag{7.5}$$

The load applied to the composite, F_c, is shared between the fiber and the matrix; therefore,

$$F_c = F_f + F_m \tag{7.6}$$

Since there is relationship among load, stress, and cross-sectional area, $P = \sigma A$, Eq. (7.6) can be written as follows:

$$\sigma_{cl} A_c = \sigma_f A_f + \sigma_m A_m \tag{7.7}$$

where A_c is the cross-sectional area of the composite, A_f is the cross-sectional area of the fiber, and A_m is the cross-sectional area of the matrix.

From Eqs. (7.4) and (7.5),

$$\sigma_{cl} A_c = \left(E_f A_f + E_m A_m \right) \varepsilon_{cl} \tag{7.8}$$

or,

$$E_{cl} = \frac{\sigma_{cl}}{\varepsilon_{cl}} = \left(E_f \frac{A_f}{A_c} + E_m \frac{A_m}{A_c} \right) \tag{7.9}$$

where E_{cl} is the longitudinal elastic modulus of the composite. Sometimes it is written as E_{11} or E_1.

The volume fraction of the fiber, V_f, can be written as the ratio of the fiber volume, LA_f, to the composite volume, LA_c:

$$V_f = \frac{LA_f}{LA_c} = \frac{A_f}{A_c} \tag{7.10}$$

In the same manner, the volume fraction of the matrix, V_m, can be written as

$$V_m = \frac{A_m}{A_c} \tag{7.11}$$

Because the volume of the unidirectional fiber-reinforced thermoplastic composite is the total volume of both the fiber and the matrix, $LA_f + LA_m = LA_c$, which can be rewritten as $A_f/A_c + A_m/A_c = 1$, or

$$V_f + V_m = 1 \tag{7.12}$$

From Eq. (7.9),

$$E_{cl} = \left(V_f E_f + V_m E_m \right) \tag{7.13}$$

Equation (7.13) can also be written as

$$E_{cl} = V_f E_f + (1 - V_f) E_m \tag{7.14}$$

or,

$$E_{cl} = (1 - V_m)E_f + V_mE_m \tag{7.15}$$

Equations (7.14) and (7.15) are the equations based on ROM for calculating the elastic modulus of a unidirectional thermoplastic composite in the longitudinal direction when the elastic modulus of the fiber and the matrix and the volume fraction of the fiber or the matrix are known. The model is also called Voigt model. Based on the equation, the elastic modulus of the composite lies between the elastic modulus of the matrix and the fiber. When there is only matrix (or $V_f = 0$), $E_{cl} = E_m$. When there is only fiber (or $V_m = 0$), $E_{cl} = E_f$.

The above-mentioned ROM equations are applicable to unidirectional fiber thermoplastic composites. A modified ROM (Eq. (7.16)) is developed for thermoplastic composites with other fiber architectures, such as biaxial fibers and quasi-isotropic fibers:

$$E_c = \eta_\theta V_f E_f + V_m E_m \tag{7.16}$$

where η_θ is the efficiency factor or Krenchel factor, $\eta_\theta = 1$ for unidirectional fiber thermoplastic composite, $\eta_\theta = 0.5$ for biaxial fiber thermoplastic composite, and $\eta_\theta = 0.375$ for quasi-isotropic thermoplastic composite.

When a unidirectional fiber thermoplastic composite is under loading, F_c, the load is shared by the fibers and the matrix. The percentage of the load shared by each constituent can be calculated. The equation for calculating the percentage of the load shared by each constituent is derived below.

Equation (7.6) can be written as

$$\frac{F_f}{F_c} = \frac{F_f}{F_f + F_m} \tag{7.17}$$

or,

$$\frac{F_f}{F_c} = \frac{\sigma_f A_f}{\sigma_f A_f + \sigma_m A_m} \tag{7.18}$$

Hence,

$$\frac{F_f}{F_c} = \frac{1}{1 + \left(\frac{\sigma_m A_m}{\sigma_f A_f}\right)} \tag{7.19}$$

After combining Eqs. (7.4), (7.5), (7.10), (7.11), and (7.19),

$$\frac{F_f}{F_c} = \frac{1}{1 + \left(\frac{E_m V_m}{E_f V_f}\right)} \tag{7.20}$$

The percentage of the load that the fiber supports can be calculated using Eq. (7.20). Because the elastic modulus of the fiber is significantly higher than that of the matrix, $E_f \gg E_m$. If the volume fraction of the fiber is not considerably lower than that of the matrix, $E_m V_m \ll E_f V_f$ and $F_f \approx F_c$. It indicates that the fiber bears most of the load applied to the composite.

ROM can also be applied to calculate the elastic modulus of the composite in the transverse direction. When the load is applied in the transverse direction (Fig. 7.3b), there is a change in the displacement thickness direction, Δt_c, which is the sum of displacements of the matrix and the fiber in the thickness direction:

$$\Delta t_c = \Delta t_m + \Delta t_f \tag{7.21}$$

or,

$$\frac{\Delta t_c}{t_c} = \frac{\Delta t_m}{t_c} + \frac{\Delta t_f}{t_c} \tag{7.22}$$

The strain of the composite in the transverse direction is calculated by dividing the change of thickness by the original thickness:

$$\varepsilon_{ct} = \frac{\Delta t_c}{t_c}$$

The strain of the matrix, ε_m, is the change in its thickness (Δt_m) over its original thickness (t_m), and it can be written as the following equation:

$$\varepsilon_m = \frac{\Delta t_m}{t_m}$$

or,

$$\Delta t_m = \varepsilon_m t_m$$

Similarly, the thickness change in the fiber is $\Delta t_f = \varepsilon_f t_f$.

Since $\varepsilon_{ct} = \dfrac{\Delta t_c}{t_c} = \dfrac{\Delta t_m}{t_m} \dfrac{t_m}{t_c} + \dfrac{\Delta t_f}{t_f} \dfrac{t_f}{t_c}$, therefore,

$$\varepsilon_{ct} = \varepsilon_m \frac{t_m}{t_c} + \varepsilon_f \frac{t_f}{t_c}$$

For a given cross-sectional area of the composite under the applied load, the volume fractions of fiber and the matrix are $V_f = \dfrac{t_f}{t_c}$ and $V_m = \dfrac{t_m}{t_c}$, respectively:

$$\varepsilon_{ct} = \varepsilon_m V_m + \varepsilon_f V_f$$

$$\frac{\sigma_{ct}}{E_{ct}} = \frac{\sigma_m}{E_m} V_m + \frac{\sigma_f}{E_f} V_f$$

Since $\sigma_{ct} = \sigma_m = \sigma_f$,

$$\frac{1}{E_{ct}} = \frac{V_m}{E_m} + \frac{V_f}{E_f} \tag{7.23}$$

Equation (7.23) is used to calculate the elastic modulus in the transverse direction of the thermoplastic composite, E_{ct}. E_{ct} is also written as E_{22} or E_2. It is generally challenging to determine the elastic modulus in the transverse direction through material testing because of the small thickness of the composite. Eq. (7.23) offers a solution to attain the value for macromechanical models and material property input for finite element analysis.

7.2.1.2 Poisson's ratio v_{xy}

A solid material, when applied with a load in one direction, will develop strains perpendicular to the direction of the load in addition to the strain parallel to the load. The strain that is parallel to the load direction is called axial strain or longitudinal strain. The strain perpendicular to the load direction is called lateral or transverse strain. The negative ratio of the transverse strain (ε_t) to the longitudinal strain (ε_l) is called Poisson's ratio (v):

$$v = -\frac{\varepsilon_t}{\varepsilon_l} \tag{7.24}$$

For an isotropic material, its Poisson's ratio remains the same regardless of the loading direction. Therefore, its Poisson's ratio is independent of the direction. Thermoplastic composites that are reinforced with particulates and nanomaterials that have low aspect ratios are isotropic materials and their Poisson's ratios are not directionally dependent. However, for an anisotropic material such as unidirectional fiber-reinforced thermoplastic composites, the strain in response to stress can differ significantly in different directions. Therefore, its Poisson's ratio is directionally dependent. Subscripts are added to v to specify the loading direction and the strain caused in other directions. The first number or letter represents the direction that the load is applied to the composite. The second number or letter specifies the direction of the strain caused by the load. If the fibers in a unidirectional fiber thermoplastic composite are aligned in the x direction and the load is applied in that direction, v_{xy} is defined by the following equation:

$$v_{xy} = -\frac{\varepsilon_y}{\varepsilon_x} \tag{7.25}$$

On the other hand, when the load is applied to the same composite in its y direction, the Poisson's ratio, v_{yx}, is defined by the following equation:

$$v_{yx} = -\frac{\varepsilon_x}{\varepsilon_y}$$

It is noted that v_{xy} is not equal to v_{yx}. Because E_x is significantly greater than E_y, v_{xy} is much greater than v_{yx}. For that reason, v_{xy} is defined as the major Poisson's ratio, while v_{yx} the minor Poisson's ratio. The major and minor Poisson's ratios are related by a reciprocal relation:

$$\frac{v_{xy}}{E_x} = \frac{v_{yx}}{E_y}$$

When the composite is loaded in the x direction, a transverse strain, ε_y, is generated. The deformation in the transverse direction, Δ_B, can be expressed in relation to ε_y and the total width of the composite, B:

$$\Delta_B = -B\varepsilon_y = Bv_{xy}\varepsilon_x$$

Because the deformation in the composite is comprised of the deformations in both the matrix and the fiber, therefore,

$$\Delta_B = \Delta_{mB} + \Delta_{fB}$$

where Δ_{mB} is the deformation of the matrix in the transverse direction and Δ_{fB} the deformation of the fibers in the transverse direction.

The deformation of the matrix in the transverse direction can be expressed by the following equation:

$$\Delta_{mB} = -B_m\varepsilon_{my}$$

Because $B_m = BV_m$ and $\varepsilon_{my} = -v_m\varepsilon_{mx}$,

$$\Delta_{mB} = BV_mv_m\varepsilon_{mx}$$

or,

$$\Delta_{mB} = BV_mv_m\varepsilon_x$$

where ε_{my} is the strain of the matrix in y direction, V_m is the volume fraction of the matrix, v_m is the Poisson's ratio of the matrix, and ε_{mx} is the strain of the matrix in x direction, which is equal to the strain of the composite in x direction, ε_x.

Similarly, the deformation of the fibers in the transverse direction can be derived below:

$$\Delta_{fB} = -B_f\varepsilon_{fy} = BV_fv_f\varepsilon_{fx} = BV_fv_f\varepsilon_x$$

where ε_{fy} is the strain of the fiber in y direction, V_f is the volume fraction of the fiber, v_f is the Poisson's ratio of the fiber, ε_{fx} is the strain of the fiber in x direction which is equal to the strain of the composite in x direction, ε_x. It is noted that if the fiber is made of isotropic materials, such as glass fiber or basalt fiber, v_f is not directionally dependent. However, if the fiber is made of anisotropic materials, such as carbon fiber or polymer fiber, v_f is directionally dependent and $v_f = v_{xy}^f$ (v_{xy}^f is the Poisson's ratio of the fiber when loaded in x direction or fiber direction) should be used in the equation.

Since $\Delta_B = Bv_{xy}\varepsilon_x$ and $\Delta_B = \Delta_{mB} + \Delta_{fB}$, the following relation can be established:

$$Bv_{xy}\varepsilon_x = BV_m v_m \varepsilon_x + BV_f v_f \varepsilon_x$$

or,

$$v_{xy} = V_m v_m + V_f v_f \tag{7.26}$$

The major Poisson's ratio, v_{xy}, of a unidirectional fiber thermoplastic composite can be calculated by Eq. (7.26) if the volume fractions of the fiber or the matrix and the Poisson's ratios of the fiber and the matrix are known.

7.2.1.3 Shear modulus G_{xy}

Shear modulus of a unidirectional fiber-reinforced thermoplastic composite can also be calculated using the ROM within its linear elastic range. When a shear stress, τ, is applied to the composite, the shear deformation of the composite δ is equal to γB, or $\delta = \gamma B$. The shear strains on the matrix and the fiber are described, respectively, in the equations below:

$$\gamma_m = \frac{\tau}{G_m} \text{ and } \gamma_f = \frac{\tau}{G_f}$$

where γ_m is the shear strain of the matrix, G_m is the shear modulus of the matrix, γ_f is the shear strain of the fiber, and G_f is the shear modulus of the fiber.

The shear deformations for the matrix and for the fiber can be written as $\delta_m = BV_m \gamma_m$, and $\delta_f = BV_f \gamma_f$, respectively. The shear deformation for the composite is the sum of both the shear deformations of the matrix and the fiber, therefore,

$$\delta = \delta_m + \delta_f = BV_m \gamma_m + BV_f \gamma_f$$

Since $\delta = \gamma B$ and $\gamma B = BV_m \gamma_m + BV_f \gamma_f$, thus, $\gamma = V_m \gamma_m + V_f \gamma_f$.

After substituting shear strains using $\gamma_m = \dfrac{\tau}{G_m}$, $\gamma_f = \dfrac{\tau}{G_f}$, and $\gamma = \dfrac{\tau}{G_{xy}}$,

$$\frac{\tau}{G_{xy}} = V_m \gamma_m + V_f \gamma_f = V_m \frac{\tau}{G_m} + V_f \frac{\tau}{G_f}$$

or,

$$\frac{1}{G_{xy}} = \frac{V_m}{G_m} + \frac{V_f}{G_f} \tag{7.27}$$

Equation (7.27) defines the shear modulus of the composite G_{xy} as a function of volume fractions of the fiber or the matrix, and the shear moduli of the fiber and the matrix.

Example question 7.1: A unidirectional glass fiber-reinforced polypropylene composite has a fiber weight percentage of 60% (60 wt%). The glass fiber has a density of 2.6 g/cm^3 and elastic modulus of 72 GPa. The density and the elastic modulus of polypropylene are 0.9 g/cm^3 and 1.2 GPa, respectively.
(a) What is the estimated elastic modulus of the composite in the fiber direction (longitudinal direction)?
(b) What is the estimated elastic modulus of the composite in the transverse direction?
(c) If an external load is applied to the composite in the fiber direction, what is the percentage of the load supported by the fibers?

Solution:
(a) The ROM can be used to estimate the elastic modulus of the composite in the longitudinal direction:

$$E_{cl} = [(1 - V_m)E_f + V_m E_m]$$

However, it is noted that fiber volume fraction (not fiber weight fraction) is used in the ROM. Because only the fiber weight fraction is given in the question, a conversion from the fiber weight fraction to fiber volume fraction is required.

It is considered that the total mass of the composite is 100 g. Therefore, the mass of the glass fibers (m_f) is 60 g and the mass of the matrix (m_m) is 40 g. Because the densities of both fiber and matrix are known, the total volume of the composite (v_c) can be calculated.
Volume of the composite is calculated below,

$$V_c = V_m + V_f = \frac{m_m}{\rho_m} + \frac{m_f}{\rho_f} = \left(\frac{40}{0.9}\right) + \left(\frac{60}{2.6}\right) = 67.5 \text{cm}^3$$

The fiber volume fraction of the composite is $V_f = \frac{V_f}{V_c} = 23.1/67.5 = 34.2$ vol%.
Therefore, the matrix volume fraction is $V_m = 1 - V_f = 65.8$ vol%.
It is noted that the fiber weight percentage can also be calculated using Eq. (6.1).
The elastic modulus of the composite in the longitudinal direction, E_{cl}, can be calculated below:

$$E_{cl} = (34.2\%)(72\text{GPa}) + (65.8\%)(1.2\text{GPa}) = 25.4\text{GPa}$$

(b) The elastic modulus of the composite in the transverse direction, E_{ct}, is calculated by Eq. (7.23):

$$\frac{1}{E_{ct}} = \frac{65.8\%}{1.2\text{GPa}} + \frac{34.2\%}{72\text{GPa}}$$

$$E_{ct} = 1.8 \ GPa$$

It can be seen that the elastic modulus in the transverse direction is much less than that in the longitudinal direction. It is slightly higher than the elastic modulus of the matrix.
(c) From Eq. (7.20), the percentage of the load supported by the fiber with respect to the total load applied to the composite is calculated below:

$$\frac{F_f}{F_c} = \frac{1}{1 + \left(\frac{E_m V_m}{E_f V_f}\right)}$$

$$\frac{F_f}{F_c} = \frac{1}{1 + \left[\frac{(1.2)(65.8\%)}{(72)(34.2\%)}\right]} = 96.9\%$$

The fibers in the composite support 96.9% of the overall load and the matrix bears the rest of the load, which is 3.1% of the overall load.

7.2.2 Halpin–Tsai equations

The Halpin–Tsai model was developed in the 1960s. It includes a set of semiempirical equations that integrate the properties of the matrix and reinforcement together with the volume fraction and geometry of the reinforcement. It can be used to calculate the properties of discontinuous fiber-reinforced composites, such as elastic modulus, shear modulus, and Poisson's ratio. The model can be expressed in the following equations:

$$\frac{P}{P_m} = \frac{(1 + \xi \eta V_f)}{(1 - \eta V_f)} \tag{7.28}$$

$$\eta = \left(\frac{P_f}{P_m} - 1\right) \bigg/ \left(\frac{P_f}{P_m} + \xi\right) \tag{7.29}$$

where P represents one of the properties of the discontinuous fiber-reinforced thermoplastic composite such as E_{11}, E_{22}, G_{12}, G_{23}, v_{12}, or v_{23}, P_m, and P_f which are the corresponding properties of the matrix (E_m, G_m, or v_m) and fiber (E_f, G_f, or v_f), respectively; V_f is the fiber volume fraction; ξ is an empirical factor that depends on conditions such as fiber geometry, fiber length, and loading condition; η is a function that is constructed in such a way that when $V_f = 0$ (neat thermoplastics), $P = P_m$, and when $V_f = 1$, $P = P_f$.

The Halpin–Tsai equations can be simplified to the ROM when ξ is considered to be extreme values. When $\xi = \infty$, the Halpin–Tsai equation can be written as $P = P_m(1 - V_f) + P_m V_f$. This equation is the ROM for the longitudinal properties of unidirectional fiber-reinforced thermoplastic composites. When $\xi = 0$, the Halpin–Tsai equations can be reduced to $\frac{1}{P} = \frac{V_m}{P_m} + \frac{V_f}{P_f}$, which is the ROM for the properties of unidirectional fiber-reinforced thermoplastic composites in the transverse direction. Therefore, the Halpin–Tsai model defines the upper and lower limit of the properties of thermoplastic composites. Any values between 0 and ∞ for ξ can define the property for thermoplastic composites.

For discontinuous fiber-reinforced thermoplastic composites, ξ varies based on the geometry and orientation of the fibers. When the fibers are below critical fiber length, $\xi = 2(l/d)$ is used for calculation of the elastic modulus of a short fiber thermoplastic composite with fiber orientation in the longitudinal direction, defined as the direction that the fibers are aligned, for example, flow direction of an injection molded composite; $\xi = 2$ is used for calculation of its elastic modulus in the transverse direction; $\xi = 1$ is used for calculation of its longitudinal shear modulus.

The Halpin–Tsai equations provide accurate prediction for thermoplastic composites at low volume fractions. Although underprediction of modulus at high volume fractions can occur, modifications can be made to compensate the underprediction by assigning ξ a function of fiber volume fraction at high volume fractions, for example, $\xi = 1 + 40v_f^{10}$ [8].

7.3 Macromechanics of thermoplastic composite

Macromechanics is the study of stress and strain behaviors of laminated composites, including laminated thermoplastic composites. It is considered that each lamina in the composite laminate, such as uni-tape, is homogeneous and possesses average properties that can be calculated through micromechanical models. The macromechanics approach has practical use in designing continuous fiber-reinforced thermoplastic composites. Although macromechanical analysis does not provide information on local failure mechanisms, it can predict material behaviors of individual lamina and the performance of the laminate.

Classical lamination theory (CLT) is a theoretical model that provides the stress and strain distribution in laminated composites based on their material properties and stacking sequences. It is also known as classical laminate theory. CLT is based on the stress–strain relationships, such as generalized Hooke's law, of the composite material. The following sections describe material anisotropy of thermoplastic composites, the generalized Hooke's law, transformation matrix, and the CLT.

7.3.1 Isotropy and anisotropy of thermoplastic composites

Solid materials can be categorized into isotropic materials and anisotropic materials. Isotropic materials have identical properties in all random directions. Examples include most bulk thermoplastics that have any random molecular chain orientations and thermoplastic composites reinforced with particulate and nanomaterials that have low aspect ratios. The mechanical properties of the isotropic material are not dependent on directions. On the other hand, anisotropic materials have directionally dependent properties. A typical example is the uni-tape. It has all the fibers aligned in one direction, which induces significantly higher mechanical properties in that direction than the properties in the other directions. Because of the direction-dependent properties, uni-tape is an anisotropic material, or more specifically, a transversely isotropic material (one type of anisotropic material). The directional dependence of the material property is called material anisotropy.

Uni-tapes or any unidirectional fiber thermoplastic composites possess high material anisotropy because of the significant difference in material properties, such as modulus and strength, in the fiber direction and in the transverse direction.

In general, continuous fiber-reinforced thermoplastic composites are anisotropic materials as their properties are directionally dependent. For example, in a thermoplastic composite laminate that is made of uni-tapes, the uni-tapes are stacked in different angles and molded into products. Figure 7.4 shows a laminate comprised of individual laminae with various fiber orientations (layup sequence of [0/45/90/−45/0]). The properties can be significantly different contingent on the orientations of the uni-tapes. It is worth noting that a thermoplastic composite laminate can have quasi-isotropic material properties. Laminae, such as uni-tapes, are arranged such that the in-plane properties are independent of direction (see Section 7.3.4). However, the quasi-isotropic thermoplastic composite is still not an isotropic material because its property in the thickness direction is much less than the in-plane directions due to lack of fibers oriented in the thickness direction.

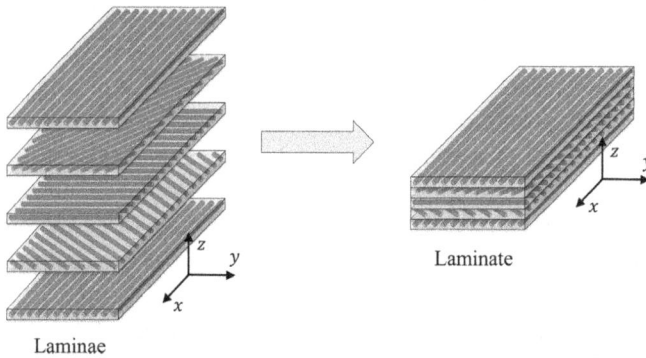

Laminae

Laminate

Fig. 7.4: Thermoplastic composite laminate comprised of individual laminae with various fiber orientations.

A special case of anisotropic materials is orthotropic materials. Those materials possess properties that are different in three mutually perpendicular directions. Because their properties depend on the direction in which they are measured, orthotropic materials are thus anisotropic materials. An example of orthotropic thermoplastic composites is the woven fabric-reinforced thermoplastic composite or cross-ply thermoplastic composite. Transversely isotropic materials are a special case of orthotropic materials. This type of material has properties that are symmetric about an axis that is normal to a plane of isotropy. A typical example is uni-tape or any unidirectional fiber-reinforced thermoplastic composites. The fiber direction is the symmetry axis and the plane of isotropy is normal to the fiber direction.

Because thermoplastic composites possess directionally dependent properties, their stress and strain behaviors are considerably more complex than those of the isotropic materials that have identical properties in all directions. The complex stress and strain response of the thermoplastic composite can significantly complicate design of thermoplastic composite structures. However, it can also offer opportunities for design engineers to use the anisotropic material characteristic to design structures with tailored properties in certain directions and efficient load-bearing capacity.

7.3.2 Generalized Hooke's law

When an isotropic material, such as particulate-filled thermoplastic composites consisting of equiaxed particles with an aspect ratio of around 1 (Fig. 2.63), undergoes stresses σ_{xx}, σ_{yy}, and σ_{zz} (Fig. 7.5), each of which causes a strain in x, y, and z directions. σ_{xx}, σ_{yy}, and σ_{zz} are often simplified to σ_x, σ_y, and σ_z, respectively.

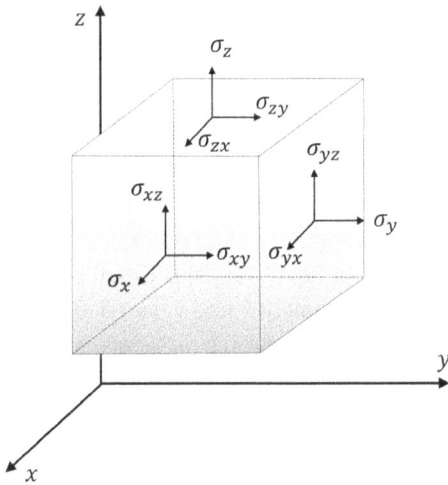

Fig. 7.5: Three-dimensional stress components showing normal stresses (σ_x, σ_y, and σ_z) and shear stresses.

The stress in the x direction, σ_x, results in a strain of (σ_x/E) in the x direction (according to $\sigma_x = E\varepsilon_x$), a strain of $(-v\sigma_x/E)$ in the y direction (based on $\sigma_x = E\varepsilon_x$ and $v = -\varepsilon_y/\varepsilon_x$), and a strain of $(-v\sigma_x/E)$ in the z direction (based on $\sigma_x = E\varepsilon_x$ and $v = -\varepsilon_z/\varepsilon_x$). Similarly, the stress in the y direction, σ_y, results in a strain of (σ_y/E) in the y direction, a strain of $(-v\sigma_y/E)$ in the x direction, and a strain of $(-v\sigma_y/E)$ in the z direction; the stress in the z direction, σ_z, results in a strain of (σ_z/E) in the z direction, a strain of $(-v\sigma_z/E)$ in the x direction, and a strain of $(-v\sigma_z/E)$ in the y direction.

The strains in the same direction can be summed together using a superposition approach when the stress and strain of the composite are still within the linear elastic range. The total strain in the x direction can be expressed by totaling all the strains in the x direction, namely, the strain (σ_x/E) caused by σ_x, the strain $(-v\sigma_y/E)$ caused by σ_y, and the strain $(-v\sigma_z/E)$ caused by σ_z:

$$\varepsilon_x = \frac{\sigma_x}{E} - v\frac{\sigma_y}{E} - v\frac{\sigma_z}{E}$$

Similarly,

$$\varepsilon_y = -v\frac{\sigma_x}{E} + \frac{\sigma_y}{E} - v\frac{\sigma_z}{E}$$

and,

$$\varepsilon_z = -v\frac{\sigma_x}{E} - v\frac{\sigma_y}{E} + \frac{\sigma_z}{E}$$

The shear stress and shear strain can be expressed by $\gamma_{xy} = \tau_{xy}/G$, $\gamma_{yz} = \tau_{yz}/G$, and $\gamma_{zx} = \tau_{zx}/G$. It is noted that there is no coupling effect in shear stresses and strains as seen in the normal stresses and strains. The shear modulus is related to the elastic modulus and the Poisson's ratio by the equation below:

$$G = \frac{E}{2(1+v)}$$

When both the normal stress–strain relationship and shear stress–strain relationship are combined, the overall stress and strain relationship of an isotropic material can be defined in a form of $\{\varepsilon\} = [s]\{\sigma\}$ as shown in Eq. (7.30). $[s]$ is called compliance matrix. Two independent material constants, namely E and v, can define the compliance matrix or the stress and strain behavior of isotropic materials, such as the thermoplastic composite filled with particles with an aspect ratio of around 1:

$$\begin{Bmatrix} \varepsilon_x \\ \varepsilon_y \\ \varepsilon_z \\ \gamma_{xy} \\ \gamma_{yz} \\ \gamma_{zx} \end{Bmatrix} = \begin{bmatrix} 1/E & -v/E & -v/E & 0 & 0 & 0 \\ -v/E & 1/E & -v/E & 0 & 0 & 0 \\ -v/E & -v/E & 1/E & 0 & 0 & 0 \\ 0 & 0 & 0 & 2(1+v)/E & 0 & 0 \\ 0 & 0 & 0 & 0 & 2(1+v)/E & 0 \\ 0 & 0 & 0 & 0 & 0 & 2(1+v)/E \end{bmatrix} \begin{Bmatrix} \sigma_x \\ \sigma_y \\ \sigma_z \\ \tau_{xy} \\ \tau_{yz} \\ \tau_{zx} \end{Bmatrix}$$

$$(7.30)$$

Certain thermoplastic composites, such as woven fabric-reinforced or cross-ply thermoplastic composites, are an orthotropic material. Their elastic moduli are directionally dependent and designated by E_x, E_y, and E_z. The Poisson's ratios of the orthotropic thermoplastic composite are also directionally dependent.

The same superposition approach can be used to define the stress and strain relationship in the orthotropic thermoplastic composite. The strains caused by the stress in the x direction (σ_x) are ε'_x, ε'_y, and ε'_z, in the x, y, and z directions, respectively. The strains are defined below:

$$\varepsilon_x' = \frac{\sigma_x}{E_x}$$

$$\varepsilon_y' = -v_{xy}\varepsilon_x' = -v_{xy}\frac{\sigma_x}{E_x}$$

$$\varepsilon_z' = -v_{xz}\varepsilon_x' = -v_{xz}\frac{\sigma_x}{E_x}$$

Similarly, there are three strains caused by σ_y and the strains are defined below:

$$\varepsilon_y'' = \frac{\sigma_y}{E_y}$$

$$\varepsilon_x'' = -v_{yx}\varepsilon_y'' = -v_{yx}\frac{\sigma_y}{E_y}$$

$$\varepsilon_z'' = -v_{yz}\varepsilon_y'' = -v_{yz}\frac{\sigma_y}{E_y}$$

Similarly, there are also three strains, ε_z''', ε_x''', and ε_y''', caused by σ_z. Those strains are defined as follows:

$$\varepsilon_z''' = \frac{\sigma_z}{E_z}$$

$$\varepsilon_x''' = -v_{zx}\varepsilon_z''' = -v_{zx}\frac{\sigma_z}{E}$$

$$\varepsilon_y''' = -v_{zy}\varepsilon_z''' = -v_{zy}\frac{\sigma_z}{E_z}$$

The strains in each direction can be summed together when the stress and strain of the composite are still within the linear elastic range, therefore,

$$\varepsilon_x = \varepsilon_x' + \varepsilon_x'' + \varepsilon_x''' = \frac{\sigma_x}{E_x} - v_{yx}\frac{\sigma_y}{E_y} - v_{zx}\frac{\sigma_z}{E_z}$$

$$\varepsilon_y = \varepsilon_y' + \varepsilon_y'' + \varepsilon_y''' = -v_{xy}\frac{\sigma_x}{E_x} + \frac{\sigma_y}{E_y} - v_{zy}\frac{\sigma_z}{E_z}$$

and,

$$\varepsilon_z = \varepsilon_z' + \varepsilon_z'' + \varepsilon_z''' = -v_{xz}\frac{\sigma_x}{E_x} - v_{yz}\frac{\sigma_y}{E_y} + \frac{\sigma_z}{E_z}$$

The normal stresses and strains for an orthotropic thermoplastic composite can be written in the following equation:

$$
\begin{Bmatrix} \varepsilon_x \\ \varepsilon_y \\ \varepsilon_z \end{Bmatrix} = \begin{bmatrix} 1/E_x & -v_{yx}/E_y & -v_{zx}/E_z \\ -v_{xy}/E_x & 1/E_y & -v_{zy}/E_z \\ -v_{xz}/E_x & -v_{yz}/E_y & 1/E_z \end{bmatrix} \begin{Bmatrix} \sigma_x \\ \sigma_y \\ \sigma_z \end{Bmatrix}
$$

After integrating shear stress and strain relationships, namely, $\gamma_{yz} = \tau_{yz}/G_{yz}$, $\gamma_{xz} = \tau_{xz}/G_{xz}$, and $\gamma_{xy} = \tau_{xy}/G_{xy}$, the strain and stress relationship for an ortho-tropic thermoplastic composite material can be written as follows:

$$
\begin{Bmatrix} \varepsilon_x \\ \varepsilon_y \\ \varepsilon_z \\ \gamma_{xy} \\ \gamma_{yz} \\ \gamma_{zx} \end{Bmatrix} = \begin{bmatrix} 1/E_x & -v_{yx}/E_y & -v_{zx}/E_z & 0 & 0 & 0 \\ -v_{xy}/E_x & 1/E_y & -v_{zy}/E_z & 0 & 0 & 0 \\ -v_{xz}/E_x & -v_{yz}/E_y & 1/E_z & 0 & 0 & 0 \\ 0 & 0 & 0 & 1/G_{xy} & 0 & 0 \\ 0 & 0 & 0 & 0 & 1/G_{yz} & 0 \\ 0 & 0 & 0 & 0 & 0 & 1/G_{zx} \end{bmatrix} \begin{Bmatrix} \sigma_x \\ \sigma_y \\ \sigma_z \\ \tau_{xy} \\ \tau_{yz} \\ \tau_{zx} \end{Bmatrix}
$$

The terms in the compliance matrix are often expressed using compliance tensors, or s_{ij} ($i, j = 1, 2,$ or 6):

$$
\begin{Bmatrix} \varepsilon_x \\ \varepsilon_y \\ \varepsilon_z \\ \gamma_{xy} \\ \gamma_{yz} \\ \gamma_{zx} \end{Bmatrix} = \begin{Bmatrix} S_{11} & S_{12} & S_{13} & 0 & 0 & 0 \\ S_{21} & S_{22} & S_{23} & 0 & 0 & 0 \\ S_{31} & S_{32} & S_{33} & 0 & 0 & 0 \\ 0 & 0 & 0 & S_{44} & 0 & 0 \\ 0 & 0 & 0 & 0 & S_{55} & 0 \\ 0 & 0 & 0 & 0 & 0 & S_{66} \end{Bmatrix} \begin{Bmatrix} \sigma_x \\ \sigma_y \\ \sigma_z \\ \tau_{xy} \\ \tau_{yz} \\ \tau_{zx} \end{Bmatrix}
$$

where $s_{11} = 1/E_x$, $s_{12} = -v_{yx}/E_y$, $s_{13} = -v_{zx}/E_z$, $s_{21} = -v_{xy}/E_x$, $s_{22} = 1/E_y$, $s_{23} = -v_{zy}/E_z$, $s_{31} = -v_{xz}/E_x$, $s_{32} = -v_{yz}/E_y$, $s_{33} = 1/E_z$, $s_{44} = 1/G_{yz}$, $s_{55} = 1/G_{xz}$, and $s_{66} = 1/G_{xy}$.

There are 12 material constants in the compliance matrix. However, the equations below, based on the Maxwell reciprocal theorem, reduce the material constants to 9 because 3 of the 12 material constants are not independent. The equations also indi-cate that $s_{12} = s_{21}$, $s_{13} = s_{31}$, and $s_{23} = s_{32}$.

$$
\frac{v_{xy}}{E_x} = \frac{v_{yx}}{E_y}; \frac{v_{yz}}{E_y} = \frac{v_{zy}}{E_z}; \text{ and } \frac{v_{xz}}{E_x} = \frac{v_{zx}}{E_z}
$$

Overall, nine independent material constants define the stress and strain relation-ship for orthotropic thermoplastic composites. Those material constants are often re-quired as the inputs for modeling structures made of orthotropic materials in finite element analysis.

The stress–strain relationship can also be written in the following equation using stiffness matrix:

$$\begin{Bmatrix} \sigma_x \\ \sigma_y \\ \sigma_z \\ \tau_{xy} \\ \tau_{yz} \\ \tau_{zx} \end{Bmatrix} = \begin{Bmatrix} C_{11} & C_{12} & C_{13} & 0 & 0 & 0 \\ C_{21} & C_{22} & C_{23} & 0 & 0 & 0 \\ C_{31} & C_{32} & C_{33} & 0 & 0 & 0 \\ 0 & 0 & 0 & C_{44} & 0 & 0 \\ 0 & 0 & 0 & 0 & C_{55} & 0 \\ 0 & 0 & 0 & 0 & 0 & C_{66} \end{Bmatrix} \begin{Bmatrix} \varepsilon_x \\ \varepsilon_y \\ \varepsilon_z \\ \gamma_{xy} \\ \gamma_{yz} \\ \gamma_{zx} \end{Bmatrix} \qquad (7.31)$$

where C_{ij} ($i, j = 1, 2,$ or 6) is defined as the stiffness tensor and $[C]$ is the stiffness matrix, inverse of the compliance matrix $[S]$.

Because the laminated thermoplastic composite is normally thin and has two dimensions (i.e., length and width) much larger than the third dimension (i.e., thickness), it is considered to be under a plane stress condition. The principal stress in the thickness direction (z direction) is much smaller than the stresses in the other directions and is negligible due to the plane stress condition. In addition, the shear stresses τ_{yz} and τ_{zx} are also negligible:

$$\sigma_z = 0; \tau_{yz} = 0; \text{ and } \tau_{zx} = 0$$

For a thermoplastic composite lamina that has its fiber aligned in the x direction in the geometry coordinate system (or global coordinate system) (Fig. 7.6), its stress and strain relationship can be described below after considering the plane stress condition:

$$\begin{Bmatrix} \varepsilon_x \\ \varepsilon_y \\ \gamma_{xy} \end{Bmatrix} = \begin{bmatrix} 1/E_x & -\nu_{xy}/E_x & 0 \\ -\nu_{xy}/E_x & 1/E_y & 0 \\ 0 & 0 & 1/G_{xy} \end{bmatrix} \begin{Bmatrix} \sigma_x \\ \sigma_y \\ \tau_{xy} \end{Bmatrix}$$

The equation can also be written as

$$\begin{Bmatrix} \sigma_x \\ \sigma_y \\ \tau_{xy} \end{Bmatrix} = \begin{bmatrix} C_{11} & C_{12} & 0 \\ C_{12} & C_{22} & 0 \\ 0 & 0 & C_{44} \end{bmatrix} \begin{Bmatrix} \varepsilon_x \\ \varepsilon_y \\ \gamma_{xy} \end{Bmatrix} \qquad (7.32)$$

where $C_{11} = \dfrac{E_x}{(1 - \nu_{xy}\nu_{yx})}$, $C_{12} = \dfrac{\nu_{yx}E_x}{(1 - \nu_{xy}\nu_{yx})}$, $C_{22} = \dfrac{E_y}{(1 - \nu_{xy}\nu_{yx})}$, and $C_{44} = G_{xy}$. It is noted that C_{12} can also be written as $\dfrac{\nu_{xy}E_y}{(1 - \nu_{xy}\nu_{yx})}$ based on the Maxwell reciprocal theorem $\left(\dfrac{\nu_{xy}}{E_x} = \dfrac{\nu_{yx}}{E_y}\right)$.

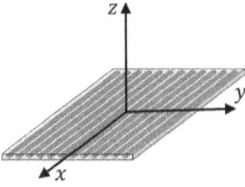

Fig. 7.6: A thermoplastic composite lamina that has all the fibers aligned in the x direction of the geometry coordinate system (global coordinate system).

7.3.3 Off-axis lamina

The schematic in Fig. 7.6 shows a lamina that has the fiber direction aligned with the global *x* direction or *x* axis. However, due to various layup sequences used in thermoplastic composite, the lamina often does not have the fiber direction aligned with the global *x* direction. Figure 7.7 shows a lamina with its fiber direction not aligned with the global *x* direction. Figure 7.7a,b shows the axes in a global coordinate system and the principal material directions (local coordinate system) of the same lamina, respectively. In the local coordinate system, the fiber direction is designated as 1-axis; the direction that is normal to the 1-axis and in the plane of the lamina is defined as 2-axis; and the direction that is normal to the 1-axis and 2-axis is 3-axis. The 1-axis has an offset angle of θ with the *x* axis in the global coordinate system and

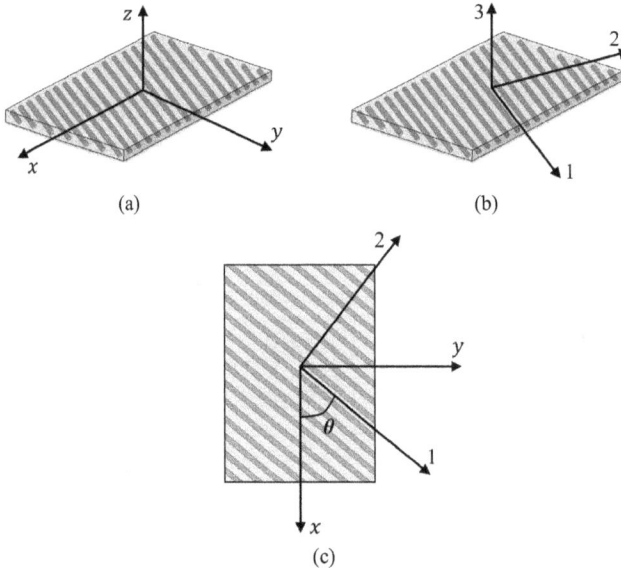

Fig. 7.7: (a) Global coordinate system for a lamina; (b) principal axes of the lamina in a local coordinate system; (c) top view of the off-axis angle θ between the global axes (*x* and *y* axes) and local axes (1 and 2 axes) of the lamina.

the 2-axis has the same angle offset. Figure 7.7c shows the superposition of the global axes (x and y axes) and local axes (1 and 2 axes) of the lamina.

The thermoplastic composite laminate may have each lamina with different offset angle θ with the global axes. Figure 7.4 illustrates a thermoplastic composite laminate comprised of five layers of laminae with various fiber orientations. The top and bottom layers have fibers oriented in the same direction as the global x axis (Fig. 7.6). The three layers in the middle, however, have fibers not oriented in the global x axis. The lamina with the fiber not oriented to the global x axis is called off-axis lamina. The angle θ is called off-axis angle.

Figure 7.8 shows five layers of glass fiber-reinforced polypropylene uni-tapes prepared for molding. Each uni-tape (or lamina) has dimensions of 152.4 mm × 152.4 mm. The uni-tapes are intentionally arranged in a cascade fashion to show the different orientation of each layer with the global x axis. The top uni-tape has all of its fibers aligned with the global x axis ($\theta = 0°$). The uni-tapes below are off-axis laminae and their off-axis angles are 22.5°, 45°, 67.5°, and 90° (from top to bottom), respectively.

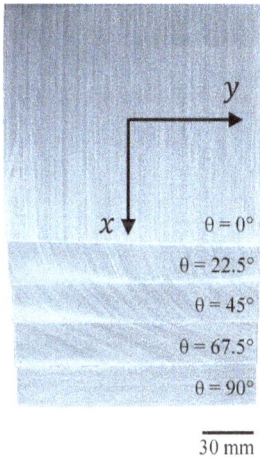

Fig. 7.8: Five layers of glass fiber-reinforced polypropylene uni-tapes that have different fiber angles with the global x axis. The top uni-tape has all of its fibers aligned with the global x axis ($\theta = 0°$) and the uni-tapes below are off-axis laminae and their off-axis angles are 22.5°, 45°, 67.5°, and 90° (from top to bottom), respectively.

For the top uni-tape layer ($\theta = 0°$), its principal material axes of symmetry coincide with the geometric axes, and its stress and strain relationship is defined by Eq. (7.33). It is considered that the lamina is in a plane stress condition because its thickness is much less than its width and length:

$$\left\{ \begin{array}{c} \sigma_1 \\ \sigma_2 \\ \tau_{12} \end{array} \right\} = \begin{bmatrix} Q_{11} & Q_{12} & 0 \\ Q_{12} & Q_{22} & 0 \\ 0 & 0 & Q_{66} \end{bmatrix} \left\{ \begin{array}{c} \varepsilon_1 \\ \varepsilon_2 \\ \gamma_{12} \end{array} \right\} \tag{7.33}$$

where Q_{ij} ($i, j = 1, 2,$ or 6) are the reduced stiffnesses of the lamina.

If the material symmetry axes of the lamina and the geometric axes do not coincide, for example, the uni-tape layer with $\theta = 45°$ in Fig. 7.8, it is a general case of orthotropy. Its stress and strain relationship is defined as follows:

$$\left\{ \begin{array}{c} \sigma_x \\ \sigma_y \\ \tau_{xy} \end{array} \right\} = \begin{bmatrix} \bar{Q}_{11} & \bar{Q}_{12} & \bar{Q}_{16} \\ \bar{Q}_{12} & \bar{Q}_{22} & \bar{Q}_{26} \\ \bar{Q}_{16} & \bar{Q}_{26} & \bar{Q}_{66} \end{bmatrix} \left\{ \begin{array}{c} \varepsilon_x \\ \varepsilon_y \\ \gamma_{xy} \end{array} \right\} \tag{7.34}$$

where $\bar{Q}_{ij}$ (i, $j = 1$, 2, or 6) are called transformed reduced stiffnesses or off-axis reduced stiffnesses. The off-axis reduced stiffnesses are calculated based on the reduced stiffnesses (Q_{ij}) and the off-axis angle (θ) using the following equations. The detailed derivation of the equations can be found in Reference [9]:

$$\bar{Q}_{11} = Q_{11}m^4 + 2(Q_{12} + 2Q_{66})n^2m^2 + Q_{22}n^4$$

$$\bar{Q}_{12} = (Q_{11} + Q_{22} - 4Q_{66})n^2m^2 + Q_{12}(n^4 + m^4)$$

$$\bar{Q}_{16} = (Q_{11} - Q_{12} - 2Q_{66})nm^3 + (Q_{12} - Q_{22} + 2Q_{66})n^3m$$

$$\bar{Q}_{22} = Q_{11}n^4 + 2(Q_{12} + 2Q_{66})n^2m^2 + Q_{22}m^4$$

$$\bar{Q}_{26} = (Q_{11} - Q_{12} - 2Q_{66})n^3m + (Q_{12} - Q_{22} + 2Q_{66})nm^3$$

$$\bar{Q}_{66} = (Q_{11} + Q_{22} - 2Q_{12} - 2Q_{66})n^2m^2 + Q_{66}(n^4 + m^4)$$

where $m = cos\theta$ and $n = sin\theta$.

7.3.4 Classical lamination theory (CLT)

CLT is a theory that describes the effect of fiber directions, stacking arrangements, layer thickness, and material properties of each lamina on the mechanical properties of laminated composites, including thermoplastic composites. It is an analytical tool that can be used to evaluate the stress, strain, and curvature in the composite as well as the complex coupling effect that can occur in the composite.

CLT can predict the properties of thermoplastic composite laminates as a function of material variables such as the properties of individual lamina, thickness of individual lamina, and orientation of individual lamina. If the material variables of each lamina in a thermoplastic composite are known, the performance of the thermoplastic composite laminate can be predicted using CLT. The properties of individual lamina can be calculated through micromechanical models such as the ROM. The properties can also be measured through materials testing. Additionally, the properties may be found from manufacturer's datasheet. Some mechanical properties of uni-tapes are listed in Tabs. 3.1 and 3.2.

Several assumptions are made in the CLT. Those assumptions are listed below:

(a) The thickness of the composite laminate is much smaller than its length and width, and the composite laminate is in a plane stress condition.

(b) Each lamina is homogeneous and orthotropic.

(c) Each lamina follows Hooke's law, or the stress–strain relation in each lamina is linear.

(d) No debonding or delamination occurs between the laminae during deformation.

(e) Displacements are continuous and small throughout the laminate.

(f) The laminate strain varies linearly through the thickness as a function of the laminate mid-plane strains and curvatures.

Figure 7.9 shows the force resultants and moment resultants in a laminated thermoplastic composite. Force resultant is defined as the normal or shear force over unit length of the composite. N_x is the normal force resultant caused by the force in the x direction. The unit length is the length in the y direction. Therefore, N_x is equal to normal force in the x direction over unit length in the y direction. Similarly, N_y is the normal force resultant in the y direction (normal force in the y direction/unit length in the x direction). N_{xy} is the shear force resultant (shear force/unit length) in the xy plane. Moment resultant is defined as the moment over unit length of the composite. M_x and M_y are the bending moment resultants (bending moment/unit length) in the yz plane and xz plane, respectively. M_{xy} is the twist moment resultant (twisting moment/unit length).

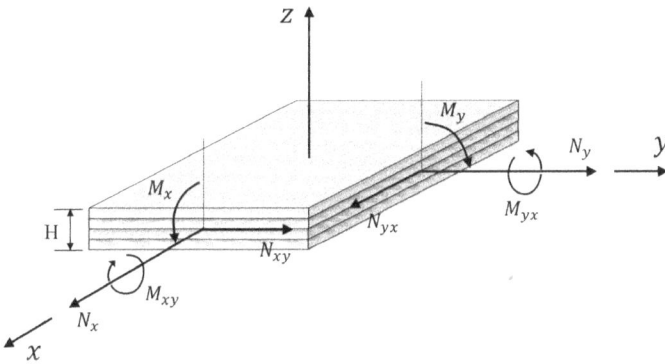

Fig. 7.9: Force resultants (N_x, N_y, and N_{xy}) and moment resultants (M_x, M_y, and M_{xy}) in a laminated thermoplastic composite.

Integration of the stresses across the thickness of each lamina (through the thickness of the laminate H) provides the force per unit length. Therefore, each force resultant can be defined by integrating the stress through the laminate thickness. The equation below shows that N_x is equal to the integration of the stress in the global x

direction, σ_x, through the thickness of the thermoplastic composite laminate from $-H/2$ to $H/2$ (Fig. 7.10):

$$N_x = \int_{-H/2}^{H/2} \sigma_x dz$$

Similarly, the other two force resultants, N_y and N_{xy}, can also be defined by integrating stresses σ_y and τ_{xy} through the laminate thickness (from $-H/2$ to $H/2$), respectively:

$$N_y = \int_{-H/2}^{H/2} \sigma_y dz$$

$$N_{xy} = \int_{-H/2}^{H/2} \tau_{xy} dz$$

The normal stresses (σ_x and σ_y) can cause bending moments, and the shear stress τ_{xy} can cause a twisting moment. By integrating the stresses through the laminate thickness, the moment resultants M_x, M_y, and M_{xy} can be defined, respectively, below:

$$M_x = \int_{-H/2}^{H/2} \sigma_x z dz$$

$$M_y = \int_{-H/2}^{H/2} \sigma_y z dz$$

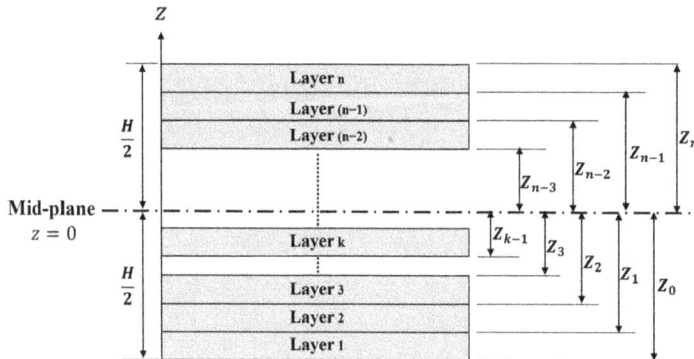

Fig. 7.10: A thermoplastic composite laminate consisting of n layers of laminae.

$$M_{xy} = \int_{-H/2}^{H/2} \tau_{xy} z \, dz$$

The force resultants, N_x, N_y, and N_{xy}, result in mid-plane strains ε_x^o, ε_y^o, and γ_{xy}^o, respectively. The moment resultants, M_x, M_y, and M_{xy}, cause bending curvatures at the mid-planes κ_x^o, κ_y^o, and κ_{xy}^o, respectively. Mid-plane is the plane when $z = 0$ (Fig. 7.10). Sometimes the superscript "o" in the mid-plane strain and curvature is omitted. The strains of the composite laminate, ε_x, ε_y, and γ_{xy}, can be written in relation to the mid-plane strains, bending curvatures at the mid-plane, and the location in the thickness z direction:

$$\left\{ \begin{array}{c} \varepsilon_x \\ \varepsilon_y \\ \gamma_{xy} \end{array} \right\} = \left\{ \begin{array}{c} \varepsilon_x^o + z\kappa_x^o \\ \varepsilon_y^o + z\kappa_y^o \\ \gamma_{xy}^o + z\kappa_{xy}^o \end{array} \right\}$$

The strains in Eq. (7.34) can be substituted, therefore,

$$\left\{ \begin{array}{c} \sigma_x \\ \sigma_y \\ \tau_{xy} \end{array} \right\} = \left[\begin{array}{ccc} \bar{Q}_{11} & \bar{Q}_{12} & \bar{Q}_{16} \\ \bar{Q}_{12} & \bar{Q}_{22} & \bar{Q}_{26} \\ \bar{Q}_{16} & \bar{Q}_{26} & \bar{Q}_{66} \end{array} \right] \left\{ \begin{array}{c} \varepsilon_x^o + z\kappa_x^o \\ \varepsilon_y^o + z\kappa_y^o \\ \gamma_{xy}^o + z\kappa_{xy}^o \end{array} \right\}$$

σ_x can be expressed in relation to off-axis reduced stiffnesses, mid-plane strains, mid-plane curvatures, and location in the thickness z direction:

$$\sigma_x = \bar{Q}_{11} \left(\varepsilon_x^o + z\kappa_x^o \right) + \bar{Q}_{12} \left(\varepsilon_y^o + z\kappa_y^o \right) + \bar{Q}_{16} \left(\gamma_{xy}^o + z\kappa_{xy}^o \right)$$

Thus,

$$N_x = \int_{-H/2}^{H/2} \sigma_x \, dz = \int_{-H/2}^{H/2} \left\{ \bar{Q}_{11} \left(\varepsilon_x^o + z\kappa_x^o \right) + \bar{Q}_{12} \left(\varepsilon_y^o + z\kappa_y^o \right) + \bar{Q}_{16} \left(\gamma_{xy}^o + z\kappa_{xy}^o \right) \right\} dz$$

$$N_x = \int_{-H/2}^{H/2} \left\{ \bar{Q}_{11}\varepsilon_x^o + \bar{Q}_{11}z\kappa_x^o + \bar{Q}_{12}\varepsilon_y^o + \bar{Q}_{12}z\kappa_y^o + \bar{Q}_{16}\gamma_{xy}^o + \bar{Q}_{16}z\kappa_{xy}^o \right\} dz$$

$$N_x = \int_{-H/2}^{H/2} \bar{Q}_{11}\varepsilon_x^o \, dz + \int_{-H/2}^{H/2} \bar{Q}_{11}z\kappa_x^o \, dz + \int_{-H/2}^{H/2} \bar{Q}_{12}\varepsilon_y^o \, dz + \int_{-H/2}^{H/2} \bar{Q}_{12}z\kappa_y^o \, dz + \int_{-H/2}^{H/2} \bar{Q}_{16}\gamma_{xy}^o \, dz +$$

$$\int_{-H/2}^{H/2} \bar{Q}_{16}z\kappa_{xy}^o \, dz$$

$$N_x = \varepsilon_x^o \int_{-H/2}^{H/2} \bar{Q}_{11} dz + \kappa_x^o \int_{-H/2}^{H/2} \bar{Q}_{11} z dz + \varepsilon_y^o \int_{-H/2}^{H/2} \bar{Q}_{12} dz + \kappa_y^o \int_{-H/2}^{H/2} \bar{Q}_{12} z dz +$$

$$\gamma_{xy}^o \int_{-H/2}^{H/2} \bar{Q}_{16} dz + \kappa_{xy}^o \int_{-H/2}^{H/2} \bar{Q}_{16} z dz$$

(7.35)

The first term in Eq. (7.35), $\varepsilon_x^o \int_{-H/2}^{H/2} \bar{Q}_{11} dz$, can be expanded by considering individual laminae (Fig. 7.10) and their off-axis reduced stiffnesses:

$$\varepsilon_x^o \int_{-H/2}^{H/2} \bar{Q}_{11} dz = \varepsilon_x^o \left(\int_{z_0}^{z_1} \bar{Q}_{11_1} dz + \int_{z_1}^{z_2} \bar{Q}_{11_2} dz + \cdots + \int_{z_{k-1}}^{z_k} \bar{Q}_{11_k} dz + \cdots + \int_{z_{N-1}}^{z_N} \bar{Q}_{11_N} dz \right)$$

where $\bar{Q}_{11_k}$ is the $\bar{Q}_{11}$ value for the kth lamina layer. For instance, $\bar{Q}_{11_1}$ is the off-axis reduced stiffness of the first lamina layer. The $\bar{Q}_{11}$ value for each layer can be calculated based on the fiber orientation and properties of that layer (elastic moduli and Poisson's ratios, see Section 7.3.3). The off-axis reduced stiffness is a constant and not a function of z, therefore,

$$\varepsilon_x^o \int_{-H/2}^{H/2} \bar{Q}_{11} dz = \varepsilon_x^o \left[\bar{Q}_{11_1}(z_1 - z_0) + \bar{Q}_{11_2}(z_2 - z_1) + \cdots + \bar{Q}_{11_k}(z_k - z_{k-1}) + \cdots + \bar{Q}_{11_N}(z_N - z_{N-1}) \right]$$

or,

$$\int_{-H/2}^{H/2} \bar{Q}_{11} dz = \sum_{k=1}^{N} \bar{Q}_{11_k}(z_k - z_{k-1})$$

The difference in the z value $(z_k - z_{k-1})$ is the thickness of the kth layer, h_k:

$$\int_{-H/2}^{H/2} \bar{Q}_{11} dz = \sum_{k=1}^{N} \bar{Q}_{11_k} h_k = A_{11}$$

Similarly,

$$\int_{-H/2}^{H/2} \bar{Q}_{12} dz = \sum_{k=1}^{N} \bar{Q}_{12_k} h_k = A_{12}$$

and,

$$\int_{-H/2}^{H/2} \bar{Q}_{16} dz = \sum_{k=1}^{N} \bar{Q}_{16_k} h_k = A_{16}$$

Therefore, the first, third, and fifth terms in Eq. (7.35) can be written as $A_{11}\varepsilon_x^o$, $A_{12}\varepsilon_y^o$, and $A_{16}\varepsilon_{xy}^o$, respectively.

The second term in Eq. (7.35) can be rewritten as follows:

$$\kappa_x^o \int_{-H/2}^{H/2} \bar{Q}_{11} z dz = \kappa_x^o \left[\frac{1}{2}\bar{Q}_{11_1}\left(z_1^2 - z_0^2\right) + \frac{1}{2}\bar{Q}_{11_2}\left(z_2^2 - z_1^2\right) + \cdots + \frac{1}{2}\bar{Q}_{11_k}\left(z_k^2 - z_{k-1}^2\right) + \cdots \right.$$
$$\left. + \frac{1}{2}\bar{Q}_{11_N}\left(z_N^2 - z_{N-1}^2\right) \right]$$

or,

$$\kappa_x^o \int_{-H/2}^{H/2} \bar{Q}_{11} z dz = \kappa_x^o \left[\frac{1}{2}\sum_{k=1}^{N} \bar{Q}_{11_k}\left(z_k^2 - z_{k-1}^2\right) \right] = B_{11}\kappa_x^o$$

Similarly, the fourth and sixth term can be rewritten, respectively, as follow:

$$\kappa_y^o \int_{-H/2}^{H/2} \bar{Q}_{12} z dz = \kappa_y^o \left[\frac{1}{2}\sum_{k=1}^{N} \bar{Q}_{12_k}\left(z_k^2 - z_{k-1}^2\right) \right] = B_{12}\kappa_y^o$$

$$\kappa_{xy}^o \int_{-H/2}^{H/2} \bar{Q}_{16} z dz = \kappa_{xy}^o \left[\frac{1}{2}\sum_{k=1}^{N} \bar{Q}_{16_k}\left(z_k^2 - z_{k-1}^2\right) \right] = B_{16}\kappa_{xy}^o$$

Therefore, Eq. (7.35) can be written as

$$N_x = A_{11}\varepsilon_x^o + A_{12}\varepsilon_y^o + A_{16}\gamma_{xy}^o + B_{11}\kappa_x^o + B_{12}\kappa_y^o + B_{16}\kappa_{xy}^o$$

Similarly, N_y, N_{xy}, M_x, M_y, and M_{xy} can be derived and described, respectively, in the following equations:

$$N_y = A_{12}\varepsilon_x^o + A_{22}\varepsilon_y^o + A_{26}\gamma_{xy}^o + B_{12}\kappa_x^o + B_{22}\kappa_y^o + B_{26}\kappa_{xy}^o$$

$$N_{xy} = A_{16}\varepsilon_x^o + A_{26}\varepsilon_y^o + A_{66}\gamma_{xy}^o + B_{16}\kappa_x^o + B_{26}\kappa_y^o + B_{66}\kappa_{xy}^o$$

$$M_x = B_{11}\varepsilon_x^o + B_{12}\varepsilon_y^o + B_{16}\gamma_{xy}^o + D_{11}\kappa_x^o + D_{12}\kappa_y^o + D_{16}\kappa_{xy}^o$$

$$M_y = B_{12}\varepsilon_x^o + B_{22}\varepsilon_y^o + B_{26}\gamma_{xy}^o + D_{12}\kappa_x^o + D_{22}\kappa_y^o + D_{26}\kappa_{xy}^o$$

$$M_{xy} = B_{16}\varepsilon_x^o + B_{26}\varepsilon_y^o + B_{66}\gamma_{xy}^o + D_{16}\kappa_x^o + D_{26}\kappa_y^o + D_{66}\kappa_{xy}^o$$

or,

$$\begin{Bmatrix} N_x \\ N_y \\ N_{xy} \\ M_x \\ M_y \\ M_{xy} \end{Bmatrix} = \begin{bmatrix} A_{11} & A_{12} & A_{16} & B_{11} & B_{12} & B_{16} \\ A_{12} & A_{22} & A_{26} & B_{12} & B_{22} & B_{26} \\ A_{16} & A_{26} & A_{66} & B_{16} & B_{26} & B_{66} \\ B_{11} & B_{12} & B_{16} & D_{11} & D_{12} & D_{16} \\ B_{12} & B_{22} & B_{26} & D_{12} & D_{22} & D_{26} \\ B_{16} & B_{26} & B_{66} & D_{16} & D_{26} & D_{66} \end{bmatrix} \begin{Bmatrix} \varepsilon_x^o \\ \varepsilon_y^o \\ \gamma_{xy}^o \\ \kappa_x^o \\ \kappa_y^o \\ \kappa_{xy}^o \end{Bmatrix} \qquad (7.36)$$

The 6×6 matrix in Eq. (7.36) is called ABD matric. The 3×3 matrix $[A]$ formed by the A_{ij} ($i, j = 1, 2,$ or 6) terms is called extensional stiffness matrix that relates the in-plane resultant forces to the in-plane strains. The 3×3 matrix $[B]$ formed by the B_{ij} ($i, j = 1, 2,$ or 6) terms is called bending–extension coupling matrix that relates the resultant forces and moments to the mid-plane strains and mid-plane bending curvatures. The 3×3 matrix $[D]$ formed by the D_{ij} ($i, j = 1, 2,$ or 6) terms is called bending stiffness matrix that relates the resultant bending moments to the bending curvatures. The A_{ij}, B_{ij}, and D_{ij} terms are defined, respectively, in the following equations:

$$A_{ij} = \sum_{k=1}^{N} \bar{Q}_{ij_k} (z_k - z_{k-1})$$

$$B_{ij} = \frac{1}{2} \sum_{k=1}^{N} \bar{Q}_{ij_k} (z_k^2 - z_{k-1}^2)$$

$$D_{ij} = \frac{1}{3} \sum_{k=1}^{N} \bar{Q}_{ij_k} (z_k^3 - z_{k-1}^3)$$

Figure 7.11 shows different couplings in Eq. (7.36) and associated stiffness terms. Those couplings are shear and extension coupling, bending and extension coupling, and bending and twisting coupling. It shows the ABD matrix with specified physical significance of stiffness terms in force and moment resultants.

Figure 7.12 compares the deformation response of thermoplastic composites with different fiber orientations without bending–twisting coupling and with bending–twisting coupling. Both composites have exaggerated thicknesses to show the difference in twist. The composite in Fig. 7.12a is a cross-ply laminate and has all the fibers in either 0 or 90°. It does not have twist–bending coupling ($D_{16} = D_{26} = 0$) and, therefore, no twist occurs when it bends. As a comparison, another thermoplastic composite (Fig. 7.12b) has off-axis fibers only that are not oriented in the bending direction. When it is under a bending load, the composite undergoes twisting due to the bending-twist coupling.

Equation (7.35) describes a general relationship among the force and moment resultants and mid-plane strains and bending curvatures at the mid-plane of laminated thermoplastic composites. The ABD matrix in the equation determines the mechanical behavior of the thermoplastic laminate. The terms in the ABD matrix are dependent on variables such as fiber directions, stacking arrangements, layer

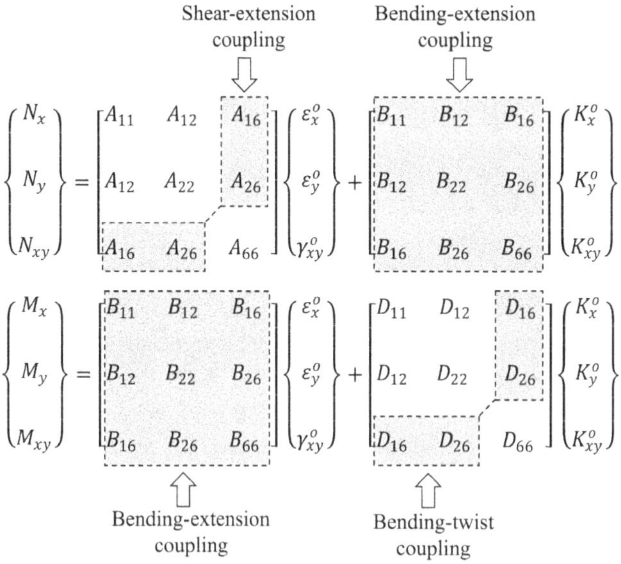

$$
\begin{Bmatrix} N_x \\ N_y \\ N_{xy} \end{Bmatrix} = \begin{bmatrix} A_{11} & A_{12} & A_{16} \\ A_{12} & A_{22} & A_{26} \\ A_{16} & A_{26} & A_{66} \end{bmatrix} \begin{Bmatrix} \varepsilon_x^o \\ \varepsilon_y^o \\ \gamma_{xy}^o \end{Bmatrix} + \begin{bmatrix} B_{11} & B_{12} & B_{16} \\ B_{12} & B_{22} & B_{26} \\ B_{16} & B_{26} & B_{66} \end{bmatrix} \begin{Bmatrix} K_x^o \\ K_y^o \\ K_{xy}^o \end{Bmatrix}
$$

Shear-extension coupling · Bending-extension coupling

$$
\begin{Bmatrix} M_x \\ M_y \\ M_{xy} \end{Bmatrix} = \begin{bmatrix} B_{11} & B_{12} & B_{16} \\ B_{12} & B_{22} & B_{26} \\ B_{16} & B_{26} & B_{66} \end{bmatrix} \begin{Bmatrix} \varepsilon_x^o \\ \varepsilon_y^o \\ \gamma_{xy}^o \end{Bmatrix} + \begin{bmatrix} D_{11} & D_{12} & D_{16} \\ D_{12} & D_{22} & D_{26} \\ D_{16} & D_{26} & D_{66} \end{bmatrix} \begin{Bmatrix} K_x^o \\ K_y^o \\ K_{xy}^o \end{Bmatrix}
$$

Bending-extension coupling · Bending-twist coupling

Fig. 7.11: Shear–extension coupling, bending–extension coupling, and bending–twisting coupling in the ABD matrix (adapted from Reference [10]).

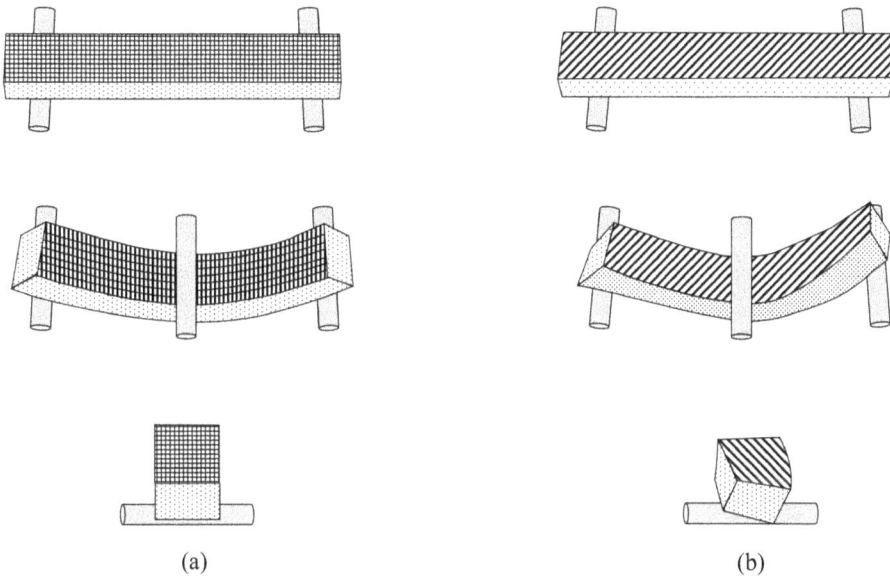

(a)   (b)

Fig. 7.12: (a) A cross-ply thermoplastic composite showing no twist; (b) twist induced by bending-twist coupling in the composite with off-axis fibers that are not oriented in the bending direction (adapted from Reference [10]).

thickness, and material properties. There are various layup sequences in thermo-plastic composite laminates and those variables, some of which are special cases, can vary. Those special cases include symmetric, balanced, symmetric and bal-anced, cross-ply, anti-symmetric, and quasi-isotropic layup sequences. The ABD matrices vary in those cases and are described below:

1. **A symmetric laminate** is defined as a laminated composite that has symmetrical layup sequence and layer thickness along its mid-plane, for example, $[90/0/45]_s$. A subscript "s," denoting "symmetry," is added to the layup sequence, and there are totally six plies in $[90/0/45]_s$, which is expanded to be $[90/0/45/45/0/90]$. For a sym-metric thermoplastic composite, which has a layup sequence symmetric to its mid-plane, there is no extension–bending coupling and $[B] = [0]$. When the symmetric laminate is subjected to forces only, its mid-plane curvature is zero. When it is sub-jected to moments only, the mid-plane strain is zero. When there is an extensional stress, no bending moment will be induced. Similarly, when there is a bending mo-ment, no extensional strain is resulted:

$$
\begin{Bmatrix} N_x \\ N_y \\ N_{xy} \\ M_x \\ M_y \\ M_{xy} \end{Bmatrix} =
\begin{bmatrix}
A_{11} & A_{12} & A_{16} & 0 & 0 & 0 \\
A_{12} & A_{22} & A_{26} & 0 & 0 & 0 \\
A_{16} & A_{26} & A_{66} & 0 & 0 & 0 \\
0 & 0 & 0 & D_{11} & D_{12} & D_{16} \\
0 & 0 & 0 & D_{12} & D_{22} & D_{26} \\
0 & 0 & 0 & D_{16} & D_{26} & D_{66}
\end{bmatrix}
\begin{Bmatrix} \varepsilon_x^o \\ \varepsilon_y^o \\ \gamma_{xy}^o \\ \kappa_x^o \\ \kappa_y^o \\ \kappa_{xy}^o \end{Bmatrix}
$$

The force and moment equations can be decoupled and written in two separate equations below:

$$
\begin{Bmatrix} N_x \\ N_y \\ N_{xy} \end{Bmatrix} =
\begin{bmatrix}
A_{11} & A_{12} & A_{16} \\
A_{12} & A_{22} & A_{26} \\
A_{16} & A_{26} & A_{66}
\end{bmatrix}
\begin{Bmatrix} \varepsilon_x^o \\ \varepsilon_y^o \\ \gamma_{xy}^o \end{Bmatrix}
$$

and,

$$
\begin{Bmatrix} M_x \\ M_y \\ M_{xy} \end{Bmatrix} =
\begin{bmatrix}
D_{11} & D_{12} & D_{16} \\
D_{12} & D_{22} & D_{26} \\
D_{16} & D_{26} & D_{66}
\end{bmatrix}
\begin{Bmatrix} \kappa_x^o \\ \kappa_y^o \\ \kappa_{xy}^o \end{Bmatrix}
$$

2. **A balanced laminate** refers to the laminate that has paired plies of the same thick-ness and material with the angles of the plies $+\theta$ and $-\theta$. For the balanced lami-nate, for example, $[0/-60/30/60/-30/0]$, there is no shear-extension coupling and $A_{16} = A_{26} = 0$. The force and moment equation can be written as follows:

$$
\begin{Bmatrix} N_x \\ N_y \\ N_{xy} \\ M_x \\ M_y \\ M_{xy} \end{Bmatrix} =
\begin{bmatrix}
A_{11} & A_{12} & 0 & B_{11} & B_{12} & B_{16} \\
A_{12} & A_{22} & 0 & B_{12} & B_{22} & B_{26} \\
0 & 0 & A_{66} & B_{16} & B_{26} & B_{66} \\
B_{11} & B_{12} & B_{16} & D_{11} & D_{12} & D_{16} \\
B_{12} & B_{22} & B_{26} & D_{12} & D_{22} & D_{26} \\
B_{16} & B_{26} & B_{66} & D_{16} & D_{26} & D_{66}
\end{bmatrix}
\begin{Bmatrix} \varepsilon_x^o \\ \varepsilon_y^o \\ \gamma_{xy}^o \\ \kappa_x^o \\ \kappa_y^o \\ \kappa_{xy}^o \end{Bmatrix}
$$

3. **A symmetrical and balanced laminate** refers a laminate with plies that are not only symmetrical but also balanced, for example, a laminate with plies arranged in $[0/\pm 45/90]_s$ which is expanded to be $[0/+45/-45/90/90/-45/+45/0]$. There is no extension–bending coupling and also [B]=0:

$$
\begin{Bmatrix} N_x \\ N_y \\ N_{xy} \end{Bmatrix} =
\begin{bmatrix}
A_{11} & A_{12} & 0 \\
A_{12} & A_{22} & 0 \\
0 & 0 & A_{66}
\end{bmatrix}
\begin{Bmatrix} \varepsilon_x^o \\ \varepsilon_y^o \\ \gamma_{xy}^o \end{Bmatrix}
$$

and,

$$
\begin{Bmatrix} M_x \\ M_y \\ M_{xy} \end{Bmatrix} =
\begin{bmatrix}
D_{11} & D_{12} & D_{16} \\
D_{12} & D_{22} & D_{26} \\
D_{16} & D_{26} & D_{66}
\end{bmatrix}
\begin{Bmatrix} \kappa_x^o \\ \kappa_y^o \\ \kappa_{xy}^o \end{Bmatrix}
$$

4. **A cross-ply laminate** is a composite with plies that have either an orientation of 0 or 90°, for example, $[0/90]_4$. The subscript "4" denotes four repeats of 0/90 in the composite, and it can be expanded to be $[0/90/0/90/0/90/0/90]$. Cross-ply is one of the common layup sequences seen in composites, including thermoplastic composites. No coupling exists between shear and extensional forces, or between the bending and twisting moments, and $A_{16} = A_{26} = 0$, $B_{16} = B_{26} = 0$, and $D_{16} = D_{26} = 0$:

$$
\begin{Bmatrix} N_x \\ N_y \\ N_{xy} \\ M_x \\ M_y \\ M_{xy} \end{Bmatrix} =
\begin{bmatrix}
A_{11} & A_{12} & 0 & B_{11} & B_{12} & 0 \\
A_{12} & A_{22} & 0 & B_{12} & B_{22} & 0 \\
0 & 0 & A_{66} & 0 & 0 & B_{66} \\
B_{11} & B_{12} & 0 & D_{11} & D_{12} & 0 \\
B_{12} & B_{22} & 0 & D_{12} & D_{22} & 0 \\
0 & 0 & B_{66} & 0 & 0 & D_{66}
\end{bmatrix}
\begin{Bmatrix} \varepsilon_x^o \\ \varepsilon_y^o \\ \gamma_{xy}^o \\ \kappa_x^o \\ \kappa_y^o \\ \kappa_{xy}^o \end{Bmatrix}
$$

When the cross-ply laminate is also symmetric, for example $[0/90]_s$, there is no coupling between the force and moment terms and $[B] = [0]$:

$$
\begin{Bmatrix} N_x \\ N_y \\ N_{xy} \end{Bmatrix} = \begin{bmatrix} A_{11} & A_{12} & 0 \\ A_{12} & A_{22} & 0 \\ 0 & 0 & A_{66} \end{bmatrix} \begin{Bmatrix} \varepsilon_x^o \\ \varepsilon_y^o \\ \gamma_{xy}^o \end{Bmatrix}
$$

and,

$$
\begin{Bmatrix} M_x \\ M_y \\ M_{xy} \end{Bmatrix} = \begin{bmatrix} D_{11} & D_{12} & 0 \\ D_{12} & D_{22} & 0 \\ 0 & 0 & D_{66} \end{bmatrix} \begin{Bmatrix} \kappa_x^o \\ \kappa_y^o \\ \kappa_{xy}^o \end{Bmatrix}
$$

5. **A quasi-isotropic laminate** refers to the laminate that consists of plies with certain arrangements such that it produces an isotropic [A] matrix. Its extensional stiffness matrix is similar to that of isotropic materials and $A_{11} = A_{22}$, $A_{16} = A_{26} = 0$ (extension and shear are uncoupled), and $A_{66} = (A_{11} - A_{12})/2$. The quasi-isotropic composite laminate has at least four plies and all the plies are arranged to have the same angle $\Delta\theta$ between adjacent ply orientations, for example, [90/45/–45/0]. It has the angle of 45° between two radially adjacent plies. For a quasi-isotropic laminate with n plies, the angle is $180/n$. The stiffnesses are independent of the angle of rotation of the laminate. For example, the laminate with a layup sequence of [–60/75/–15/30] remains quasi-isotropic after all of the plies in [90/45/–45/0] reorient by a rotation angle of +30° because the angle between adjacent ply orientations remains at 45°. The material isotropy in quasi-isotropic composite laminates is in-plane only. The force and moment equation can be written as follows:

$$
\begin{Bmatrix} N_x \\ N_y \\ N_{xy} \\ M_x \\ M_y \\ M_{xy} \end{Bmatrix} = \begin{bmatrix} A_{11} & A_{12} & 0 & B_{11} & B_{12} & B_{16} \\ A_{12} & A_{11} & 0 & B_{12} & B_{22} & B_{26} \\ 0 & 0 & (A_{11} - A_{12})/2 & B_{16} & B_{26} & B_{66} \\ B_{11} & B_{12} & B_{16} & D_{11} & D_{12} & D_{16} \\ B_{12} & B_{22} & B_{26} & D_{12} & D_{22} & D_{26} \\ B_{16} & B_{26} & B_{66} & D_{16} & D_{26} & D_{66} \end{bmatrix} \begin{Bmatrix} \varepsilon_x^o \\ \varepsilon_y^o \\ \gamma_{xy}^o \\ \kappa_x^o \\ \kappa_y^o \\ \kappa_{xy}^o \end{Bmatrix}
$$

6. **An antisymmetric laminate** has paired plies with the same thickness and material above and below the mid-plane but opposite sign for the ply orientation. For every lamina of $+\theta$ orientation above the mid-plane in the anti-symmetric laminate, there is an identical lamina (with the same thickness and material) of $-\theta$ orientation at the same distance below the mid-plane, for example, [– 45/15/30/– 30/– 15/45]. There is no bending–twisting coupling and $A_{16} = A_{26} = 0$, $D_{16} = D_{26} = 0$:

$$
\begin{Bmatrix}
N_x \\
N_y \\
N_{xy} \\
M_x \\
M_y \\
M_{xy}
\end{Bmatrix}
=
\begin{bmatrix}
A_{11} & A_{12} & 0 & B_{11} & B_{12} & B_{16} \\
A_{12} & A_{22} & 0 & B_{12} & B_{22} & B_{26} \\
0 & 0 & A_{66} & B_{16} & B_{26} & B_{66} \\
B_{11} & B_{12} & B_{16} & D_{11} & D_{12} & 0 \\
B_{12} & B_{22} & B_{26} & D_{12} & D_{22} & 0 \\
B_{16} & B_{26} & B_{66} & 0 & 0 & D_{66}
\end{bmatrix}
\begin{Bmatrix}
\varepsilon_x^o \\
\varepsilon_y^o \\
\gamma_{xy}^o \\
\kappa_x^o \\
\kappa_y^o \\
\kappa_{xy}^o
\end{Bmatrix}
$$

7.4 Summary

- Micromechanics refers to analysis of composites, including thermoplastic composites, at the level of their individual constituents, that is, the fiber and the matrix.
- The ROM is a mathematical method to calculate the properties of a mixture based on the volume fraction and property of each constituent in the mixture.
- Modified ROM equations can be applied to continuous fiber thermoplastic composites with different fiber architectures, such as biaxial fiber and quasi-isotropic fiber architectures.
- The Halpin–Tsai model is a semiempirical model that takes consideration of not only the properties of the matrix and reinforcement but also their percentages and reinforcement geometry.
- Thermoplastic composites reinforced with particulate and nanomaterials with low aspect ratios have isotropic material properties, and their properties are identical in all random directions. Two independent material constants, namely E and v, can define the compliance matrix or the stress and strain relationship of the isotropic thermoplastic composite material.
- Force resultant is defined as the normal or shear force over unit length of thermoplastic composites, and moment resultant is defined as the moment over unit length of the composite.
- CLT is a macromechanical method that describes the effect of fiber directions, stacking arrangements, layer thickness, and material properties of each lamina on the mechanical properties of laminated composites, including thermoplastic composites.

References

[1]   Voigt W. Ueber die Beziehung zwischen den beiden Elasticitätsconstanten isotroper Körper. Annalen der physik. 1889;274(12):573–87.

[2]   Reuss A. Calculation of the flow limits of mixed crystals on the basis of the plasticity of monocrystals. Zeitschrift für Angewandte Mathematik und Mechanik. 1929;9:49–58.

[3]   Hashin Z, Rosen BW. The elastic moduli of fiber-reinforced materials. Journal of Applied Mechanics. 1964;31:223–32.

[4]   Hill R. Theory of mechanical properties of fibre-strengthened materials – III. Self-consistent model. Journal of the Mechanics and Physics of Solids. 1965;13(4):189–98.

[5]   Budiansky B. On the elastic moduli of some heterogeneous materials. Journal of the Mechanics and Physics of Solids. 1965;13(4):223–7.

[6]   Mori T, Tanaka K. Average stress in matrix and average elastic energy of materials with misfitting inclusions. Acta Metallurgica. 1973;21(5):571–4.

[7]   Aboudi J. Mechanics of Composite Materials: a Unified Micromechanical Approach. Elsevier: 2013.

[8]   Hewitt R, De Malherbe M. An approximation for the longitudinal shear modulus of continuous fibre composites. Journal of Composite Materials. 1970;4(2):280–2.

[9]   Chawla KK. Composite Materials: Science and Engineering. Springer Science & Business Media: 2012.

[10] Jones RM. Mechanics of Composite Materials. CRC press: 1998.

Chapter 8
Recycling of thermoplastic composites

8.1 Introduction

Recycling refers to reusing waste materials and converting them into useful resources. It is gaining more and more attention because of the strong need for an eco-friendly and sustainable environment. From the viewpoint of environmental protection, economic impact, and ethics, it is an urgent and pressing need to increase the recycling rate for solid waste materials including thermoplastics and thermoplastic composites. The benefits from recycling can be significant. Firstly, it can reduce the amount of waste materials sent to landfills and diminish the pollution in air, soil, and water caused by the waste material in landfills. Secondly, recycling is economically rewarding. Recycled materials are generally at a lower cost than virgin materials, which can reduce the overall material cost and benefit the manufacturing industry. In addition, recycling can reduce the exploitation of new resources and lead to better preservation of natural resources for the future generations to come.

The thermoplastic composite is fully recyclable and can be repurposed and reused via cost-effective recycling methods such as remelting and remolding. It is one of the advantages that thermoplastic composites possess over thermoset composites. There has been a challenge to fully recycle thermoset composites because the thermoset cannot be remelted or remolded due to its cross-linking structure. Although the fibers in the thermoset composites can be reclaimed through different approaches (see Case Study 8.1), it is extremely difficult to fully recycle the thermoset matrix as it is. The thermoset matrix is normally converted to heat, hydrocarbons, gases, and other by-products by thermally or chemically breaking down its cross-linked molecular structure. The difficulty in dealing with thermoset composite scraps or end-of-life (EoL) components further highlights the significance of full recyclability that the thermoplastic composite can provide. Thermoplastics generally do not have a cross-linking structure and their polymer chains can slide across each other when heated. This characteristic allows the thermoplastic matrix to be remelted and remolded after being heated to its processing temperatures and, therefore, full recyclability for thermoplastic composites. Considering that there will be more stringent regulations on recycling, thermoplastic composites possess a great advantage because of their full recyclability and cost-effective recycling processes.

There are several sources of thermoplastic composite scraps for recycling. Those thermoplastic composite scraps include the waste generated at different production stages, namely, waste generated from the preform production stage, trim-offs from thermoplastic composite preforms, scrap from machining or finishing operation, and rejected parts during molding. In addition, many components made of thermoplastic

https://doi.org/10.1515/9781501519055-008

composites that reach their end of service life, for example, automotive components, are another source of thermoplastic composites for recycling. All of these sources are described below.

1. **Scrap generated from the preform production stage.** Preforms are produced at the first step (pre-impregnation step) of the processing of thermoplastic composite (see Section 4.1). Waste materials can be generated from this step. Those waste materials include fibers, thermoplastics, as well as preforms that do not meet quality requirements, for example, the preforms produced before processing conditions stabilize.

2. **Trim-offs from thermoplastic composite preforms.** Trim-offs can be generated from the second step of the processing of thermoplastic composites, namely, consolidation (molding) stage. Preforms, for example, uni-tapes, are firstly cut to desired dimensions before being stacked for molding. Trim-offs are often generated due to the irregular geometry of parts to be produced.

3. **Scrap from finishing operation on molded parts.** The scrap from the finishing operations is another source of recycled thermoplastic composites. Some of the molded discontinuous fiber thermoplastic composite products have near-net shape. However, some processes can produce parts with excess materials, such as flash or injection gates, that need to be trimmed off. Other finishing operations, such as drilling and grinding, have to be done on certain thermoplastic composite parts. Figure 8.1a shows discontinuous glass fiber nylon composite scrap from machining. In addition, continuous fiber thermoplastic composite parts molded from preforms can have edges that are outside of the part and need to be trimmed. Figure 8.1b shows woven glass fiber PET composite scrap trimmed from a molded truck trailer part.

4. **Rejected part.** Rejected part refers to the part that does not meet production specifications and is not accepted for use. It is one of the scrap sources for recycling. The failure of meeting the specification can be resulted from change of processing parameters or malfunction of the machinery in the production line. The rejected part normally has defects such as incomplete filling, excessive warpage, sink marks, weld lines, and burn marks.

5. **EoL components.** When thermoplastic composite components do not function as intended and are beyond repair, the components are said to reach the EoL. The components are normally decommissioned and are another source of thermoplastic composite scrap for recycling.

Figure 8.2 shows the recycling methods for recycling thermoplastic composites, namely, mechanical recycling, chemical recycling, and thermal recycling methods. Those methods vary in their approaches, purposes, and products generated. Among those methods, mechanical recycling method is dominantly used for recycling thermoplastic composites.

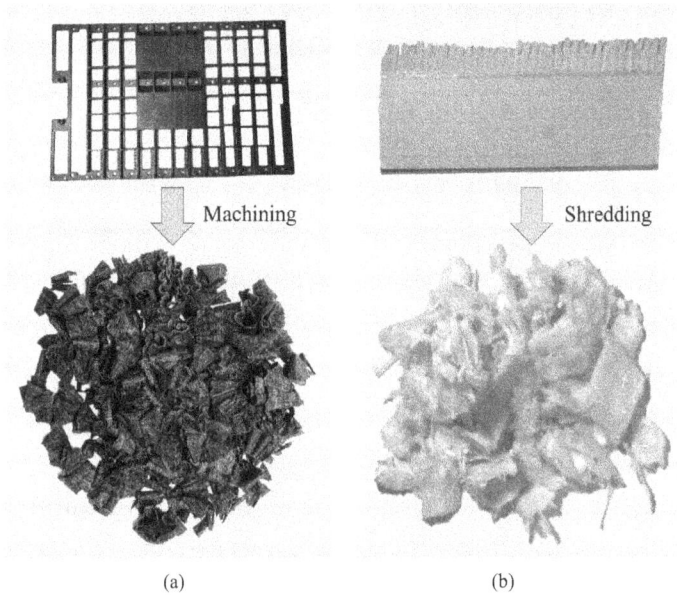

Fig. 8.1: (a) Long glass fiber-reinforced nylon composite part after machining (top) and machining chips for recycling (bottom) and (b) woven glass fiber-reinforced PET composite trim-off (top) and shredded pieces from the trim-off (bottom).

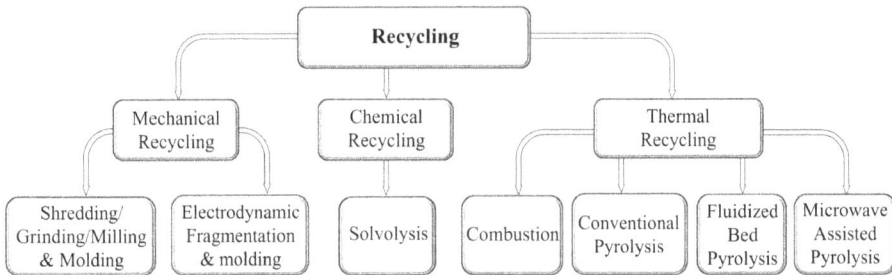

Fig. 8.2: Different methods for recycling thermoplastic composites. Mechanical recycling method is dominantly used for recycling thermoplastic composites.

The thermoplastic composite is mostly recycled through mechanical recycling and reprocessed without the need of separating the fiber from the matrix. The mechanical recycling is efficient and does not use any hazardous chemicals; therefore, it is dominantly used in recycling thermoplastic composites. The other methods, such as chemical or thermal recycling methods, are primarily used for recycling thermoset composites, and not often used for recycling thermoplastic composites. Some of the methods are able to reclaim fibers by removing the matrix in the composite scrap while some of the methods only harvest the thermal energy. Table 8.1 summarizes the product, by-product, advantage, and disadvantage of each recycling method.

Tab. 8.1: Various methods for recycling thermoplastic composites.

Recycling methods		Products reclaimed	By-products	Advantages	Disadvantages
Mechanical recycling	Shredding/ milling/ grinding and molding	Composite	Dust	– Cost-effectiveness – No hazardous materials	– Fiber length attrition – High machinery wear
	Electrodynamic fragmentation and molding	Composite	Wastewater and gas	– High disintegration rate	– Severe mass loss – High voltage required – Under development
Chemical recycling	Solvolysis	Fibers and matrix	Waste chemical	– Fiber length retention – Fiber and matrix separation	– Hazardous waste chemical – Limited scalability
Thermal recycling	Combustion and incineration	Fibers and heat	Fumes	– Energy efficiency	– Hazardous off gases – Fiber property degradation
	Conventional pyrolysis	Fibers	Char and hydrocarbons	– Cost-effectiveness – Fibers reclaimed	– Char deposition on fiber – Hazardous off gases – Fiber property degradation
	Fluidized bed pyrolysis	Fibers			
	Microwave-assisted pyrolysis	Fibers			

8.2 Mechanical recycling

Mechanical recycling is the most commonly used method to recycle thermoplastic composites. It mainly involves physically reducing the size of thermoplastic composite scraps and molding the scrap. The size of the scrap is firstly reduced through forces such as shear, impact, or electrical shock forces. It should be adequately small such that the scrap can be fed into extruders for consequent melting and molding process. This process does not separate the fiber and the matrix and takes full advantage of the remelting and reshaping capability of the thermoplastic matrix in the composite. It is considered to be most economic and least impactful to the environment. The mechanical recycling steps include shredding, grinding, milling, or electrodynamic fragmentation of the composite scrap followed by melting and molding.

Physical and chemical change normally occur to the thermoplastic composite during the recycling processes. Firstly, the fiber length is reduced during shredding (or other size reduction processes). Consequent melting and molding of the composite lead to more attrition of the fibers because of the shear force and interaction among fibers involved during those processes. The matrix also undergoes another thermal cycle that can result in degradation and deterioration of mechanical properties.

8.2.1 Shredding, grinding, or milling

The recycled thermoplastic composites from different sources (see Section 8.1) can significantly vary in sizes. For example, the chips or residue from machining normally have small sizes and can be directly fed into extruders for melting. However, some scraps, such as rejected parts from production or EoL composite parts, are too large to be fed into extruders for melting. Those thermoplastic composite scraps have to be cut into smaller pieces prior to melting.

Shredders, grinders, and mills are equipment that use mechanical forces, such as shear forces or impact forces, to break down large thermoplastic composite scraps or parts into pieces with desired sizes. Shredders and grinders are two terms that are often used interchangeably although there is difference in their designs and minimal achievable size. Shredding normally yields centimeter-scale composite pieces while grinding results in pieces at a millimeter scale. The shredder normally has a design with a low speed and high torque for shredding a large piece into small pieces while the grinder has a design for grinding off small pieces from a large composite piece. Milling is commonly used for reducing the length of recycled fibers by cutting mills or hammer mills. It is a downgrading process that can be used to reduce the recycled thermoplastic composite to a much smaller size compared to shredding and grinding. The shredding, grinding, and milling processes have high tool wear because of the high strength and the abrasive characteristic of the fibers, such as glass fibers, in the recycled composite.

A screen or mesh is normally used to control the size of the final composite pieces by allowing the composite pieces with a size no more than the aperture size to go through the screen or the mesh. The larger pieces that cannot go through the screen or the mesh are repeatedly shredded, ground, or milled until a desired size is reached.

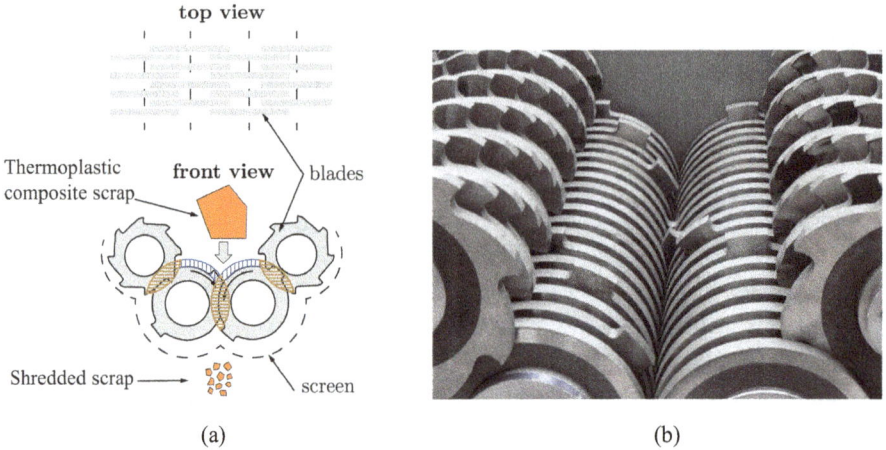

Fig. 8.3: (a) Schematic of a shredder (top and front views) and (b) inside view of a four-shaft shredder for shredding thermoplastic composites (reprinted from Reference [1] with permission).

The main by-product created from shredding, grinding, and milling is dust. The dust includes fibers that are separated from the matrix. It is a health hazard, and safety measure should be taken. If the recycled thermoplastic composite contains carbon fibers, loose carbon fibers can be generated and drift in air, which can cause issues such as short circuits.

8.2.2 Electrodynamical fragmentation

Electrodynamical fragmentation process was developed in the 1960s for disintegrating rocks in the mining industry using an electrical discharge. This process was recently adopted in an attempt to disintegrate thermoplastic composites that need to be recycled. The process involves passing the electric discharge through a liquid media, such as water, oil, or liquid nitrogen, where the thermoplastic composite is placed. The electric discharge is generated at a high voltage (i.e., 180 kV) and creates a plasma channel in the composite that induces extremely high temperatures and pressures in a confined area [2]. The shock wave resulted from the plasma channel fractures and disintegrates the composite. Figure 8.4a schematically shows the electrodynamical fragmentation process.

This process is able to disintegrate thermoplastic composites at a high rate because of its high energy intensity. However, its efficiency dramatically decreases when the number of fragments is reduced in the liquid media after the fragments with desired sizes are removed. In addition, there are a certain amount of material loss from the process due to material sublimation and pyrolysis of the thermoplastic matrix under extremely high temperatures. A material loss of 13% has been reported [2].

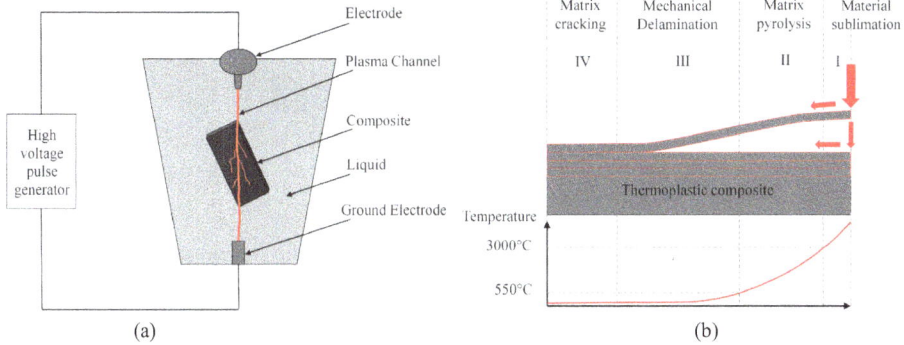

Fig. 8.4: (a) A typical setup of the electrodynamical fragmentation process for disintegrating recycled thermoplastic composites and (b) a schematic describing the mechanisms of disintegrating the composite in the fragmentation process (adapted from Reference [2]).

Figure 8.4b shows the material sublimation and matrix pyrolysis adjacent to the plasma channel that lead to the material loss. Delamination and matrix cracking resulted from the shock wave are also shown in the figure [2].

8.2.3 Remolding

The recycled thermoplastic composite after size reduction is typically processed via injection molding, compression molding, or extrusion. Those processes are covered in Section 4.5. The thermoplastic composite pieces with desired sizes are firstly fed into an extruder for melting. The recycled thermoplastic composite is often mixed with virgin thermoplastic composites with a certain ratio to maintain the balance between cost saving and performance. The recycled and virgin thermoplastic composites normally have the same fiber percentage and the same type of thermoplastic matrix; however, virgin thermoplastic composites with different fiber contents or even neat thermoplastic can also be added. The composite melt is then molded into products through different molding processes. Figure 8.5 shows a schematic of shredding, melting, and compression molding of recycled thermoplastic composites. As mentioned in Section 4.6.2, the compression molding process retains fiber length well and the mechanical properties of the recycled composites can be maintained after multiple iterations of shredding, melting, and molding processes [3]. However, recycling from injection molding or extrusion process can induce degradation of material properties. The properties such as tensile strength decrease with the number of injection molding or extrusion cycles mainly due to fiber attrition [4–6].

The thermoplastic matrix in the recycled thermoplastic composite experiences another melting cycle which can cause degradation of the thermoplastic, such as chain scissoring, oxidation, and generation of volatile products. The chain scissoring

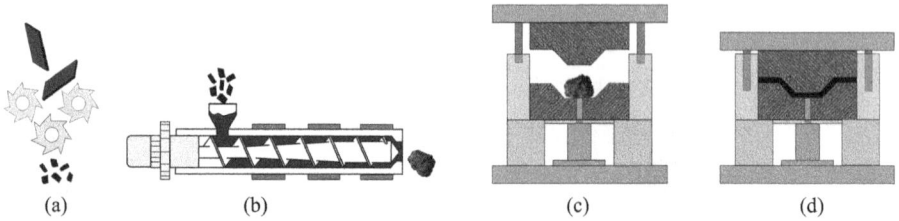

Fig. 8.5: Typical processes for recycling thermoplastic composites: (a) shredding of thermoplastic composite scrap; (b) remelting of the shredded composite; (c) and (d) compression molding of the composite melt.

results in smaller molecular weight of the thermoplastic matrix and lower mechanical properties.

8.3 Chemical recycling

Chemical recycling is another method used to recycle polymer matrix composites. It uses chemicals that are able to attack and degrade the molecular chains in the matrix. The main purpose of this method is to separate the fiber from the matrix, normally carbon fibers from carbon fiber-reinforced composites. Although it is currently used mainly for thermoset composites, it can also be used in recycling thermoplastic composites.

Solvolysis refers to the reaction that breaks the chemical bonds in the polymer matrix using a solvent and produces monomers and oligomers. A high pressure is also used in the meantime to achieve efficient solvolysis. Figure 8.6 shows the setup for solvolysis. Composite scraps are fed into a reactor in which high pressure and elevated temperature are applied. Separation of fibers from the composite scrap is achieved after the solvent dissolves the matrix into monomers and oligomers. The challenge of using solvolysis is to find suitable solvents for the polymer, especially high-performance thermoplastics. For example, polyphenylenesulfide (PPS) and polyether ether ketone (PEEK) have no solvent at room temperature and complete dissolution of those thermoplastics can only happen in high concentration chemicals at elevated temperatures.

Dissolution of the matrix in polymer matrix composites including thermoplastic composites can also be achieved by using supercritical fluids. Those fluids have a temperature and pressure high than their critical-point values and possess a great combination of liquid-like density, dissolving capability, gas-like viscosity, and diffusivity. The supercritical fluid can decompose the thermoplastic into relatively large oligomers while the fibers remain unaffected. The liquids used as supercritical fluids include water, methanol, acetone, alcohol, and propanol. Figure 8.7a illustrates the supercritical water resulted from a high pressure (22.1 MPa) and a

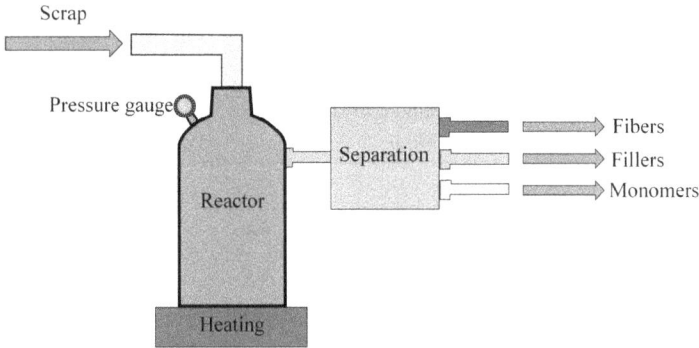

Fig. 8.6: A schematic of the solvolysis setup for separating the fibers in composites (adapted from Reference [7]).

high temperature (374 °C). Figure 8.7b illustrates the supercritical ethanol under a high pressure (6.2 MPa) and at a high temperature (241 °C). When the fluid is at its critical point, there is no change in state when pressure is increased or temperature is raised.

Supercritical fluids have been normally used in removing epoxy from its composite and reclaim carbon fibers. The fluid is also used in thermoplastic composites. For example, a mixture of ethanol and water has been used to decompose PEEK in its carbon fiber composite at 350 °C within 30 min with a catalyst to reclaim the carbon fiber [8]. The oligomers dissolved in the supercritical fluid can also be collected.

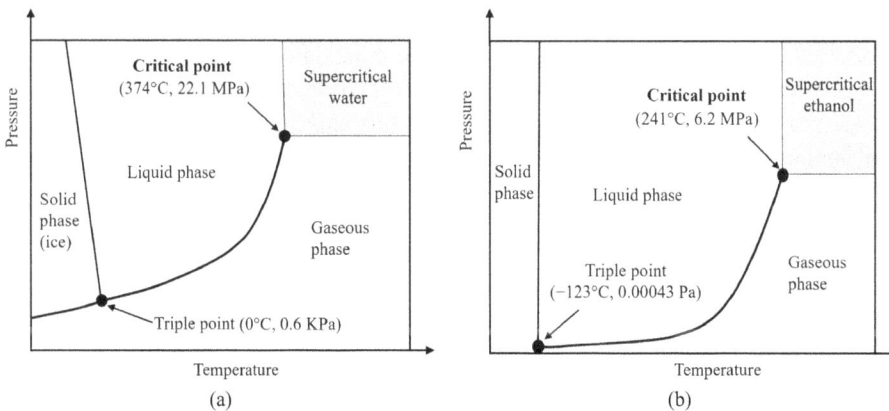

Fig. 8.7: (a) Phase diagram of water showing its critical point at 374 °C and 22.1 MPa and (b) phase diagram of ethanol showing its critical point at 241 °C and 6.2 MPa.

Tab. 8.2: Temperature and pressure required to reach supercritical conditions for typical compounds used to separate fibers from matrix in composite scraps [9–15].

Solvent	Supercritical temperature (°C)	Supercritical pressure (MPa)	Supercritical density (g/cm^3)
Acetone	235	4.7	0.235
Benzyl alcohol	323	4.5	0.447
1-Butanol	287	4.9	0.270
Ethanol	241–243	6.14–6.38	0.276
Methanol	240	8.2	0.282
Propanol	235–264	4.76–5.17	0.274
Water	368–380	22.1–24	0.322–0.368

8.4 Thermal recycling

Thermal recycling is another method for recycling polymer matrix composites. It involves thermally removing the matrix in the composite and reclaiming the fibers. The matrix is burned off or carbonized at high temperatures ranging from 300 to 1,000 °C. It is a method mainly used for recycling thermoset composites; however, it can also be used for thermoplastic composites.

Thermal recycling methods can be classified into pyrolysis, gasification, combustion, and incineration depending on the oxygen amount (Fig. 8.8). Those methods result in different outputs. Due to no oxygen involved in pyrolysis, the products generated from pyrolysis include tars, hydrocarbons, and char in addition to fibers and fillers. When excessive oxygen is used, incineration occurs and gases and ashes are the main products. When limited oxygen is introduced to the composite, the matrix is mainly converted into gases and a small amount of char.

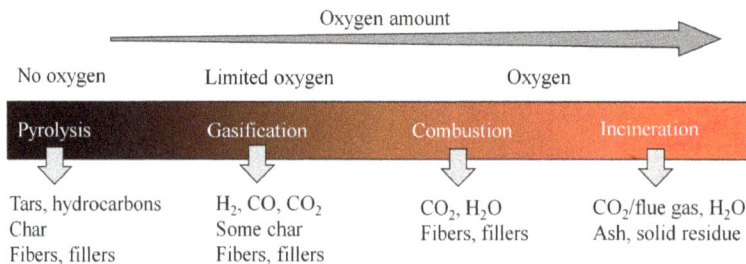

Fig. 8.8: Pyrolysis, gasification, combustion, and incineration for thermally recycling polymer matrix composites (adapted from Reference [7]).

Combustion and incineration take place at a relatively high temperature with the presence of oxygen. Their purpose is to oxidize the matrix in the composite scrap. A tremendous amount of heat can be generated and harvested.

Pyrolysis is a common method to recycle thermoset composites for their fibers. According to their setup, there are conventional pyrolysis, fluidized bed pyrolysis, and microwave-assisted pyrolysis. Those pyrolysis methods can be adopted to reclaim the fibers from recycled thermoplastic composites.

1. **Conventional pyrolysis** is often used to recycle the carbon fiber from carbon fiber-reinforced polymer matrix composites. The composite is heated to above 300 °C in an inert atmosphere. Thermochemical reactions occur to the matrix and create chars and tars. The fibers in the composite, such as carbon fiber, are protected from oxidation because of the inert atmosphere. This approach has also been used to characterize the carbon fiber content in a thermoplastic composite [16].

2. **Fluidized bed pyrolysis** is attracting more attention as an efficient pyrolysis method to reclaim fibers from polymer matrix composites. A fluidized bed is a mixture of a particulate solid material and a fluid that behaves as a fluid under appropriate conditions. In the fluidized bed pyrolysis process, shredded composites are fed into the fluidized bed, which offers an extremely high surface contact area with the recycled composite because of its fluidic characteristic. The matrix can be heated to its degradation temperature at a high rate and decompose into gases, tar, and char. The gas can be collected and used as the heat source for the pyrolysis process.

3. **Microwave-assisted pyrolysis** is another variation of the pyrolysis process. It uses microwave as the energy source to heat the composite, decompose the matrix, and separate the fibers. Microwave is a form of electromagnetic radiation that causes the polar molecules in the polymer matrix to rotate and produce heat. The method is not applicable to composites with nonpolar polymer matrices. Because the microwave can penetrate through the composite and therefore the radiation is absorbed throughout the composite and efficient heating is resulted. The pyrolysis process is normally done under an inert atmosphere to prevent burning of the matrix and oxidation of carbon fibers. The matrix can eventually be converted into by-products, such as tars, hydrocarbons, and gases, and separation of the fibers from the matrix is achieved.

8.5 Properties of recycled fiber

One major purpose in chemically and thermally recycling composites is to reclaim the fibers. It is beneficial to retain fiber properties as much as possible during the recycling process. However, the mechanical properties of the fibers can suffer from the harsh thermal and chemical treatment in addition to possible reduction in fiber length, both of which can deteriorate their reinforcing performance.

The residual tensile strength of the reclaimed glass fiber can be significantly affected by the high temperature. The fiber strength starts to drop significantly when the temperature reaches to a certain level. Figure 8.9 illustrates the effect of temperature on the residual tensile strength of E-glass fibers [17]. The fibers are held at each temperature for approximately 2 h. The decrease in their tensile strength is mainly due to the growth of preexisting surface flaws and generation of new flaws. However, the elastic modulus of the recycled glass fiber is minimally affected by the temperature exposure.

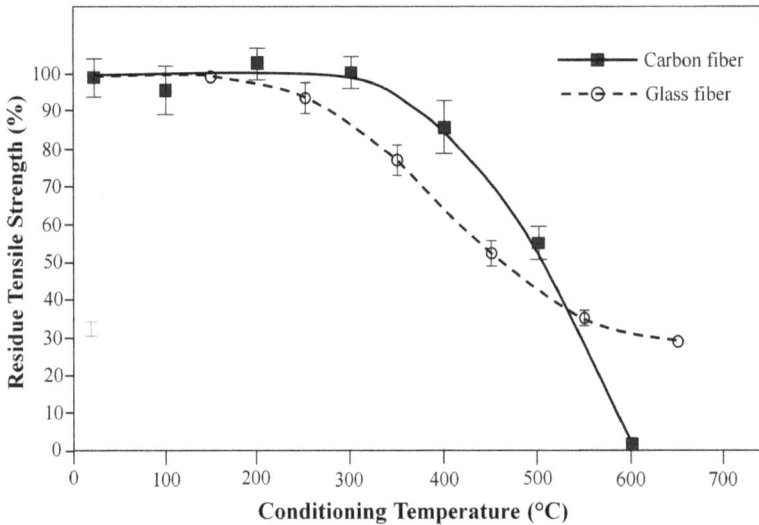

Fig. 8.9: Decrease in the residual tensile strength of glass fibers (reprinted from Reference [17] with permission) and carbon fibers (adapted from Reference [18]) with temperature.

Figure 8.9 also shows the decrease of residual tensile strength for carbon fibers after exposure to elevated temperatures. There is a drastic drop of the residual tensile strength after exposure at 400 °C. The decrease in tensile strength after exposure at a high temperature is resulted from the oxidation of the carbon fiber surface and generation of defects on the fiber surface. Because the fiber strength is highly sensitive to the defect, the fiber shows decreased residual tensile strength. The effect of the temperature exposure on elastic modulus of the carbon fiber is negligible [18].

Case study 8.1: Recycling of carbon fibers: from thermoset composites to thermoplastic composites

Carbon fibers possess superior specific modulus and strength as mentioned in Section 2.3.1. However, their high cost has limited their use to mainly high-end applications, such as aircraft, aerospace, and sports. One major factor that contributes to the high cost is the tremendous energy consumption during production of virgin carbon fibers. The energy used to manufacture virgin high strength carbon fibers ranges from 186 to 595 MJ/kg [19–23]. The energy for producing ultra-high modulus fibers can even reach 911–1,010 MJ/kg [24]. Those values vary considerably due to different precursor materials used, varying carbon fiber grades, and so on. The production process also produces a large amount of greenhouse gasses such as CO_2. For example, there is 19.3–21.3 kg CO_2 created for producing 1 kg carbon fibers [24].

The need for low-cost carbon fibers has mainly driven the tremendous interest in recycling of carbon fibers. The main material source for recycled carbon fibers is carbon fiber-reinforced epoxy composite. There are different carbon fiber epoxy composite scrap sources, including scraps generated from manufacture of carbon fiber epoxy prepregs and carbon fiber epoxy composite components after reaching their EoL. For example, one main source for recycled carbon fibers is carbon fiber thermoset composites used in aircrafts. It is estimated that thousands of aircrafts are in graveyards, and several hundreds of aircraft are withdrawn from service annually. Those aircrafts, especially the newer models, used a large amount of carbon fibers. It is estimated that carbon fiber composites make up 40% of the total weight of an aircraft. Some aircraft models have even a higher carbon fiber composite percentage up to 50%. The composites are mainly carbon fiber epoxy composite laminates as well as carbon fiber epoxy composite skins from sandwich structures.

It has been reported that 15–182 MJ is required to reclaim 1 kg carbon fibers from carbon fiber-reinforced epoxy composites [20, 25, 26]. The energy consumption for recycling carbon fibers can vary tremendously because of the variety of scraps used, various methods involved in the recycling process, the difference in the batch size, as well as the final forms (continuous or discontinuous carbon fibers) of the recycled carbon fibers. Overall, the average energy consumption for producing 1 kg recycled carbon fibers is approximately one fourth of the average energy required to produce the same amount of virgin carbon fibers. As a result, the recycled carbon fibers are 20–40% less expensive than virgin carbon fibers. Therefore, the recycling approach can provide a valuable solution to the increasing demand of low-cost carbon fibers for stiffness-driven applications, such as automotive. The market size of recycled carbon fibers is continuously growing because of the increasing demand of low-cost carbon fibers. It is estimated that the global market size for recycled carbon fibers grows from USD 109 million in 2020 to USD 193 million by 2025 at a compound annual growth rate of 12.0% [27].

As aforementioned, carbon fibers are recycled from two main epoxy composite scrap sources such as prepregs and end-products that do not meet specifications or have reached EoL. Carbon fibers are reclaimed from those composites through pyrolysis or solvolysis. The partially cured epoxy resin in the prepreg scraps is generally removed by solvolysis. The end-products from cured thermoset composites are normally shredded, and the epoxy matrix is removed through pyrolysis. The fiber length attrition can occur during the recycling process, especially during shredding that converts continuous fibers to discontinuous fibers (generally long fibers).

Figure 8.10 illustrates typical steps for recycling carbon fibers from thermoset composites and combining them with thermoplastics, including collecting carbon fiber thermoset composite wastes, cleaning and size reduction, removal of thermoset matrix, combining recycled carbon fibers with thermoplastics, and molding carbon fiber thermoplastic composites. The thermoset composite waste is firstly cleaned and reduced in size for the consequent matrix removing process. The thermoset matrix is normally removed through solvolysis (for prepregs with partially cured thermoset matrix) or pyrolysis (for end-products with fully cured thermoset matrix). Figure 8.11a shows recycled woven carbon fabrics through solvolysis and Fig. 8.11b shows discontinuous carbon fibers recycled through pyrolysis. The original sizing on carbon fibers is removed after the pyrolysis process (thermally) or solvolysis process (chemically), which facilitates application of new sizing desired for thermoplastic matrices.

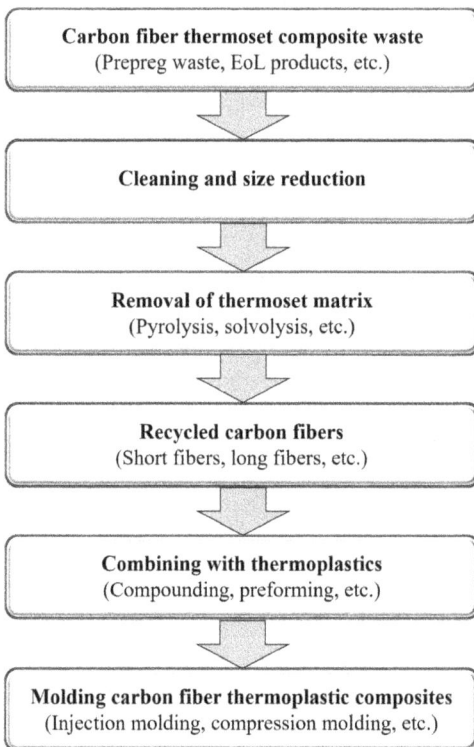

Carbon fiber thermoset composite waste
(Prepreg waste, EoL products, etc.)

↓

Cleaning and size reduction

↓

Removal of thermoset matrix
(Pyrolysis, solvolysis, etc.)

↓

Recycled carbon fibers
(Short fibers, long fibers, etc.)

↓

Combining with thermoplastics
(Compounding, preforming, etc.)

↓

Molding carbon fiber thermoplastic composites
(Injection molding, compression molding, etc.)

Fig. 8.10: Typical procedures for recycling carbon fibers from thermoset composites for producing thermoplastic composites.

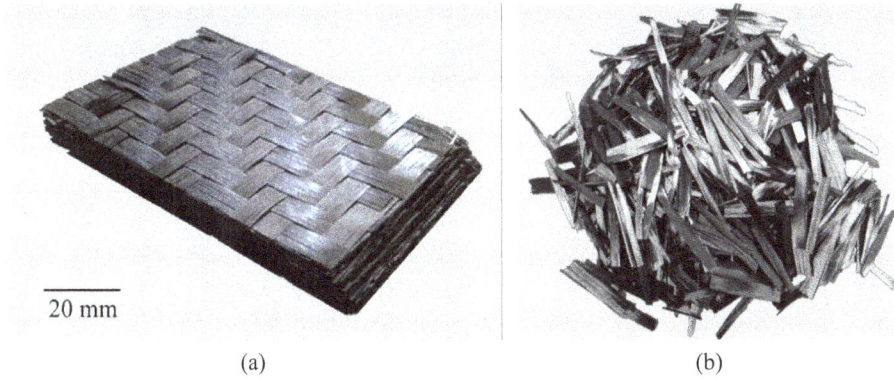

Fig. 8.11: (a) Recycled woven carbon fabrics through solvolysis (reprinted from Reference [28] with permission) and (b) recycled discontinuous carbon fibers through pyrolysis.

The recycled carbon fibers are normally prepared into different forms, such as "fluffy" fiber and fiber mat, and combined with thermoplastics (Fig. 8.12(a,b)). A typical approach involves compounding the fluffy carbon fibers with thermoplastic in different forms (such as powder and pellets) for processes such as compression molding or injection molding. Figure 8.12a shows the recycled carbon fibers mixed with PPS powders for compression molding. Another typical approach is to produce mat preforms that consist of recycled carbon fibers and thermoplastic fibers. The mat preform is then heated and molded through thermostamping, compression molding, and so on. This approach can minimize fiber length attrition and therefore maximize the load-bearing capacity of the carbon fibers and their composites. Figure 8.12b shows a mat consisting of recycled carbon fibers and nylon fibers. The recycled carbon fiber can also be produced into a fiber mat consisting of recycled carbon fibers only (Fig. 8.12c) for combining with thermosets or thermoplastics.

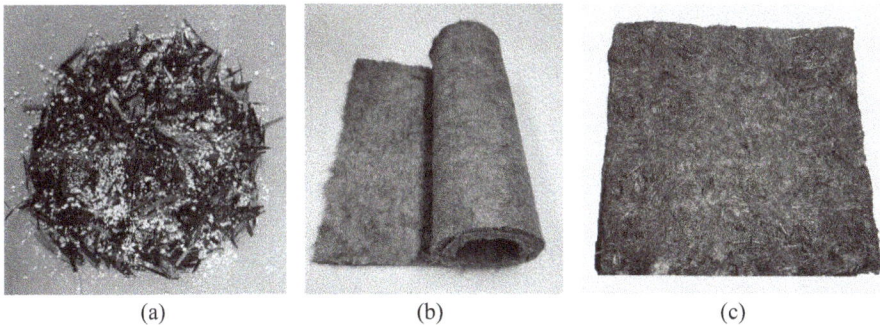

Fig. 8.12: Different forms of recycled carbon fibers for molding processes: (a) recycled carbon fibers mixed with PPS powders; (b) fiber mat consisting of recycled carbon fibers and nylon fibers; and (c) fiber mat consisting of recycled carbon fibers only.

Figure 8.13(a,b) show thermoplastic composite components made of the carbon fiber recycled from EoL aircraft components. A rocker panel demonstrator shown in Fig. 8.13a consists of 20 wt% carbon fiber and 80 wt% PET. PET fibers are firstly mixed with carbon fibers recycled from carbon fiber thermoset composites, and flat preforms of nonwoven fiber mats are developed through a vacuum-based, waterborne deposition process. The preform undergoes a thermostamping process and is molded into panels. A 25% weight reduction is achieved over their acrylonitrile butadiene styrene plastic counterpart. The recycled carbon fiber composite component also provides improved stiffness and cold impact performance [29, 30]. Figure 8.13b shows thermoplastic composite drum sticks made of carbon fibers recycled from carbon fiber thermoset composites. The recycled carbon fibers are compounded with nylon pellets and injection molded into drum sticks. Besides those components, recycled carbon fibers have also been used to produce other thermoplastic composite products such as automotive components, athletic footwear, hockey skates and blades, surf board, and kayak paddles.

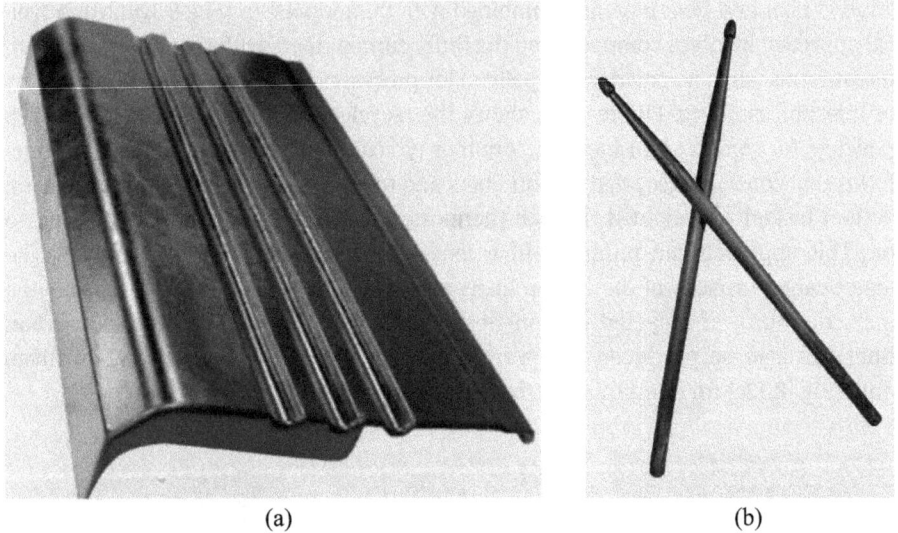

(a)   (b)

Fig. 8.13: Thermoplastic composite components consisting of carbon fibers that are recycled from thermoset composites: (a) recycled carbon fiber-reinforced PET rocker panel for transportation application (adapted from References [29] and [30]) and (b) drum sticks made of recycled carbon fiber-reinforced nylon composite.

8.6 Summary

- Thermoplastic composites can be fully recycled, which is one of the advantages that thermoplastic composites possess over thermoset composites.
- Thermoplastic composite scraps are generated from different production steps. Those scraps include the waste from preform production, trim-offs from thermoplastic composite preforms, scrap from finishing operation on molded parts, rejected parts, and EoL components.
- Methods for recycling polymer matrix composites mainly include mechanical recycling, chemical recycling, and thermal recycling. Mechanical recycling is dominantly used in recycling thermoplastic composites.
- Both carbon fibers and glass fibers recycled through pyrolysis show minimal reduction in elastic modulus but a significant drop in tensile strength.
- Mechanical recycling of thermoplastic composites does not require separation between the fiber and the matrix. It involves size reduction using shredding/grinding/milling, remelting, and remolding. Fiber length attrition and thermal degradation of the matrix can occur during the processes.
- Carbon fibers recycled from carbon fiber thermoset composites can be combined with thermoplastics to produce value-added carbon fiber-reinforced thermoplastic composites.

References

[1] Vincent GA, de Bruijn TA, Wijskamp S, Rasheed MIA, van Drongelen M, Akkerman R. Shredding and sieving thermoplastic composite scrap: Method development and analyses of the fibre length distributions. Composites Part B: Engineering. 2019;176:107197.
[2] Roux M, Eguémann N, Dransfeld C, Thiébaud F, Perreux D. Thermoplastic carbon fibre-reinforced polymer recycling with electrodynamical fragmentation: From cradle to cradle. Journal of Thermoplastic Composite Materials. 2017;30(3):381–403.
[3] Mark Janney UV, Ryan Sutton, and Haibin Ning. Re-Grind Study of PPS-Based Long Fiber Thermoplastic Composites. SAMPE: 2014.
[4] Eriksson PA, Albertsson AC, Boydell P, Prautzsch G, Månson JA. Prediction of mechanical properties of recycled fiberglass reinforced polyamide 66. Polymer Composites. 1996;17(6):830–9.
[5] Lee KH, Lim SJ, Kim WN. Rheological and thermal properties of polyamide 6 and polyamide 6/glass fiber composite with repeated extrusion. Macromolecular Research. 2014;22(6):624–31.
[6] Casado JA, Carrascal I, Diego S, Polanco JA, Gutiérrez-Solana F, García A. Mechanical behavior of recycled reinforced polyamide railway fasteners. Polymer Composites. 2010;31(7):1142–9.
[7] van Oudheusden A. Recycling of composite materials. 2019.
[8] Dandy L, Oliveux G, Wood J, Jenkins M, Leeke G. Accelerated degradation of Polyetheretherketone (PEEK) composite materials for recycling applications. Polymer Degradation and Stability. 2015;112:52–62.

[9]   Dauguet M, Mantaux O, Perry N, Zhao YF. Recycling of CFRP for high value applications: Effect of sizing removal and environmental analysis of the SuperCritical Fluid Solvolysis. Procedia Cirp. 2015;29:734–9.

[10] Postorino P, Tromp R, Ricci M, Soper A, Neilson G. The interatomic structure of water at supercritical temperatures. Nature. 1993;366(6456):668–70.

[11] Wagner W, Pruß A. The IAPWS formulation 1995 for the thermodynamic properties of ordinary water substance for general and scientific use. Journal of Physical and Chemical Reference Data. 2002;31(2):387–535.

[12] Zhao L, Hou Z, Liu C, Wang Y, Dai L. A catalyst-free novel synthesis of diethyl carbonate from ethyl carbamate in supercritical ethanol. Chinese Chemical Letters. 2014;25(10):1395–8.

[13] Kamitanaka T, Matsuda T, Harada T. Mechanism for the reduction of ketones to the corresponding alcohols using supercritical 2-propanol. Tetrahedron. 2007;63(6):1429–34.

[14] Sun Y, Ponnusamy S, Muppaneni T, Reddy HK, Patil PD, Li C, et al. Optimization of high-energy density biodiesel production from Camelina sativa oil under supercritical 1-butanol conditions. Fuel. 2014;135:522–9.

[15] Mukai S-a, Koyama T, Tsujii K, Deguchi S. Anomalous long-range repulsion between silica surfaces induced by density inhomogeneities in supercritical ethanol. Soft matter. 2014;10 (35):6645–50.

[16] Wang Q, Ning H, Vaidya U, Pillay S, Nolen L-A. Fiber content measurement for carbon fiber–reinforced thermoplastic composites using carbonization-in-nitrogen method. Journal of Thermoplastic Composite Materials. 2018;31(1):79–90.

[17] Feih S, Manatpon K, Mathys Z, Gibson A, Mouritz A. Strength degradation of glass fibers at high temperatures. Journal of Materials Science. 2009;44(2):392–400.

[18] Kessler E, Gadow R, Straub J. Basalt, glass and carbon fibers and their fiber reinforced polymer composites under thermal and mechanical load. AIMS Materials Science. 2016;3:1561–76.

[19] Meng F, Olivetti EA, Zhao Y, Chang JC, Pickering SJ, McKechnie J. Comparing life cycle energy and global warming potential of carbon fiber composite recycling technologies and waste management options. ACS Sustainable Chemistry & Engineering. 2018;6(8):9854–65.

[20] Suzuki T, Takahashi J, editors. Prediction of energy intensity of carbon fiber reinforced plastics for mass-produced passenger cars. Proceedings of 9th Japan International SAMPE Symposium; 2005.

[21] Takahashi J, Zushi H, Suzuki T, Nagai H, Kageyama K, Yoshinari H, editors. Life cycle assessment of ultra lightweight vehicles using CFRP. 5th International Conference on EcoBalance Tsukuba, Japan; 2002.

[22] Van Acker K, Verpoest I, De Moor J, Duflou J-R, Dewulf W. Lightweight materials for the automotive: environmental impact analysis of the use of composites. Revue de Métallurgie. 2009;106(12):541–6.

[23] Johnson M, Sullivan J. Lightweight Materials for Automotive Application: An Assessment of Material Production Data for Magnesium and Carbon Fiber. Argonne National Lab. (ANL), Argonne, IL (United States): 2014.

[24] CES EduPack. Granta Design: http://www.grantadesign.com/education/. Accessed: Feb 2021.

[25] Keith MJ, Oliveux G, Leeke GA. Optimisation of solvolysis for recycling carbon fibre reinforced composites. 2016.

[26] Shibata M, Nakagawa K. CFRP recycling technology using depolymerization under ordinary pressure. HItachi Chemical. 2014:6.

[27] https://www.marketsandmarkets.com/PressReleases/recycled-carbon-fiber.asp. Accessed: Feb 2021.

[28] Oliveux G, Bailleul J-L, Gillet A, Mantaux O, Leeke GA. Recovery and reuse of discontinuous carbon fibres by solvolysis: Realignment and properties of remanufactured materials. Composites Science and Technology. 2017;139:99–108.

[29] Mauhar M. IV. 5. Low Cost Carbon Fiber Composites for Lightweight Vehicle Parts—Materials Innovation Technologies LLC. Lightweight Materials R&D Program.

[30] Gardiner G. Recycled carbon fiber update: closing the CFRP lifecycle loop. Composites Technology. 2014;20(6):28–33.

Index

https://doi.org/10.1515/9781501519055-009

www.ingramcontent.com/pod-product-compliance
Lightning Source LLC
Chambersburg PA
CBHW080915220326
41598CB00034B/5574